BUILDING A POPULAR SCIENCE LIBRARY COLLECTION FOR HIGH SCHOOL TO ADULT LEARNERS

Building a Popular Science Library Collection for High School to Adult Learners

ISSUES AND RECOMMENDED RESOURCES

Gregg Sapp

GREENWOOD PRESS
Westport, Connecticut • London

Library of Congress Cataloging-in-Publication Data

Sapp, Gregg.
    Building a popular science library collection for high school to
adult learners : issues and recommended resources / Gregg Sapp.
       p.   cm.
    Includes bibliographical references and index.
    ISBN 0–313–28936–0
    1. Libraries—United States—Special collections—Science.
I. Title.
Z688.S3S27   1995
025.2'75—dc20   94–46939

British Library Cataloguing in Publication Data is available.

Library of Congress Catalog Card Number: 94–46939
ISBN: 0–313–28936–0

First published in 1995

Greenwood Press, 88 Post Road West, Westport, CT 06881
An imprint of Greenwood Publishing Group, Inc.

Printed in the United States of America

The paper used in this book complies with the
Permanent Paper Standard issued by the National
Information Standards Organization (Z39.48–1984).

10 9 8 7 6 5 4 3 2 1

To Kelsey and Keegan, with love,
I hope that you never stop learning.

Contents

viii   Contents

Preface

In 1985 I accepted my first professional library position as a science librarian at Idaho State University. In addition to the normal pressure associated with starting a new job, I was additionally concerned about being a *science librarian* per se and working with technical information from disciplines in which I had little if any academic preparation. Although I had worked in various technical libraries in the past, like most librarians my educational strengths were in the arts, humanities, and social sciences. This apprehension led me to undertake voracious, self-directed reading in popular science literature, and that activity, over time, has become an ongoing personal learning strategy. Professionally, I have become so committed to popular scientific literature that I have spent the last five years researching topics related to science literacy, science education, and information resources. This book is a product of that work.

Popular science information is important because it provides the only vehicle for curious nonspecialists to learn about myriad scientific and technical issues, short of going back to school to do so. Further, for those who are in school, popular science can be integrated into the college or high school general education curriculum in ways that stimulate greater understanding of and interest in subjects that often are quickly forgotten. Several recent surveys have found American science education and the public understanding of science to be deficient—perhaps dangerously so. When scientists, educators, and public servants discuss reforms in these areas, the need for more and better popular scientific information is almost always stated.

Even so, the nature of popular science information resources is not well understood. Even those who call for increased popularization of scientific knowledge sometimes do so guardedly. The reason is essentially that, while primary scientific information is submitted to peer scrutiny and assessment, there are no comparable quality checks where popular science is concerned. The fear is that scientific speculation might be presented as fact, unrealistic promises might be made based upon faulty research findings, or information might be used selectively, even insidiously, by persons with partisan agendas. These are legitimate concerns.

As information professionals, librarians who use or select popular science resources must possess the necessary skills by which to evaluate that literature. Where such evaluation is not done appropriately, the library can become the institutional means by which defective science information reaches the public. Part I of this book provides public, high school, and college librarians with a basic understanding of popular science information so that they can make better choices on behalf of their library patrons. Part II, organized by subject area, cites specific books, periodicals, and other resources that could constitute a core science literacy collection. Between the two parts, theory and practice are melded.

ACKNOWLEDGMENTS

Thanks are due to many. Barbara Rader at Greenwood Press proposed the idea for this book. Drs. Pearce Mullen and Robert Fellencz of Montana State University offered suggestions and encouragement. George Suttle was a compatriot and occasional sounding board. Paul Wylie helped keep me on track. I owe thanks to Dave Michael, because I forgot him in my last book. My staff in Access Services Department at the University of Miami Richter Library were exceedingly patient and tolerant. Thanks also to Paul Michel, on whom the irony of this book will not be lost, and Bill Dobbins, ditto. Great thanks are due my family: my wife, Beatrice, who suffered through those endless weekends while I pounded the word processor; my children, Kelsey and Keegan; and my parents and siblings, especially Laura.

Part I

Scientific Information, Popular Science, and Lifelong Learning

Chapter One

Science Literacy, Science Education, and Public Life

Science and technology impinge on our daily lives in countless ways. Electronic gadgets that a few years ago would have seemed futuristic are now common in our homes and workplaces. Just a decade ago, for example, cellular phones, microwave ovens, ATM machines, video recorders, and compact discs were new commercial technologies. Today the frontiers have been pushed back, and now virtual reality and high definition television are among the emerging high-tech products. Frequently, however, everyday technologies can seem like black boxes, the inner workings of which are totally inscrutable to us. This is often true of even the most mundane tools and products. For example, how many of us know how even the ordinary light bulb works? Without a basic understanding of how these devices work, we are "blinded" by science—incapable of understanding or appreciating many of the things that we encounter virtually every day.

Further, those who lack an intellectual grounding in science are missing out on a fascinating cultural experience. Science has changed how we think and perceive the world in which we live, and how we express ourselves. To give just one example, most of us are familiar with the word "hacker" and the kind of person it describes, but relatively few know of the flourishing subculture that hackers have developed. It is so rich, in fact, that Eric Raymond has written a reference book, *The New Hacker's Dictionary* (MIT Press, 1993), to describe and standardize its language.

Most crucially, however, as citizens we are often called upon, either directly or indirectly, to assess complex issues that are founded in science. At the national level, it is estimated that two-thirds of the bills currently pending in

Congress involve science and technology in some way.[1] Numerous scientific issues have broad public relevance; for example:

- Billions of taxpayer dollars were spent on such projects as the Space Shuttle Program, the Strategic Defense Initiative, the Hubble Space Telescope, and the Superconducting Supercollider. Although quite different in purpose and conception, all these enterprises encountered serious financial and/or technical problems, for which taxpayers footed the bill.
- Genetically and biologically engineered products, including foods, are rapidly becoming commercially available. Many people, however, fear that such products are unsafe.
- Research into the development of many new drugs that are of potentially great benefit, including AIDS treatments, is continually hampered by economic and political pressures. Meanwhile, millions suffer from possibly treatable diseases.
- As some legislators seek to pass laws designed to safeguard threatened environments and preserve biodiversity, others are more concerned about the economic impacts of such measures.
- The Human Genome Project, a government funded effort to map the entire human genetic code, raises both legislative and ethical questions.
- Proposed computer networks—"information superhighways"—hold the promise of democratizing information, but could also become commercialized and accomplish the opposite.
- In the 1970s, research into the development of sustainable energy resources was a political priority. Today, despite dwindling resources, the urgency of those days seems to have faded from public memory.
- There is a clamorous public outcry to reform health care, but there are serious doubts as to how this can be accomplished when medical technology and procedures are becoming more sophisticated and, inevitably, more expensive.

And there are scores of other issues that demand political action and citizen involvement. Especially in local versions of these national debates, public mobilization can make a difference. How well does the public understand the scientific underpinnings of matters such as these?

The term "science literacy" has been used to describe the general public's understanding of scientific and technical concepts. Recently, writers in popular, professional, and scholarly publications alike have expressed concern regarding the state of America's science literacy. Poor science literacy begins with the science education that children receive in elementary school, where, according to criticism leveled in a recent *Newsweek* article, "American science education serves not to nurture children's natural curiosity but to extinguish it with catalogs of dreary facts and terms."[2] These problems continue into high school and then into college, where decreasing numbers of students have chosen to seek degrees in science and engineering (S/E). The final result is an adult population that lacks a basic grasp of the scientific and technological foundations of modern life. Educators argue that true science literacy must be cultivated and maintained over a lifetime.[3] This book accepts that thesis and will focus, in particular, on information resources that are appropriate tools for stimulating a deeper appreciation and understanding of science among high school, undergraduate, and general adult populations.

DEFINITIONS OF SCIENCE LITERACY

The term "science literacy" is easier to use than to define. Perhaps the simplest definition is offered by James Trefil and Robert Hazen in their book *Science Matters*: "For us, science literacy constitutes the knowledge you need to understand public issues."[4] This definition is an example of what might be called "civic science literacy,"[5] which focuses on the importance of the general public's role in furthering political dialogue on the many sci-tech issues currently confronting us. Similarly, various educators have written about the necessity of instilling within students not only a better understanding of these issues, but also the motivation to participate in such a dialogue. Historian of science Michael Shortland expresses this view succinctly: "In a word, to become scientifically literate is to become an effective citizen."[6]

While the democratic aspect of science literacy is very important, it is only one facet. Jon Miller of the International Center for the Advancement of Science Literacy, who has conducted the most extensive surveys on the science literacy of the American public,[7] tested for it in three distinct categories: (1) a basic factual knowledge of core scientific concepts and a minimal vocabulary of technical terms; (2) an understanding of the scientific method and the ability to distinguish between legitimate and spurious science; and (3) an appreciation of the role of science in society. In this formulation, science literacy begins to assume the form of a holistic personality—the sum of what one knows, how one obtains and evaluates new information, and what one does with it.

Science literacy is perhaps best defined by the attributes and attitudes of those who possess it. It is cultivated rather than learned, and it requires maintenance over the course of a lifetime. Shortland cites the following components of science literacy:

- An appreciation of the nature and aims of science and technology, including their historical origins and the epistemological and practical values they embody.
- A knowledge of the ways in which science and technology actually work, including the funding of research, the conventions of scientific practice, and the application of new discoveries.
- A basic grasp of how to interpret numerical data, especially relating to probability and statistics.
- A general grounding in selected areas of science, including a number of key interdisciplinary areas.

Figure 1.1
The Almond Model of Public Attitudes

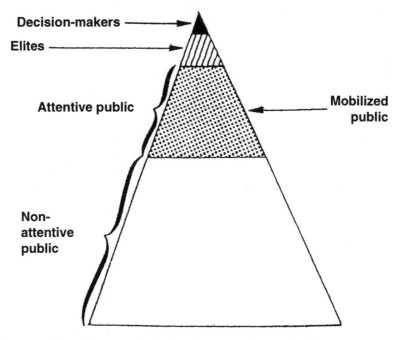

- An appreciation of the interrelationships between science and technology and society, including the role of scientists and technicians as experts in society.
- An ability to acquire new scientific information in the future.[8]

Because the word "literacy" can imply the ability to *read* technical literature, the term "science literacy" itself has been challenged. Kenneth Prewitt, a president of the Social Science Research Council, prefers the term "science savvy": "My understanding of the scientifically savvy citizen is a person who understands how science and technology impinge upon public lives. Although this understanding would be enriched by substantive knowledge of science, it is not coterminous with it."[9]

Miller describes science literacy as occurring within an "attentive public," a "self-selected group that has a high level of interest in, and a functional knowledge about a given issue area."[10] The accompanying model, originally developed by G. A. Almond in the 1950s,[11] depicts a stratified pyramid wherein the attentive public, which in this case is science literate, occupies a block near the top of the pyramid, just below the decision makers and policy leaders (see Figure 1.1). A basic characteristic of the attentive public is its desire to seek information on science. By contrast, the nonattentive public, which in this case is science illiterate, resides at the bottom of the pyramid and is either unwilling or unable to keep up with and use scientific information.

For the purposes of this book, it is particularly important to note that science literacy is built on an infrastructure of information; it is the result of successful, specialized, and ongoing information seeking behavior. The following definition of science literacy emphasizes this information centrality:

Science literacy is an active understanding of scientific methods and of the social, economic, and cultural roles of science as they are conveyed through various media and is thus built on an ability to acquire, update, and use relevant information about science.[12]

The advantage of this definition is that it addresses equally the behaviors and the resources by which science literacy is achieved. Science literacy is not just knowledge of facts. The French mathematician and philosopher Jules-Henri Poincaré said, "Science is built of fact just as a house is made of bricks, but facts do not make science any more than a pile of bricks make a house." A person can know virtually nothing of quantum physics and still be scientifically literate. What matters is that one possesses the motivation and the ability to seek information on those topics about which one becomes curious or has a need to know. These are developmental traits and aptitudes that fit into the model of lifetime learning, which begins in primary education, is internalized during the critical young and early adult years, then continues as a matter of lifestyle throughout adulthood. This book explores some of the information resources that can make science literacy possible among those in the young adult to mature years of the lifetime learning continuum.

PARALLEL DIMENSIONS OF SCIENCE LITERACY

Science literacy thus has two distinct but interrelated dimensions: (1) the science that our students learn in the schools and (2) the science that is understood by adults in our society. Recent studies have found both to be sorely lacking.

Several private institutions and governmental agencies have produced extensive statistical and demographic profiles of the weaknesses in American science education. The National Assessment of Educational Progress project, which was formed by Congress following the Reagan administration's scathing report *A Nation at Risk* (1983),[13] studied trends in science education throughout the 1980s and presented them in *The 1990 Science Report Card*,[14] a document frequently cited by many concerned with science education and literacy.

In the 1990 report, researchers from the National Center for Education Statistics used an item response method to survey 20,000 American students in grades four, eight, and twelve. Questions were taken from the life, physical, earth, space, and general sciences. The multiple choice and inferential questionnaires were designed to test for science skills at progressive levels from zero to 500, with significant benchmarks established at the following levels of aptitude:

- Level 200: Understands Simple Scientific Principles. Students at this level are developing some understanding of simple scientific principles, particularly in the life sciences.

- Level 250: Applies General Scientific Information. Students at this level can interpret data from simple tables and make inferences about the outcomes of experimental procedures.

- Level 300: Analyzes Scientific Procedures and Data. Students at this level can evaluate the appropriateness of the design of an experiment and have the skills to apply their knowledge in interpreting information from text and graphs.

- Level 350: Integrates Specialized Scientific Information. Students at this level can infer relationships and draw conclusions using detailed scientific knowledge from the physical sciences.

Student performance at all levels involves the mastery of facts, concepts, and principles, but at the higher levels performance becomes increasingly related to the ability to apply this knowledge in evaluating problems and making reasonable inferences based on scientific principles. In short, students at these levels are capable of "doing science."

Among the high school seniors tested, nearly all (99%) had demonstrated the basic knowledge characteristic of Level 200, and 84 percent performed at Level 250. Beyond this point, however, there was a dramatic plunge. Fewer than half (45%) of the students reached Level 300, and a mere 9 percent scored at Level 350 or above. From these statistics, it would seem that most graduating high school students have learned the rudiments of science—the basic core concepts that are generally known by a science literate person—but they do not exhibit the ability to put this knowledge to practical use, nor is there evidence that they have internalized these concepts in a way that might engender within them the behavioral traits, like issue attentiveness and active information seeking, that characterize the higher orders of science literacy. This is troubling in that some educators have suggested that the skills associated with Levels 300 and 350 are prerequisites to success in college science studies.[15] Further, the average overall score of twelfth graders registered in 1990, which was 290.4, was down significantly from the 1970 average of 304.8.[16]

Since the majority of high school students do not perform at test levels that would indicate that they have assimilated scientific principles, it should not be surprising that neither do they show the behavioral traits that are associated with science literacy. Thirty-five percent of graduating high school seniors state unambiguously that they do not like science.[17] Most students (89%) have earned course credit in biology, but fewer than half (45.4%) have done so in chemistry, and barely one in five (20.1%) have braved physics.[18] (Perhaps not coincidentally, the same study also found that many high school instructors lack formal education in the physical sciences.) Good study habits are often found wanting. Figure 1.2 illustrates the relationships between science proficiency and various home behaviors. The positive correlation between high science scores and in-

Figure 1.2
Distribution of Students and Average Science Proficiencies by Factors Related to
the Home

	GRADE 4		GRADE 8		GRADE 12	
	Percent of Students	Average Proficiency	Percent of Students	Average Proficiency	Percent of Students	Average Proficiency
Types of Reading Materials in the Home						
Zero to two types	34 (1.0)	222 (1.1)	20 (0.7)	241 (1.7)	14 (0.7)	272 (1.9)
Three types	34 (0.7)	235 (1.1)	30 (0.8)	260 (1.2)	27 (0.8)	289 (1.5)
Four types	33 (0.9)	244 (1.0)	50 (0.9)	274 (1.5)	59 (1.0)	301 (1.3)
Daily Amount of Time Spent on Homework — All Subjects						
None assigned	21 (1.3)	237 (1.5)	6 (0.5)	249 (3.6)	12 (0.7)	273 (1.6)
Did not do it	5 (0.3)	215 (1.7)	7 (0.5)	245 (2.3)	9 (0.4)	292 (2.8)
One-half hour or less	34 (1.1)	233 (1.2)	20 (0.7)	261 (1.9)	21 (0.6)	295 (1.6)
One hour	24 (0.7)	237 (1.3)	40 (0.8)	268 (1.3)	32 (0.7)	296 (1.3)
Two hours	17 (0.8)	228 (1.4)	19 (0.6)	268 (1.7)	18 (0.8)	299 (1.8)
More than two hours	—	—	8 (0.5)	265 (3.0)	9 (0.5)	303 (2.4)
Daily Pages Read for School and Homework — All Subjects						
Five or fewer pages	26 (0.8)	226 (1.1)	32 (0.9)	253 (1.6)	33 (1.1)	281 (1.5)
Six to 10 pages	21 (0.7)	234 (1.3)	28 (0.6)	266 (1.5)	24 (0.8)	292 (1.3)
More than 10 pages	53 (1.1)	236 (1.1)	40 (1.0)	271 (1.6)	43 (1.3)	304 (1.5)
Days of School Missed last Month						
None	—	—	44 (0.8)	269 (1.1)	30 (0.9)	300 (1.4)
One or two days	—	—	33 (0.6)	268 (1.6)	38 (0.6)	297 (1.7)
Three days or more	—	—	23 (0.6)	249 (1.8)	32 (0.8)	285 (1.6)
Parents Living in Home						
Both parents	75 (0.8)	238 (1.0)	77 (0.6)	270 (1.3)	75 (0.7)	300 (1.1)
Single parent	20 (0.7)	222 (1.3)	19 (0.6)	251 (1.7)	20 (0.7)	286 (1.9)
Neither parent	5 (0.4)	205 (2.2)	4 (0.3)	231(3.9)	5 (0.3)	274 (3.6)
Daily Hours of Television Viewing						
Zero to one hour	18 (0.7)	235 (1.3)	14 (0.6)	273 (2.7)	34 (1.1)	304 (1.7)
Two hours	19 (0.6)	242 (1.4)	22 (0.8)	271 (2.1)	27 (0.6)	296 (1.5)
Three hours	16 (0.5)	238 (1.4)	22 (0.5)	267 (1.2)	19 (0.6)	291 (1.5)
Four to five hours	21 (0.6)	236 (1.2)	28 (0.8)	260 (1.3)	15 (0.7)	279 (1.7)
Six hours or more	26 (0.7)	219 (1.2)	14 (0.5)	241 (1.9)	5 (0.4)	263 (2.7)

The standard errors of the estimated percentages and proficiencies appear in parentheses. It can be said with 95 percent
certainty that for each population of interest, the value for the whole population is within plus or minus two standard errors of
the estimate for the sample.

The 1990 Science Report Card: NAEP's Assessment of Fourth, Eighth, and Twelfth Graders (National Center for Education Statistics,
U.S. Department of Education, 1992).

Figure 1.3
Distribution of Students and Average Science Proficiency by Race/Ethnicity

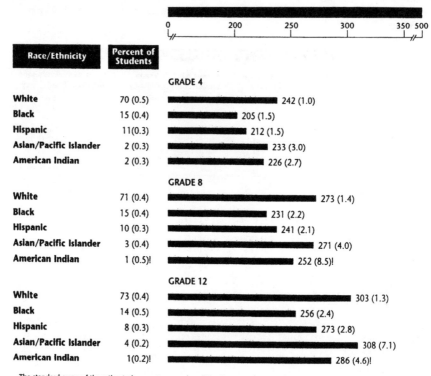

The standard errors of the estimated percentages and proficiencies appear in parentheses. It can be said with 95 percent certainty that for each population of interest, the value for the whole population is within plus or minus two standard errors of the estimate for the sample.

! Interpret with caution — the nature of the sample does not allow accurate determination of the variability of these estimated statistics.

The 1990 Science Report Card: NAEP's Assessment of Fourth, Eighth, and Twelfth Graders (National Center for Education Statistics, U.S. Department of Education, 1992).

creasing amounts of time spent reading or doing homework is not surprising, nor is the inverse relationship between time spent watching television and proficiency ratings.[19] Again, in the jargon of science literacy, these negatives are nonattentive behaviors.

These global averages are useful for discerning trends and identifying areas for remediation, but they can also mask some serious cultural inequalities in science education. Gender, race, and socioeconomic status affect students' science achievement in schools and, later, who among them goes on to become scientists and science literate citizens. Forty-three percent of twelfth grade females asked if they "like science" responded negatively, compared to just 26 percent of males. The average science proficiency score for twelfth grade males was 299, while that for females was 289; further, males scored higher in all four content areas tested.[20]

Figure 1.4
Distribution of Students and Average Science Proficiency by Type of Community

	Percent of Students	Average Proficiency
GRADE 4		
Advantaged Urban	11 (1.7)	252 (2.4)
Disadvantaged Urban	9 (1.1)	209 (2.6)
Extreme Rural	11 (1.8)	235 (2.6)
Other	69 (2.8)	233 (1.0)
GRADE 8		
Advantaged Urban	10 (2.2)!	283 (4.1)!
Disadvantaged Urban	9 (1.7)	242 (4.2)
Extreme Rural	11 (2.1)	257 (3.2)
Other	69 (2.8)	264 (1.5)
GRADE 12		
Advantaged Urban	10 (2.4)!	304 (4.4)!
Disadvantaged Urban	12 (2.5)	273 (5.3)
Extreme Rural	11 (2.7)!	291 (3.9)!
Other	67 (3.5)	296 (1.6)

The standard errors of the estimated percentages and proficiencies appear in parentheses. It can be said with 95 percent certainty that for each population of interest, the value for the whole population is within plus or minus two standard errors of the estimate for the sample.

! Interpret with caution — the nature of the sample does not allow accurate determrination of the variability of these estimated statistics.

The 1990 Science Report Card: NAEP's Assessment of Fourth, Eighth, and Twelfth Graders (National Center for Education Statistics, U.S. Department of Education, 1992).

Figures 1.3 and 1.4 summarize ethnic and socioeconomic disparities between the science achievement rates of various populations. In particular, note the large discrepancy between the average science proficiency of whites and blacks (nearly a fifty point difference!) and that between advantaged and disadvantaged urban students. Finally, females and minorities consistently score lower than white males on the technical sections of the SAT tests.[21] These statistics reveal that science has not been an equal opportunity discipline, and, while the reasons are complex, reforms must address not only the broad problem of defective science education, but also the specific gender and cultural factors that apparently discourage interest in the sciences among some groups.

Comparisons with student science achievement levels recorded in other countries help put the U.S. science literacy problem into perspective. In a 1988 school science achievement test conducted under the auspices of the International Association for the Evaluation of Educational Achievement (and with sponsorship from the National Science Foundation), American high school students placed ninth out of thirteen countries in physics scores, eleventh out of thirteen in chemistry, and last of thirteen in biological sciences. Among the areas that placed ahead of the United States in all categories were Australia, Britain, Hungary, Hong Kong, Japan, Poland, Sweden, and Singapore.[22] U.S. students fared somewhat better in the mean scores they recorded in 1989's Second International Mathematics Study,[23] in which they placed twelfth of twenty countries for algebra, but they were just sixteenth of twenty in geometry. Further, it is disturbing that the United States trailed many less developed nations, such as Nigeria, Swaziland, and Thailand.

Considering these grim statistics on the state of American science education at the secondary level, it is hardly surprising that the deficiencies become even more pronounced at the college level. Science educators refer to a "pipeline" (see Figure 1.5) through which students pass as they pursue careers in science and engineering. This analogy illustrates the funneling effect that occurs during the years of secondary and postsecondary education, at each stage of which fewer and fewer students remain in the pipeline that eventually leads to a career in some technical field. According to a 1987 National Science Foundation projection, the pipeline is becoming increasingly narrow; it is expected that of the more than 4 million high school sophomores entering the pipeline in 1977, approximately 9,700, or just 0.2 percent of the initial sample, will eventually obtain Ph.D.s in science or engineering.[24]

The relative underachievement of females and minorities that was documented at the high school level persists at the college level. In 1990 women made up just over 30 percent of the total number of S/E college students and less than 15 percent of those in engineering. Also in that year, more than twice as many men received S/E baccalaureate degrees as women (307,580 to 123,793).[25] Compounding the problem, the dropout rate for women in S/E majors is very high; as few as 3 percent of those originally seeking degrees in certain S/E subjects eventually enter careers in these fields.[26]

Among minorities, there is a similar lack of representation. Although blacks currently constitute about 12 percent of the nation's overall college population, in recent years they have received just 3 percent of the baccalaureate degrees awarded in S/E fields.[27] The figure for science degrees awarded to Hispanics is even lower, and that for Native Americans lower still. The only ethnic minority with significant per capita representation among science majors is Asian students; 70 percent of the foreign students receiving Ph.D.s in engineering belong to this group. Although preliminary figures recorded in the early 1990s show improvements in the numbers of women and minorities entering the sciences, they are still vastly underrepresented in S/E compared to their presence in the general college population. This further constricts the pipeline.

Major losses of students from the S/E pool occur during the late high school and early college years, and this phenomenon has become progressively problematic in some fields over the years. For example, between 1966 and 1988 the proportion of college freshmen planning to major in mathematics declined by half, from 11.5 percent to 5.8 percent. Incoming freshmen plans for a major rate business (25.7%) as a first choice, arts and humanities (11.3%) second, and social sciences (10.1%) third. As they progress in college, many more students convert to these disciplines as they abandon their science majors. The attrition rate in some disciplines is shockingly high, as much as 40 percent in physics.[28]

The second dimension of science literacy reflects the problems that occur when scientifically undereducated students become citizens. Because of these

Figure 1.5
The Science and Engineering Pipeline

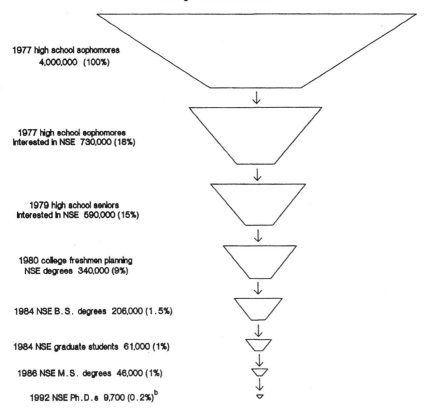

—Natural Science/Engineering[a] Pipeline: Following a Class From High School
Through Graduate School

1977 high school sophomores
4,000,000 (100%)

1977 high school sophomores
interested in NSE 730,000 (18%)

1979 high school seniors
interested in NSE 590,000 (15%)

1980 college freshmen planning
NSE degrees 340,000 (9%)

1984 NSE B.S. degrees 206,000 (1.5%)

1984 NSE graduate students 61,000 (1%)

1986 NSE M.S. degrees 46,000 (1%)

1992 NSE Ph.D.s 9,700 (0.2%)[b]

[a]Natural science/engineering (NSE) includes physical, mathematical, and life sciences, and engineering, but not the social sciences.

[b]National Science Foundation estimate, based on the historical rate in NSE of 5 percent of B.S.
graduates going on for Ph.D.s (using an 8-year average lag time from B.S. to Ph.D.) If market
conditions increase demand for Ph.D.s, then this estimate may understate future production of
NSE Ph.D.s. The number of NSE Ph.D.s in 1986 was about 12,000, or over 7 percent of NSE
B.S. graduates in 1978 (Center for Education Statistics degree data). Assuming 7 percent of
1986 B.S. graduates rather than 5 percent go on for Ph.D.s would project 14,400 Ph.D.s in
1992 rather than 9,700. Other methods of prediction (for instance, estimating Ph.D.s as a
percent of the 30-year-old age cohort) show similar responsiveness to changing participation
rates and assumptions. The Ph.D. population is very small and responds to changing conditions
in academia and the job market, so that population-based estimates should be taken as rough
indicators or warning signals rather than as solid predictions.

Note: These National Science Foundation estimates indicate the general pattern of the NSE pipeline,
but are not actual numbers of students in the pipeline. (For instance, actual natural science/
engineering B.S. production was 209,000 in 1986, Center for Education Statistics data.) The
estimates are based on data from the U.S. Department of Education-sponsored National Longitudinal Study of 1972 Seniors (for the high school senior through graduate school transitions)
and High School and Beyond Study of 1980 Seniors (for the high school sophomore to high
school senior transition). Since the National Longitudinal Study was conducted, student interest
in NSE majors has risen, but it is not yet clear whether trends in student interest with time
will follow the pattern of 1972 high school seniors revealed by the National Longitudinal Study.

Source: National Science Foundation, The Science and Engineering Pipeline, PRA Report 87–2,
April 1987, p. 3; and personal communication with National Science Foundation staff.

well documented inadequacies of American science education and the large numbers of defections from S/E programs during the critical secondary and college years, many scientists and educators—and the latter category should include librarians—have grave concerns about the poor state of science literacy among adult Americans and what it could mean to our society in an increasingly competitive, high-tech global economy.

Jon D. Miller has characterized the American public's understanding of science as "deplorably low." Miller conducted his first surveys, commissioned by the National Science Foundation, in 1979. Since that time, there has been little overall improvement and continued decline in some areas. For example, in 1979 Miller found that about 22 percent of adult Americans knew what DNA is; a subsequent survey in 1987 found that 16 percent did. An astonishing 45 percent of Americans were found to believe that the sun revolves around the Earth; less than half of the population believe in evolution. These statistics have practical consequences: in a later survey sponsored by the National Institute of Health, Miller determined that 80 percent of Americans lack the knowledge to make informed decisions on their own health care. Further, the pseudosciences are more popular than ever: 39 percent of Americans would characterize astrology as a science. Finally, the image of scientists themselves is somewhat tainted: 53 percent of Americans believe that, because of their esoteric knowledge, scientists possess "dangerous powers."[29] All in all, Miller estimates that a mere 18 percent of the adult American population are attentive to science, and of them just 5 percent qualify as true science literates. Numerous other recent surveys on specific aspects of science knowledge have yielded similarly discouraging results.[30]

Recall the two complementary models of the pyramid of issue attentiveness and the science education pipeline. Juxtapose those models and they seem to fit together like interlocking puzzle pieces. It could very well be that those same students who fall out of the pipeline at some point during the high school and college years are the adults who reside in the wide foundation of the pyramid of issue attentiveness and who, in this case, are science illiterates. Thus, the abandonment of science at some point during a student's education can lead to adult behaviors that are not conducive toward the cultivation of science literacy. Again, acquiring and maintaining science literacy is a lifetime proposition.

To improve America's science literacy, reforms are needed that address the problem in both its dimensions: reforms that aim to recruit and retain students with an active interest in the sciences, and reforms with the goal of elevating the science literacy of the general public. Inasmuch as information seeking skills are vital to science literacy, the ability to obtain and internalize appropriate information is central. The emphasis here is on "appropriate," because, in its primary literature, much published science is thoroughly incomprehensible to all but a few highly educated specialists. That is where works of popular science become extremely important to the crusade for science literacy.

THE IMPORTANCE OF POPULAR SCIENCE

Research into science education and the public understanding of science indicates widespread ignorance and misconceptions about science. It is thought to be esoteric, incomprehensible, and ivory towerish. As a result, students often tend to view science disciplines as undesirable, and that attitude can later become manifest in adult feelings of detachment from, and even a twinge of distrust toward, the work of scientists. Simultaneously, since many scientists work within very small, insular groups and achieve professional recognition by publishing their findings in specialized journals read only by their colleagues, they have sometimes had little interest in popularizing their ideas for the general public. There is a communication gap between the scientists, who often fail—or do not try—to convey the nature and significance of their work to a broader audience, and the public at large, who do not understand science or are inattentive to issues and trends in science. This gap can be bridged with the right kind of information.

Science popularization has been suggested as a way to promote science literacy. This has been recognized by the American Association for the Advancement of Science and by its British counterpart, the Royal Society,[31] both of which have called upon their constituencies to produce more and better works of popular science. Sigma Xi, the honorary scientific society, recently held an international symposium on how scientists can work to improve science literacy. Additionally, over the last decade editorials have appeared in various professional scientific journals, including *American Scientist, Bioscience, Environmental Science and Technology, Chemical and Engineering News, Journal of Geological Education, American Journal of Physics*, and *Science*, in support of popular science.[32] In one of these editorials, Carl Sagan describes the scientist's need for a healthier symbiosis with the public:

Support for scientific research derives mainly from public funds. This provides an obvious parochial reason for scientists to explain to the taxpayer what [it] is they do, and for this reason alone it is surprising more scientists are not engaged in the popularization of science. On a larger scale, there is an enormous range of grave social issues where science is central to any solution—from greenhouse warming and ozonospheric depletion to the nuclear arms race and the AIDS pandemic. . . . We are in real danger of having constructed a society fundamentally dependent on science and technology in which hardly anybody understands science and technology. This is a clear prescription for disaster.[33]

Writers in professional publications for the fields of journalism and education have also jumped on the popular science bandwagon.[34] Thus, at the level of professional organizations and societies, there is broad recognition of the need to make science accessible to nonscientists.

Increasingly, individual scientists have accepted the challenge to popularize their work. Several renowned scientists, such as Steven Hawking, Stephen Jay

Gould, Roger Penrose, Lynn Margulis, George Smoot, Freeman Dyson, Richard
Feynman, and Jane Goodall, have written nontechnical books based on their
research. Among them, Stephen Jay Gould is perhaps the most widely read; he
summarizes his reasons for writing popular science thus:

In America, for reasons that I do not understand (and they are truly perverse), such
writing for non-scientists lies immersed in deprecations—"adulteration," "simplification," "distortion for effect," "grandstanding," whiz-bang." . . . I deeply deplore the
equation of popular writing with pap and distortion for two main reasons. First, such a
designation imposes a crushing professional burden on scientists (particularly young scientists without tenure) who might like to try their hand at this expansive style. Second,
it denigrates the intelligence of millions of Americans eager for intellectual stimulation
without patronization. If we writers assume a crushing mean of mediocrity and incom-
prehension, then not only do we have contempt for our neighbors, but we also extinguish
the light of excellence.[35]

In his book *Innumeracy* mathematician John Allen Paulos echoes that sentiment
more pithily: "Mathematicians who don't deign to communicate their subject
to a wider audience are a little like multimillionaires who don't contribute anything to charity."[36]

As previously argued, science literacy is most effectively cultivated as part
of a lifetime educational process. While the origins of interests and aptitudes in
science can be traced to early childhood and include several personal and social
elements,[37] it is also clear that many students become disenchanted with the
sciences during the high school and early college years. Many educators have
advocated popular science as a way to recruit students to S/E disciplines and to
stimulate a better appreciation of S/E among those students who enter other
fields.[38] Along those lines, the following comments, made by cosmologist Marc
Davis when asked what books had been influential in steering him in his even-
tual career direction, are illustrative:

I read a lot of science books at the eighth and ninth grade level—books on relativity and
on physics and chemistry. . . . I read Bertrand Russell's *The ABC of Relativity* . . . in high
school. That was pretty amazing. I recall very vividly that later in high school, George
Gamow's books, *One Two Three . . . Infinity* in particular, were really influential. And I
was quite intrigued by *The Creation of the Universe*. But particularly *One Two Three
. . . Infinity* was just spectacularly great for me. It really made me decide that I wanted
to study physics.[39]

Typically, these books appeal to young adult readers at two levels: they introduce profound ideas that can stimulate the imagination, pique the curiosity, and
trigger an interest in learning more, and they also depict how science is done,
thereby showing the scientific process to be a human endeavor, characterized
by the setbacks and triumphs of any human work.

Popular science is a unique kind of information resource with a rich heritage.

Science literacy is built on a foundation of information retrieval, use, and assimilation. These facts suggest clearly that librarians, as information experts and gatekeepers, can make major contributions to America's crusade to achieve science literacy. As librarian Beth Clewis notes:

Scientific literacy research offers an opportunity for librarians to collect data of use and interest to other fields, especially communications and education. At its most ambitious, such research can contribute to cross-disciplinary discourse, and in doing so accomplish the dual goal of providing a theoretical basis for library policy and opening up library research to a wide audience.[40]

Librarians are trained to do the kinds of research, such as user studies and needs assessments, that can be useful in future research in science literacy. Further, since many of the survey instruments used to gauge science literacy have been criticized because they test knowledge of facts rather than the ability to find needed information, librarians can bring to the debate an insight that could redefine the very means by which science literacy is measured. Librarians should strive to provide the bibliographic and reference services by which patrons can keep up with important scientific developments and make sense of the big picture. Finally, librarians possess considerable evaluative skills that can be applied to works of science popularization. These and other ideas will be developed more fully in subsequent chapters of this book.

That having been stated, it must also be acknowledged that many librarians, whose academic backgrounds tend to be in the social sciences, arts, and humanities, are not intimately familiar with popular scientific writing and other media.[41] A major goal of this book is to better acquaint librarians with this expansive genre. Toward that end, the next chapter offers a brief history of popular science.

NOTES

1. *Science, Technology and Government for a Changing World: The Concluding Report of the Carnegie Commission on Science, Technology & Government* (New York: Carnegie Commission, 1993).

2. "Not Just for Nerds," *Newsweek* 115 (April 9, 1990): 52–64.

3. The "lifetime learning" concept is implicit in many of the works produced by the American Association for the Advancement of Science's Project 2061. See in particular *Science for All Americans* (Washington, DC: AAAS, 1989) and *Scientific Literacy*, edited by Audrey Champagne et al. (Washington, DC: AAAS, 1989); James Krieger, "Promoting Science Literacy: AAAS Sets Benchmarks for Students," *Chemical and Engineering News* (November 1, 1993): 4–5; "Science Books for a Lifetime of Reading," *Science Books and Films* 27 (May 1991): 97. See also several of the articles contained in section 4 of "Striving for Excellence: The National Education Goals" (Rockville, MD: Access ERIC, 1990 [ED 334713]).

4. James Trefil and Robert Hazen, *Science Matters: Achieving Science Literacy* (New York: Doubleday, 1991), xii.

5. Lewis Wolpert describes the point of view of those advocating "civic science literacy" in his chapter "Science and the Public" in *The Unnatural Nature of Science* (Cambridge, MA: Harvard University Press); see also similar ideas expressed in Jeremy Bernstein, "Science Education for the Non-scientist," chap. 16 in *Quarks, Cranks and the Cosmos* (New York: Basic, 1993).

6. Michael Shortland, "Advocating Science: Literacy and Public Understanding," *Impact of Science on Society* 152 (Fall 1988): 305–316.

7. Jon D. Miller, Kenneth Prewitt, and R. Pearson, *The Attitude of the U.S. Public Toward Science and Technology: A Report to the National Science Foundation* (Chicago: National Opinion Research Center, 1980); Jon D. Miller, *The Public Understanding of Science and Technology in the U.S.* (Dekalb, IL: Public Opinion Laboratory, 1991); Jon D. Miller, "The Scientifically Illiterates," *American Demographics* 9 (June 1987): 26–31; Jon D. Miller, "The Five Percent Problem," *American Scientist* 76 (March-April 1988): 166.

8. Shortland, "Advocating Science"; see also A. B. Arons, "Achieving Wider Science Literacy," *Daedalus* 112 (Spring 1983): 91–122.

9. Kenneth Prewitt, "Civil Education and Science Illiteracy," *Journal of Teacher Education* 34 (November-December 1983): 17–20.

10. Jon D. Miller, *The American People and Science Policy* (New York: Pergamon, 1983).

11. G. A. Almond, *The American People and Foreign Policy* (New York: Harcourt, 1950), 226–242.

12. Gregg Sapp, "Science Literacy: A Discussion and Information Based Definition," *College and Research Libraries* (January 1992): 21–30.

13. David Pierpoint Gardner, *A Nation at Risk: The Imperative for Educational Reform* (Washington, DC: National Commission on Excellence in Education, 1983).

14. Lee R. Jones et al., *The 1990 Science Report Card* (Washington, DC: National Center for Education Statistics, 1992). Other statistics of interest have also appeared in *Digest of Education Statistics, 1993* (Washington, DC: National Center for Education Statistics, 1993).

15. *The Liberal Art of Science* (Washington, DC: AAAS, 1990). Describes innovative undergraduate programs and outcome-based curricula at several institutions.

16. National Science Board, *Science and Engineering Indicators, 1991* (Washington, DC: National Science Board, 1993).

17. Jones et al., *1990 Science Report Card*, 81.

18. National Science Board, *Science and Engineering Indicators, 1991*, 26.

19. Jones et al., *1990 Science Report Card*, 17.

20. Ibid., chap. 1.

21. Jerilee Grandy, *Ten Year Trends in SAT Scores* . . . (Princeton, NJ: Educational Testing Service, 1985), 1–28; Office of Technology Assessment *Educating Scientists and Engineers* (Washington, DC: Government Printing Office, 1989), 32–36, 53; J. Mestre and Jack Lockhead, *Academic Preparedness in Science*, 2nd ed. (New York: College Board Publishers, 1990).

22. International Association for the Evaluation of Educational Achievement, *Science Achievement in Seventeen Countries* (New York: Pergamon, 1988); see also M. S. Klein

and F. James Rutherford, *Science Education in Global Perspective* (Washington, DC: AAAS, 1985).

23. Elliot Medrich, et al., *International Mathematics and Science Assessments* (Berkeley, CA: MPR Associates, 1992).

24. Office of Technology Assessment, *Educating Scientists and Engineers*, 12.

25. National Science Board, *Science and Engineering Indicators, 1991*, Chap. 2, p. 235.

26. Congress of the United States, *Higher Education for Science and Engineering* (Washington, DC: Government Printing Office, 1988), 34–42.

27. National Science Board, *Science and Engineering Indicators, 1991*, Chap. 2.

28. Kenneth Green, "A Profile of Undergraduates in the Sciences," *American Scientist* 7 (September-October 1989): 475–480.

29. Jon D. Miller, "Science Literacy: A Conceptual and Empirical Review," *Daedalus* 112 (Spring 1983): 26–31; Miller, Prewitt, and Pearson, *Attitude of the U.S. Public Toward Science*; Miller, "Scientifically Illiterates"; Miller, "Five Percent Problem."

30. Thomas R. Lord and Clint Rauscher, "A Sampling of Basic Life Science Literacy in a College Population," *American Biology Teacher* 53 (October 1991): 419–424; Arthur Lucas, "Public Knowledge of Biology," *Journal of Biological Education* 21 (Spring 1987): 41–45; A. M. Lucas, "Public Knowledge of Elementary Physics," *Physics Education* 23 (1988): 10–16.

31. AAAS, *Science for All Americans*; Royal Society, *The Public Understanding of Science* (London: Royal Society, 1985).

32. "Annual Meeting Hosts Symposium on the Public Understanding of Science," *American Scientist* 77 (January-February 1989): 12–15; Roald Hoffman, "Plainly Speaking," *American Scientist* 75 (July-August 1987): 418–420; Rustrum Roy, "Science Must Change or Self Destruct," *Bioscience* 36 (November 1986): 660–661; "Reporting Science for the Public," *Environmental Science and Technology* 23 (November 5, 1989): 491; Edward Jefferson, "Communicating Science," *Chemical and Engineering News* 63 (September 3, 1984): 48; E-an Zen, "Science Literacy and Why It Is Important," *Journal of Geological Education* 38 (November 1990): 463; Carl Sagan, "Why Scientists Should Popularize Science," *American Journal of Physics* 57 (April 1, 1989): 295; Robert Pool, "Science Literacy: The Enemy Is Us," *Science* (January 1991): 266–267.

33. Sagan, "Why Scientists Should Popularize Science."

34. Jeffrey Weld, "Scientific Literacy," *Educational Leadership* (December 1991-January 1992): 83–84; Tim Beardsley, "Teaching Real Science," *Scientific American* 267 (October 1992): 99–108; Lynn Arthur Steen, "Reaching for Science Literacy," *Change* (July/August 1991): 12–19.

35. Stephen Jay Gould, *Bully for Brontosaurus* (New York: Norton, 1991), 11–12.

36. John Allen Paulos, *Innumeracy: Mathematical Illiteracy and Its Consequences* (Hill and Wang, 1980), 80.

37. Jon D. Miller, *The Origins of Interest in Science and Mathematics* (Washington, DC: ERIC, 1988 [ED 303337]).

38. Librarian J. L. Cramner expressed this connection well in his article "The Popularization of Science Through Cheap Books," *Illinois Libraries* 31 (November 1949): 385–395.

39. Marc Davis, quoted in Alan Lightman and Roberta Brawer, *Origins: The Lives and Times of Modern Cosmologists* (Cambridge, MA: Harvard University Press, 1990), 341–358.

40. Beth Clewis, "Scientific Literacy: A Review and Implications for Librarianship," *Collection Management* (Fall 1990): 101–112.

41. Tony Stankus has advocated workshops as a means of improving science collection development skills among librarians; see his "Building Confidence and Competence: A Workshop in Science Journals for Beginning Librarians Without a Science Background," *Serials Librarian* 18 (Fall 1990): 28–45.

Chapter Two

A Brief History of Scientific Communication

In a sense, science is a circular process of information discovery, dissemination, and use. Communication of information is centrally important in the practice of science today. Since the rise of "modern" science—a drama which, in the West, might be said to have begun with the astronomy of Copernicus—the style, media, methods, and purposes of scientific communication have changed dramatically. Today, original scientific information is communicated by and for specialists, in ways that exclude all but their peers. This is not bad—it works perfectly well for members of the intended audience; but it also has the dual effects of fragmenting science as a whole and making scientific information unintelligible to the vast majority of people. These factors, in turn, contribute to science illiteracy.[1]

The first entry in the revised edition of Hellmans and Bunch's *The Timetables of Science* (Touchstone/Simon and Schuster, 1992) is 2,400,000 B.C. when paleolithic hominids in Africa first learned to manufacture stone tools. Thus, the very earliest emergence of what might be called science shows how, as ancient humans sought and gained understanding of the natural world, they applied this knowledge to create technologies that bettered their lives. The motivation behind science is essentially human—the desire to understand the world and our place in it—and essentially practical—the need to manipulate our world to our advantage.

Still, throughout history, as science and technology have advanced, they have become increasingly fragmented, and this process naturally excludes a large proportion of the general population. In antiquity, certain things were taught to, and thus known by, only a few. Often, as in ancient Egypt and Babylonia, scientific knowledge was embedded within ritual and myth, revealed only to priests. Later, in the Hellenistic world, a rigorous form of learning based on Aristotle's *inductive method*, which provides the foundation for what is today called the *scientific method*, took place in academies to which the majority did not have access. At another level of society, specialized crafts and technologies developed that were learned, then practiced, only by dedicated artisans. Thus, long before anything like true science existed, certain types of technical skills and knowledge were already reserved to a small group.

Today's science literacy crisis is something totally different, though. While an advanced understanding of any field will always be the private realm of specialists, in the past any educated layperson could learn about and keep up with basic scientific and technological topics. This is no longer the case, in part because of the exclusive communications system that science has developed. Remember from Chapter One that information is a key ingredient in science literacy. If the information available makes no sense to a person, it is of no use. Because the nature of scientific communication has changed so dramatically since the time of Copernicus, looking at the history of those changes can help us understand today's deficiencies in science literacy.[2]

450 YEARS OF SCIENTIFIC INFORMATION

Copernicus addressed the First Book of his historic 1543 tome *De Revolutionibus Orbium Caelestium* (On the Revolutions of the Heavenly Spheres) to the educated nonscientist of his day. In it, he described his heliocentric theory of the universe in style and language specifically intended to be read and understood by anybody with a general academic education. That he did so in a book that was otherwise extremely technical and mathematical, and which could be understood fully by just a handful, reveals much about the practice of science in that era. These educated nonscientists were the gentry, the scholars, the landowners, the nobility, and, not least of all, the ecclesiastical authorities of the day. While they were not scientists, by virtue of their learning they were in the general intellectual mainstream of sixteenth century Europe, and, further, they had money, power, and influence. Copernicus knew that the success of his theory depended in no small part on its appeal to these groups.

Although Copernicus's theory was not generally accepted until after his death, it was certainly known and discussed by many public factions. The idea that the Earth might not occupy the center of the universe was much more than a narrow scientific dispute—it was a major debate throughout all of late medieval civilization.[3] According to the noted historian of science Thomas Kuhn, "The significance of *De Revolutionibus* lies less in what it says itself than in what it caused others to say."[4] Much of the controversy centered around its theological ramifications, and at that level it was expressed through Church tractates and philosophical discourses, but it also forced an examination of religious doctrines that penetrated to all levels of society. The central issue was whether an astronomer could legitimately speculate about what Johannes Kepler called "the innermost form of nature." The revolution initiated by Copernicus was social as well as scientific. Science was on its way to becoming an institution, and that had a major impact on civilization as a whole. Nobody would be untouched by it.

Galileo's dispute with the Catholic Church symbolized the tensions between religious dogma and these new scientific theories. In 1633 an aged Galileo bowed before Pope Urban VIII and recanted his teachings of a sun-centered universe; this image has endured as a symbol of the conflict between rational science and religious faith. Although scholars continue to study and interpret the complex events that led up to the famous recantation, one factor that clearly concerned the Church was Galileo's enormous public popularity, and thus the degree to which his theories had reached the minds of the populace at large. His teachings were all the more dangerous for that very reason.

Galileo was a supreme popularizer. He was dedicated to the promotion of the experimental method in general and the Copernican system in particular. He ardently wished for all who could read to test his ideas. Instead of writing his major scientific treatises in jargon and mathematics (as he did with other works), he used his considerable literary skills to produce works that could be widely perused and discussed. Consider the popular appeal of the following passage from *Siderius Nuncius* (The Starry Messenger), where he reports the discovery of the moons of Jupiter, which he had beheld through his telescope:

There remains the matter which in my opinion deserves to be considered the most important of all—the disclosure of four PLANETS never seen from the Creation of the world until our own time. . . . Here we have a fine and elegant argument for quieting the doubts of those who, while accepting with tranquil mind the revolutions of the planets about the sun in the Copernican system, are mightily disturbed to have the moon alone revolve around the earth and accompany it in an annual rotation around the sun. . . . Now we have not just one planet rotating about another while both run through a great orbit around the sun; our own eyes show us four stars which wander around Jupiter as does the moon.[5]

Galileo recognized that this discovery was too significant to be communicated in the technical and typically understated language of science. Instead, he trumpeted the discovery. Popularizing his theories led to his greatness, and perhaps also his downfall.

A mere fifty years passed between Galileo's last published work and Isaac Newton's first, but during the interim science itself had become more institutionalized, and its means of communication had changed. Constitutionally, Galileo and Newton were worlds apart. While Galileo was robust and festive, Newton was frail and aloof. Moreover, their approaches to science were decidedly different. While Galileo presented his findings to all educated laypersons, Newton quite consciously wrote his master work, *The Mathematical Principles of Natural Philosophy* (or the *Principia*), in which he showed how calculus could be used to compute motion and physical forces, in a dense, heavily mathematicized prose that no nonphysicist could understand. He explained his motives for doing so:

> To avoid being baited by little smatterers in mathematics, I designedly made the *Principia* abstract; but yet so as to be understood by able mathematicians who, I imagine, by comprehending my demonstration would concur with my theory.[6]

These comments foreshadow the development of a new style of scientific writing, one that spoke solely and exclusively to peers. It thus fell to others to describe and interpret Newton's revolutionary work to the masses.

Although Newton himself disdained popularization, his work nonetheless had great public impact. Within his lifetime, popular renderings of the central ideas of the *Principia* began to be published, and these works, along with very popular public demonstrations of scientific experiments and new technologies, fomented an intellectual climate in which science came to be regarded as an authoritative way of knowing. British society first, then Europe and America, embraced a so-called natural philosophy. Newton's success at mathematizing the forces of push, pull, and acceleration inspired a belief that through the application of scientific principles all of the workings of nature could be predicted, understood, and thus harnessed. This, in turn, led to fervent entrepreneurship and technological development.[7]

Thus, the seventeenth century marked a period of tumultuous transition in how science was done and communicated. In his classic three-volume history of science, René Taton contended that "in less than a century—from Gilbert's *De Magnete* to Newton's *Principia*—the face of science had changed almost beyond recognition."[8] At around the turn of that century, any educated person could keep up with and comprehend virtually all published scientific works of the era. There was really no science per se, just natural philosophy, and it subsumed all disciplines and specialties. Over time, however, several trends and developments began to erode this monolithic, philosophical science. First were the great works and discoveries of the likes of Galileo, René Descartes, Francis Bacon, and Newton, whose ideas challenged the authority of beliefs that had gone unquestioned since antiquity, and in doing so gave scientists new license to speculate. This led to the development of the scientific method, a new way of thinking about and doing science. Further, new technologies, such as the telescope and Leeuwenhoek's microscope, opened entirely new vistas to experimental scrutiny. Finally, as previously mentioned, advances in mathematics, especially Newton's calculus, led to its being increasingly adopted as "the language of science." Science had become a profession; symbolic of this professionalization, the Royal Society was founded in 1660.[9]

The specialization of science meant that new, original information often could not be communicated directly to a lay audience. The scientific journal—a format designed for currency and specialization—emerged as the primary vehicle for conveying the results of new research, which were usually couched in technical language. The growing number of journals paralleled increases in the amount of research being done—a harbinger of today's "information explosion." (It has been suggested that a contributing factor to the great nineteenth century physiologist Johannes Muller's mental breakdown was his despair over being unable to keep abreast of the proliferating literature of his field.) In order to organize this information, the first indexes and review publications appeared in the early eighteenth century.[10]

Still, throughout the eighteenth century and into the nineteenth, much science remained intelligible to laypeople. Charles Darwin, for instance, wrote *The Origin of Species* with the idea that it would be read by biologists and nonbiologists alike. In the early nineteenth century armchair scientists still made major contributions in observational sciences, such as biology and earth sciences, but lacked the intellectual knowledge or access to the necessary technological equipment to make similar contributions to such fields as physics or chemistry. This created a widening information gap between scientists and the public. Popularizers such as Mary Somerville, who wrote the widely read *On the Connection of the Physical Sciences* (1846), appeared in increasing numbers to fill this gap. Further, several scientists took it upon themselves to popularize their own fields. Michael Faraday, for example, gave a series of public lectures, one entitled "The Chemistry of the Candle."[11]

In America, where a strong sense of science nationalism and a faith in technological progress had developed by the mid-nineteenth century, concentrated efforts were launched to debunk misconceptions and superstitions that impeded this progress. New publications such as *Scientific American* (1845) were inaugurated to further this cause.[12] There were significant substantive differences between many of these popular works and the primary works whose contents they reported. For example, although *Scientific American* featured popularizations written by scientists, many other forums contained material by journalists, educators, and civic leaders, who all too often had ulterior motives for writing, such as promoting a point of view or a political cause, or simply selling magazines.

This phenomenon is the subject of John Burnham's study *How Superstition Won and Science Lost*.[13] Nineteenth and early twentieth century scientists such as Faraday, Asa Gray, John Wesley Powell, John M. Coulter, F. W. Clarke, and Louis Agassiz were among a group Burnham called classic "men of science" (in this era, there were few women in science). These individuals maintained a perspective on science that transcended disciplinary fragmentation and, because of their almost evangelical enthusiasm for the scientific method, promoted the "religion of science" among a broad public. Their self-perceived role was to correct the errors of superstition. From the mid to late 1800s, these men and others like them were active popularizers; but with ever increasing specialization among the sciences, later generations were to produce fewer and fewer such figures. The task of presenting science to the public thus passed largely to those who lacked the technical education to do an adequate job.

Burnham traces four stages in the early popularization of science in the United States:

- Diffusion—when science did not need condensation, simplification and translation;
- Popularization—when men of science tried to share their vision of the religion of science.
- Dilution—when popularization passed into the hands of educators, who represented science only at second hand, and, simultaneously, journalists;
- Trivialization—when popular science consisted of impotent snippets of news, the product of authority figures.

The final stages of the process, he contends, had been reached by the middle of the twentieth century, and the result is evident in the public's poor science literacy today.

In the twentieth century, science has been popularized by an ever-increasing variety of popular media, the domain of "secondhand" science popularizers. Marcel LaFollette's study *Making Science Our Own* describes in detail the content of science features published in general interest magazines from 1910 to 1955, demonstrating how many contemporary images of science were formed.[14] Finally, science fiction emerged as a distinct literary genre. In part because of its popularity and in part because of a lack of information from more authoritative sources, it contributed much to the public's perceptions of what science could and could not do. Technology created numerous new vehicles for popular science, such as radio, motion pictures, and television.

In the years immediately after World War II, the new science of nuclear physics, which had abundantly displayed its potency at Hiroshima, stimulated a broad, media-based wave of popularization. Many of these images contained contradictions. On one hand, the accomplishments of the Manhattan Project were depicted as a triumph of human ingenuity, but on the other hand, the image of scientists suffered due to a public perception that, by intruding upon the affairs of God, they had unleashed a terrible force on humanity.[15] Another study explores in detail the multifarious images—some good, some bad, some frivolous—of the definitive scientist of the era, Albert Einstein.[16] He was widely admired for his genius, but also incorrectly faulted for having unwittingly set in motion the chain of events that led to the development of the Bomb. The time lent itself to wild speculation and fears, which were expressed in all varieties of media.

In *The New Priesthood* (1965) Ralph Lapp identified, for perhaps the first time, the potential danger to American democracy of a general public that lacked a basic understanding of science. Lapp suggested that Americans tended to regard scientists as latter-day Sadducean priests endowed with esoteric knowledge and decision-making authority.[17] These were to become major concerns in the post-Sputnik era, and political initiatives were launched in an attempt to reassert America's scientific ascendancy and competitiveness. Those initiatives do not seem to have had a long-term impact, however. The very same concerns were echoed in Rustrum Roy's 1986 editorial in *Bioscience*: "Science will die as a vital culture shaping force for the same reason that theology (not religion) died . . . a few centuries ago—it became too precious, the province of an elite priesthood."[18]

More recently, the commercial enterprise of science popularization has seen some ups and downs. The apex of the boom cycle may have come in the late 1970s and early 1980s with the inception of twenty new general science magazines (including *Omni*, *Discover*, *Science 80*, and a revamped *Science Digest*), seventeen new television shows (including *Nova*, *Omni*, *Walter Cronkite's Universe*, and such PBS series as *Cosmos* and *The Ascent of Man*), and more than sixty newspaper sections dedicated to popular science.[19] A *Time* magazine cover story on *Cosmos* creator Carl Sagan declared that "ennui" about popular science "has turned into enthusiasm."[20] Former Fermilab director Robert Wilson called popularizations "the new literature of science" that would integrate a "technology of humanism into the common culture."[21] Some of these ventures were short-lived because of market saturation and lack of advertising revenue. Nevertheless, if science popularization did not emerge as a blockbuster industry, it did prove that it can attract and sustain an audience.[22]

Recognizing how the general public learns about science is essential for understanding why science literacy in the United States is low (and, also, perhaps, why so many students eschew science studies). The general public acquires meaningful information (or *mis*information) about science through various media. Invariably, whenever professional scientists speak of the need to improve science literacy, they call for more and better popularization. Thus, America's science literacy can only be as good as the quality of the information that is available and the means by which it is sought and used. This phenomenon is described by Oscar Handlin in his historical study of science and popular culture:

Since the explanation of the scientists was remote and incomprehensible, a large part of the population satisfied its need for knowing in its own way. Side by side with the formally defined science there appeared a popular science, vague, undisciplined, unordered, and yet extremely influential. It touched upon the science of scientists, but did not accept its limits. And it more than adequately met the requirements of the people because it could more easily accommodate the traditional knowledge to which they clung.[23]

Clearly, popular science should not be presented as science per se—it is an *interpretation* of science. As with any interpretation, it can be more or less accurate and authoritative. Where it is less so, distortions and misunderstandings arise, and science literacy suffers. However, information presented at an incomprehensibly technical level cannot enlighten any but the initiated and has no direct public impact. The ability to distinguish the good from the bad, the appropriate from the inappropriate in works of popular science is a prerequisite for achieving science literacy. This suggests a major role for librarians in carefully evaluating works of popular science just as they would evaluate any other information resources.

EFFECTIVE SCIENCE POPULARIZATION

This brief historical discussion of scientific communications shows why science popularizations are needed and wanted, and how they can influence society and culture. It also addresses some of the problems that can occur when scientific information is translated and interpreted for the general public. Plain English is not the language of science. Today original scientific information is conveyed through vehicles and in a language that are practically meaningless to the layperson. When this information is popularized, it undergoes a fundamental change, and in the process distortions and errors can be introduced. Science literacy suffers.

Popularizing science requires some unique and extraordinary communications skills. Popularizing science is akin to translating information from one language to another; as anybody with even a smattering of knowledge of another language knows, some concepts simply do not have exact matches in English. This is also true with scientific information. Physicists, for example, speak of "quarks," which have "color," "flavor," and "spin," but to the nonphysicist the word "quark" sounds like nonsense, and the adjectives used to describe its properties have common meanings that have nothing to do with what the physicist is trying to convey.[24]

The task of the science popularizer is to take information that has no parallels in everyday life and convey it in a way that can be readily understood without corrupting the integrity of the original information. Effective popular science strikes a delicate balance between technical accuracy and easy comprehension. In his "How to Write a Popular Scientific Article," J.B.S. Haldane, a brilliant geneticist and a prolific popularizer, wrote: "You must . . . know a great deal more about your subject than you put on paper. . . . It means that you must constantly return from the unfamiliar facts of science to the familiar facts of everyday experience."[25] He might have added that in doing so you must take care not to misrepresent or trivialize the information. Stephen Jay Gould explains: "The rules are simple: no compromises with conceptual richness; no bypassing of ambiguity or ignorance; and removal of jargon, of course, but no dumbing down of ideas."[26] In a nutshell, effective popular science must adhere to Haldane's demand for clarity and Gould's "rules" on accuracy. This is not an easy prescription.

Effective popular science, in any medium, results from the skilled application of communications techniques. The librarian who is familiar with the genre and who recognizes these techniques is better able to evaluate the quality of a work of popular science than one who is not. Knowledge of science is not a prerequisite to making sound collection development decisions for these materials. Knowledge of the characteristics of effective popular science is, however. The next chapter discusses the nature of popular science and the means by which it can be most successfully communicated.

NOTES

1. Oscar Handlin, "Science and Technology in Popular Culture," *Daedalus* 94 (Winter 1965): 156–170; Annette Woodlief, "Science," in Thomas Inge, ed., *Concise Histories of American Popular Culture* (Westport, CT: Greenwood Press, 1982), 354–362.

2. Noriss S. Hetherington, "The History of Science and the Teaching of Science Literacy," *Journal of Thought* 17 (Summer 1982): 53–66.

3. Toby C. Huff, *The Rise of Modern Science* (New York: Cambridge University Press, 1993), chap. 9.

4. Thomas Kuhn, *The Copernican Revolution* (Cambridge, MA: Harvard University Press, 1957), chap. 6.

5. Translated by Stillman Drake in *Discoveries and Opinions of Galileo* (New York: Anchor, 1957), 50.

6. Newton is cited in Joseph Schwartz, *The Creative Moment: How Science Made Itself Alien to Modern Culture* (New York: HarperCollins, 1992), chap. 1.

7. Larry Stewart, *The Rise of Public Science* (New York: Cambridge University Press, 1992).

8. René Taton, ed. *History of Science: The Beginnings of Modern Science, from 1450 to 1800*, vol. 2. 3 vols. (New York: Basic, 1964), 393.

9. Michael Hunter, *Establishing the New Science: The Experiences of the Early Royal Society* (Woodbridge, England: Boydell, 1989), 1–27.

10. A. A. Marten, "Developments of European Scientific Journal Publishing Before 1850," in A. J. Meadows, ed., *Development of Science Publishing in Europe* (Amsterdam: Elsevier, 1980); 1–23; David Kronick, *A History of Scientific and Technical Periodicals* (New York: Scarecrow, 1962).

11. Jack Meadows, "The Growth of Science Popularization," *Impact of Science on Society* 36 (Winter 1986): 341–346.

12. Matthew D. Whalen, "Science, the Public and American Culture," *Journal of American Culture* 4 (1981): 14–26.

13. John C. Burnham, *How Superstition Won and Science Lost: Popularizing Science and Health in the United States* (New Brunswick, NJ: Rutgers University Press, 1987), 226–262.

14. Marcel C. LaFollette, *Making Science Our Own* (Chicago: University of Chicago Press, 1990).

15. Paul Boyer, *By the Bomb's Early Light: American Thought and Culture at the Dawn of the Atomic Age* (New York: Pantheon, 1983).

16. Alan J. Friedman, and Carol C. Donley, *Einstein in Myth and Muse* (Cambridge: Cambridge University Press, 1983).

17. Ralph Lapp, *The New Priesthood: The Scientific Elite and the Uses of Power* (New York: Harper, 1965).

18. Rustrum Roy, "Science Must Change or Self Destruct," *Bioscience* 36 (November 1986): 660–661.

19. Bruce V. Lewenstein, "Was There Really a Popular Science Boom?," *Science, Technology and Human Values* 12 (Spring 1987): 29–41.

20. "The Cosmic Explainer," *Time* 116 (October 20, 1980): 62–69.

21. Robert Wilson, "The New Literature of Science," *Bulletin of the Atomic Scientists* 37 (April 1981): 1–2.

22. Bruce V. Lewenstein, "Why Isn't Popular Science More Popular?," *American Scientist* 76 (September/October 1988): 447–449.

23. Handlin, "Science and Technology in Popular Culture."

24. Gregory, Bruce. *Inventing Reality: Physics as Language* (New York: Wiley, 1988).

25. J.B.S. Haldane, "How to Write a Popular Scientific Article," in *On Being the Right Size and Other Essays* (Oxford: Oxford University Press, 1985), 154–160.

26. Gould, *Bully for Brontosaurus*, preface, 11.

Chapter Three

Understanding Popular Science
Information Resources

People get information about science from many resources. Librarians are quite familiar with most of these resources, which include books, periodicals, newspapers, and videotapes. Likewise, librarians are generally knowledgeable about the electronic and networked resources that will become increasingly important in the promotion of science literacy in the future. Substantively, however, the stuff of popular science—the information content itself and how it is presented—is unique and often unfamiliar to librarians. That makes evaluating these resources as part of collection development decisions difficult and sometimes confusing.

This chapter discusses what makes popular science information unique and how to evaluate its accuracy, effectiveness, readability, and usefulness. The discussion moves from the general and theoretical to the specific and concrete. Throughout, it focuses on how the principles described here can be applied in building library collections in support of lifetime—that is, high school to adult— science literacy. Examples are given where appropriate.

An effective work of popular science must inform, to be sure, but it must also entertain and stimulate the imagination. It must be honest and accurate, but it must also make some concessions to render technical material comprehensible to the nonscientist. It deals with facts and theories, which may sometimes seem impersonal, but its human side must come through. Although scientific information reaches the public in many formats, each with unique characteristics, they share many of the same stylistic and substantive issues whether the format be hard copy, video, electronic, or other. While these principles are generally descriptive of written language, they apply to other media as well.

AN INTRODUCTION TO THE THEORY OF POPULAR SCIENCE

In 1959 British chemist C. P. Snow observed in *The Two Cultures* that a chasm was widening between the culture of the humanities and that of the sciences.[1] In academe he saw a "gulf of incomprehensibility" between literary humanists and mathematical scientists. The two groups didn't speak the same language and thus failed to understand, and sometimes distrusted, each other. He also found this rift in society at large. Today, probably the clearest evidence of the polarization of the two cultures is the crisis in science literacy.

When scientific information is popularized, it is translated from the language of the scientific culture into that of the humanistic culture. In order to understand what happens during this process, it is first useful to examine some of the peculiar characteristics of primary scientific literature.

The voice of science might be said to speak in monotone. That statement is not meant to be disparaging: it is a well-known and purposeful phenomenon. The prototype of the modern scientific article contains several prescribed components and is written in a deliberately neutral style. A recent handbook on writing science for professional publication unabashedly recommends adhering to the structural "recipe" of a science paper.[2] Its ingredients include an *introduction*, which states the problem and elucidates its context in the form of a literature review; an account of *methods*, an explicit description of how the researchers performed their investigation of the problem; a section describing the *results* of the experiment; and finally a concluding *discussion*, where the significance of these results is expounded and, almost always, an appeal for future research is issued. Throughout, maximum clarity is sought in the writing style, and for that reason rhetorical devices such as metaphors or figures of speech are seldom used. Mathematics, which is by its nature unencumbered by the nuances of language, is especially valued in this regard. The aim is to provide a replicable, positivistic account of something that occurred in nature. The product is a form of writing that one critic has called "literary novocain."[3]

When scientists refer to the "literature" of their fields, they do not use the word in its ordinary sense. Few read scientific research for an aesthetic experience, nor is it studied for its literary merit. Instead, scientific literature refers to a body of knowledge, which exists *as* knowledge, and the written words that convey new information are meant to do so clearly and precisely, without embellishment on the part of the author. Many literary critics feel that this is as it should be.[4] Their views are basically revisitations of Snow's "two cultures" theory, although, unlike Snow, they do not lament the chasm between the arts and sciences, but rather see it as a natural cleavage that has occurred as the poles of human knowledge have widened.

Recently, however, a countertradition has arisen in the literary analysis of scientific writing. These critics have argued that despite its consciously neutral voice, science writing still exhibits many of the characteristics of humanistic literature. For example, in his book *Science as Writing*, David Locke gives examples from many primary sources, including articles he has published in professional chemistry journals, to show how science writing can be rhetorical, expressionistic, and even artful.[5] In conclusion, he wonders if, in an era in which science acknowledges the existence of quantum uncertainties and confidently searches for "beautiful" theories, the cookbook approach to writing can be adequate for communicating these very abstract concepts.

Another trend in the study of scientific literature examines the historical and literary processes by which the current prototype of the scientific paper became accepted. Bazerman divides scientific writing into genres, such as letters, literature reviews, theoretical articles, handbooks and reference works, and various forms of pedagogical writings. From his analyses of articles appearing in the *Transactions of the Royal Society* from 1665 to 1800, he sees one genre in particular, the experimental report, gaining ascendancy during that period.[6] Within other genres, opinion, interpretation, and speculation were legitimate, and the rhetorical devices by which such perspectives could be communicated were commonly employed. Conversely, the experimental report developed as "representational" writing, a sequential and factual account of what was done and what was observed, using language that is as direct and unambiguous as possible. The phenomena being reported were what was important, and literary ornamentation on the part of the author was at best superfluous, at worst obstructive.

The style of writing that became favored by experimentalists and which has evolved into the standard voice of science today has certain limitations. It is the style of literary self-denial; as Brazerman suggests, it "hides." In its neutrality, it might be "representational," in the sense of providing an unbiased chronicle of natural phenomena, but it is entirely inadequate as an honest representation of the humanity of the scientific process. The reliance on jargon and specialized mathematics, both of which are inscrutable to nonscientists (and, because of scientific specialization, often even to scientists from other fields), further limits the range and scope of scientific writing. It can have no impact beyond a small, insular group.

Popular science theorists distinguish between "knowledge producers" (research scientists), "knowledge intermediaries" (popularizers), and "knowledge acquirers" (any attentive public).[7] The communication of information from the producers to the acquirers—a process called *knowledge transfer*—almost always requires an intermediary action. This is the case even when the scientist is his/her own popularizer, because, except in a few cases, the information is presented in popular form only after it has been published in technical journals.

The intermediary may or may not be a knowledge producer. For example, the intermediary may be a scientist seeking to popularize general knowledge within the discipline, or, conversely, a journalist or an educator with little or no formal academic training in the subject area. In the same way, the intermediary may or may not be a knowledge acquirer. Where physics is concerned, for example, the physicist cannot be merely a curious layperson. If the intermediary is a nonscientist, however, there is likely to be much similarity between his or her intellectual interests and those of the intended audience. In any case, the role of the intermediary is critical, because breakdowns can occur in the integrity of the information when it is transmitted.

The knowledge acquirers are extremely heterogeneous. Some are high school or college students. Some are intelligent, curious laypersons. Some are concerned citizens or consumers who are seeking information for practical purposes. Although these audiences are distinct and differently motivated, at a basic level many popular science books will serve all their needs equally well. These books are consciously written at a level appropriate for the broadest possible audience; they assume no or little prior knowledge of the subject area. Such books are said to have low degrees of "formalisation and technical precision."[8]

Sometimes, however, the knowledge acquirer might be a high school student seeking admission to a prestigious college science program, or a teacher with a vocational interest in a subject, or even a scientist from another discipline. These audiences are more knowledgeable; while probably unable to make much sense of the primary literature in the field, they will still seek information at their own somewhat more advanced technical levels. Many popular science books are written for these audiences. The concept of *levels of relevance* is important for evaluating works of popular science and can be used productively in promoting lifetime science literacy.

Finally, it is helpful to consider just what "knowledge" is being transferred. All information exists within a context. In the primary journals, the context is established by the references cited, which reveal the author's influences and where this new piece of research stands in respect to existing knowledge in the field. Having thus established a context, the author describes the research methods employed, then generally expounds upon the significance of the research in subsequent discussion or conclusion sections of the paper. It is for peer reviewers to assess the quality of the research. In works of popular science, however, direct references are not always made to the primary literature (although there may be bibliographies and reading lists), and there is nothing so rigid as a peer review process. The context, significance, and authority of the information must be established in other ways.

The preceding issues are among the most conspicuous discussed by popular science theorists. In practice, writers and producers of popular science materials confront these issues in numerous ways. Librarians familiar with the workings of popular science can make more informed collection development decisions. The following sections examine some of these factors by considering issues related to (1) the communication techniques of popular science, and (2) the content of a work of popular science.

COMMUNICATIONS TECHNIQUES FOR POPULAR SCIENCE

Popularizing science is a process of knowledge transfer from the primary, technical literature of a field written for fellow knowledge producers into the language of an entirely different group, the knowledge acquirers. By necessity, this information changes during the process. Such changes can take many forms, depending, generally, upon the intent of the author and the knowledge level of the target audience. Invariably, the greater the knowledge gap between the knowledge producers and the knowledge acquirers, the more the information must be refashioned in order to make sense. (In much the same way, it is generally more difficult to conceptually translate non-Western languages into English.) Thus, the science popularizer must first assess who comprises the intended audiences and how the information can best be presented to be meaningful to them.

The authors of many popular science books identify their target audience in the preface, using terms like "generally intelligent laypersons," "curious non-specialists," or "readers with no prior knowledge." In the introduction to his book *Why the Reckless Survive* psychologist Melvin Konner puts it nicely:

This book . . . is designed to stand up to academic scrutiny without letting the general reader see the effort involved in gaining that legitimacy. Who is this general reader? Someone who appreciates the subtleties of language but does not want to keep getting up to go to the dictionary. Someone who knows perfectly well that complex ideas can usually be expressed without gobbledygook. Someone who wants to feel that reading is conversation, not lecture or catechism. In retrospect, it seems odd that I should have doubted in the slightest the existence of such people, since I myself have been one since childhood.[9]

He intends his book to be accessible to anybody, from young adult to senior citizen, who is willing to spend some time with it. The underlying assumption, simplistic as it might sound, is that these people are *motivated* to read the book; they are legitimately interested in and want to learn about its subject.

This is not, incidentally, the most basic level of science popularization. The level of the least common denominator, represented by some of the writing that appears in newspapers and general interest magazines, does not assume that the readers are necessarily motivated to read the material. Thus, in order to lure readers away from the comics or entertainment sections, writers sometimes give their features attention-grabbing headlines and write in styles that range from whimsical to melodramatic. This is not to demean the role of magazines and newspapers in popularizing science; they can be very important and effective in that regard. The point, rather, is to identify yet another level at which popular science materials can be presented, so that librarians who are developing collections that include these materials can be aware of their strengths, weaknesses, and appropriate uses.

An almost universal concession that the author must make for nonspecialist audiences involves the handling of jargon and mathematics. Where jargon is concerned, look for statements to the effect that "technical terms will be kept to a minimum" or "scientific terms will be explained as they occur." An index or, better, a glossary enhances any science popularization.

Mathematics can be even more problematic, since vocabulary can be learned, but mathematical notation, essential to science, is often alien to a general audience. In the preface to his bestseller *A Brief History of Time*, Stephen Hawking remarked somewhat wryly that his publisher had warned him that the book's sales would decrease by half for every mathematical equation he included; but in the second chapter he devoted several pages to describing, in prose, the revolution in physics that Newton started by his invention of calculus.[10] The point is that mathematical notation may be used sparingly, if at all, but at the same time the author must pay mathematics its due in order to give readers a true sense of how science is done. Any mathematical constructs that are used frequently—such as the convention for denoting powers of ten or measurements from the Kelvin scale—should be explained explicitly and early in the text.

The terms "scispeak" and "technobabble" have been coined to describe how scientific information can look and sound to a layperson.[11] Scispeak is characterized not only by a high jargon content, but also by an impersonal style, long sentences, and a preponderance of polysyllables. Take an issue of any recent scientific research journal, open it to a randomly selected article, and read the title, the abstract, and perhaps a few sentences. What you are reading might be valuable information to the scientist, but to a novice it will probably seem like just so much scispeak.

The scispeak found in research articles must be dismantled, polished, and rephrased for general readers. In a 1993 article, Jeanne Fahnestock examined and compared original scientific writing with subsequent popular versions of the same information.[12] An example from her work is illustrative. The original article read:

A similar analysis performed on the DNA taken from either the patient's normal bladder adjacent to the tumor, or from peripheral blood leukocytes, showed the same two bands at 410 and 355 nucleotides, indicating the presence of the same two alleles as were present in the patient's carcinoma. . . . Thus, the alteration identified in this gene at the Nae I or Msp I site by restriction enzyme cleavage appears to be in the germ line and must have existed before development of the bladder carcinoma. . . . Thus, it is tempting to speculate that there is an association between this point mutation in the $c\text{-ras}_1^H$ gene and the bladder carcinoma. Although we have no information at present regarding the frequency of the mutant $c\text{-ras}_1^H$ gene in bladder tumors, we do know that this change is infrequent in the general population since analysis of DNA from 34 individuals revealed the presence of Msp I/Hpa II site. (Muschel et al. *Science* 219:855)

Here is the popularized account of the same information:

Researchers from the National Cancer Institute and Yale University Medical School believe they have found, in both normal and diseased cells of a bladder cancer patient, a mutant gene that may have caused his malignancy. The findings indicate that people may inherit certain genes that predispose them to developing some kinds of cancer. ("A Cancer Gene" *Science 83* April, 1983: 10)

Fahnestock calls the obvious translation that has taken place here a "genre shift." She also notes that the conclusions of these two articles are not in absolute agreement, which raises another issue (to be discussed in the next section) about how to assess the authority of popular science, but for now consider just the stylistic differences.

In their handbook *Communicating Science*, Michael Shortland and Jane Gregory state unambiguously, "In popular science writing, the top priority is communication with the readers."[13] The language must be clear, vivid, and logically ordered. Thus, somewhat shorter sentences are favored, especially when making a strong point, but this should not be overdone, since studies have shown that variability of sentence length within a paragraph aids overall comprehension. A good rule of thumb is that, when reading a sentence aloud, you should reach the end (or at least some logical and grammatical break, such as a comma) before you need to take a breath. Likewise, words and phrases should be chosen for brevity; "elucidate" can become "tell," "endeavor" can become "try," "due to the fact that" can become "because," and "at some time in the future" can become "later." Generally, the active voice is preferred.

Shortland and Gregory cite the following two passages, both from a 1926 popularization by the notable British physicist Arthur Eddington, as exemplifying the effects of sentence length on comprehension. The first is an original draft; the second is his revision:

1. In some of the hottest stars a series of lines known as the Pickering Series was discovered in 1896, which is spaced on precisely the same regular plan but the lines fall half way between the lines of the Balmer Series—not exactly half way because of the gradually diminishing intervals from right and left, but just where one would interpolate lines in order to double their number whilst keeping the spacing regular, though unlike the Balmer Series, the Pickering Series has never been produced in any laboratory: so what element was causing it?

2. In some of the hottest stars a series of lines known as the Pickering Series was discovered in 1896. This is spaced on precisely the same regular plan, but the lines fall half way between the lines of the Balmer Series—not exactly half way because of the gradually diminishing intervals from right and left, but just where one would interpolate lines in order to double their number whilst keeping the spacing regular. Unlike the Balmer Series, the Pickering Series has never been produced in any laboratory. What element was causing it?[14]

In the first example, the entire passage is in a single sentence. The revised version contains four sentences that vary in length from fifty-four to five words. Notice, in particular, how the shorter sentences in the second draft add emphasis.

George Gopen and Judith Swan have studied mechanical factors that affect reader understanding of scientific information, and they found that confusion often results when the sentences are not built according to reader expectations.[15] For example, the subject of a sentence should be placed at a natural "stress point,'' which is generally near the beginning, but might follow a comma or be anyplace where the reader perceives a mental break in the text. A big problem with overlong sentences is that there might be several stress points, and the main one can get lost. A second area where confusion arises is when there are in-adequate linkages between concepts, so that the context is lost from one sentence or paragraph to another.

Gopen and Swan suggest the following simple structural principles for pop-ular science writing:

1. Follow a grammatical subject as soon as possible with its verb.
2. Place in the stress position the ''new information'' you want the reader to empha-size.
3. Place the person or thing whose ''story'' a sentence is telling at the beginning of the sentence, in the topic position.
4. Place appropriate ''old information'' in the topic position for linkage backward and contextualization forward.
5. Articulate the action of every clause or sentence in its verb.
6. In general, provide context for your reader before asking that reader to consider anything new.
7. In general, try to ensure that the relative emphases of the substance coincide with the relative expectations for emphasis raised by the structure.

While they give these guidelines, the authors emphatically state that these are by no means *rules*—in fact, there are none, and some of the most creative stylists of popular science take grammatical liberties when there are good reasons for doing so. That, too, is a difference between primary and popular scientific writ-ing; the former is rigid and prescriptive, while the latter is fluid and open-ended.

Readers also have a right to expect a writer of a popular science book or article to demonstrate a certain degree of literary flair. This kind of writing must engage readers at a personal level. For example, while analogies and metaphors are rare in the primary literature, they can be extremely effective in helping readers to envision the meanings of certain scientific concepts. J.B.S. Haldane wrote:

It is good to start from a known fact, say a bomb explosion, a bird's song, or a cheese. This will enable you to illustrate some scientific principle. But here again take a familiar analogy. Compare the production of hot gas in the bomb to that of steam in a kettle, the changes which occur in the bird each year to those which take place in men once in a lifetime at puberty, the precipitation of casein by calcium salts to the formation of soap suds. If you know enough, you will be able to proceed to your goal in a series of hops rather than a single long jump.[16]

Another famous and useful analogy explains why all galaxies in the expanding universe are retreating from one another by comparing them to raisins in a rising loaf of bread. Literary constructs such as these not only make concepts easier to understand, but are vivid, lively, and entertaining.

Some writers are quite witty and employ all sorts of quirky but uniquely effective rhetorical devices to get their point across. In the following excerpt from *Dreams of a Final Theory*, Nobelist Stephen Weinberg instructs his readers in the niceties of quantum mechanics by imagining the following exchange between Tiny Tim and Ebenezer Scrooge:

Tiny Tim: I think quantum mechanics is just wonderful. I never did like the way that in Newtonian mechanics if you knew the position and velocity of every particle at one moment you could predict everything about the future, with no room for free will and no special role for humans at all. . . .

Scrooge: Bah! I may have changed my mind about Christmas, but I still know humbug when I hear it. It is true enough that the electron does not have a definite position and momentum at the same time, but this just means that these are not appropriate quantities to use when describing the electron. What an electron or any collection of particles does have at any time is a wave function. . . . The evolution of wave function is just as deterministic as the orbits of particles in Newtonian mechanics. . . . Where's your free will now![17]

Weinberg's mischievous enthusiasm for the subject is not unique among scientists; some, in fact, are noted eccentrics, and the more colorful sides of their personalities emerge to good effect in their popular writings.

Another technique commonly used by science popularizers, especially when writing for an audience with no prior knowledge of the subject, is "staging" or "pacing." The idea is simple: present technical material sequentially, in some ordered manner, and provide breaks, interludes of relatively easier discourse, between the more difficult passages. Commonly, the sequence is chronological, with new discoveries depicted in the order that they were made. Another typical way to organize a popular science exposition is by increasing levels of technical complexity. (Fortunately, since knowledge in science generally progresses over time, these two approaches are complementary.) Finally, as the material becomes more difficult, breaks become more and more important. Anecdotes can be effective, especially those that show a human side to scientific work. (Use of the first person, seldom acceptable in primary science writing, is common and welcome in popular science.) Frequent section breaks are also useful.

The following passage, from a section called "Jello-O Waves" in A. Zee's *An Old Man's Toy*, a popularization of gravity's influence in the universe, illustrates several of the techniques and principles heretofore described. It is also fun to read:

In Einstein's theory, gravity is the manifestation of curved space time. Space time can be squeezed and stretched. The Soviet physicist and humanitarian Andrei Sakharov once likened space to an elastic medium that bends and warps in response to the presence of matter, much like a piece of Jell-O in the hands of a three year old. One characteristic of the elastic media is that waves can propagate in them. Poke a piece of Jell-O, and it will shake and roll in indignation. Imagine a very long piece of Jell-O. By poking and tapping the Jell-O at one end, you can send a wave of Jell-O vibration travelling down to the other end. I suppose you can communicate with a friend in that way.

Similarly, waves can propagate in the elastic medium that is space. . . . How would we know when a gravity wave comes by? Well, how would a bacterium on the Jell-O know that a Jell-O wave has come by? As the wave comes by, the Jell-O shakes and rolls and the bacterium moves accordingly. In general, this is how you would detect any kind of wave. Sitting in a small boat on a calm day, when you bob up and down you know that a wave from a passing speedboat has come by. . . .

Similarly, when a gravitational wave comes by, space will warp and unwarp and things will move in response. Sitting on a small planet on a calm day, you know that a gravity wave has come by when your planet bobs up and down.[18]

Zee's description of gravity waves is an example of writing suitable for just about any generally intelligent layperson. Suppose, however, that a reader has just finished Zee's book and, with curiosity stimulated, wishes to proceed to a somewhat more technical treatment. Scientific information is communicated along a spectrum that ranges from the extremely technical jargon of the primary research journal to the comfortable vernacular of a newspaper article. There are many levels of relevance in between. Consumers of this information can thus move from the basic levels to progressively more advanced ones as their curiosity grows. At the more technical levels, readers can become more challenged and enriched by the information they receive. To librarians, the phenomenon of levels of relevance suggests possibilities for reference services and reading programs.

Obviously, in a science popularization written for a more advanced readership, the author might take greater liberties with jargon, mathematics, sentence structure, and literary style. All of the previously mentioned guidelines still apply, but at a higher level of sophistication. The following description of Einstein's gravitational theory comes from John Barrow's book *Theories of Everything*; because the subjects are similar, it is useful to compare it to Zee's description as an example of the next step on the popular science ladder:

There may be a good deal more to the universe than meets the eye, even the eye of faith of the cosmologist. Einstein's theory of gravitation has taught us that the notion of force may be nothing more than a convenient anthropomorphism. The classical picture of physical laws sees them as sets of rules which dictate how particles respond to the action of certain "forces" between them when the particles are set down in the traditional space whose geometry was laid bare by Euclid. Einstein's general theory of relativity provided us with a picture of gravitation that was altogether more sophisticated. The presence of the particles of matter, and their motion, determine the local topography of the space in which they sit. No longer are there mysterious forces acting between neighboring bodies. Each now moves along the most economical path available to it on the undulating space created by all particles in the universe. Thus the Sun creates a large ditch in the space near the Earth, and the Earth moves around the inside surface of that ditch. This path we call its orbit. There are not gravitational "forces" acting between distant objects. Everything takes its marching orders from the spatial topography of its immediate locale.[19]

There is nothing terribly difficult to comprehend here. Still, the style and structure of the language are a bit more complex. The third person is used. Some prior knowledge (e.g., who Euclid was) is expected. Some figures of speech are used (e.g., the Sun creates a gravity "ditch"), but the analogies are not as playful as Zee's "Jell-O waves." In short, Barrow's is the kind of book to which a reader might turn after reading Zee's. That, in turn, is a good reason why many science popularizations include lists of further readings, and why librarians should pay attention to them.

Finally, before turning to a discussion of the content of effective science popularizations, mention should be made of illustrations. These can be used effectively in three ways. First, they can clarify: some of the concepts of modern science are difficult to capture in words, and a well-conceived drawing or schematic can do much to help the reader visualize an elusive idea. When the purpose of an illustration is to clarify, it should make intuitive sense on first glance, and, on reading the caption (also essential), its meaning should be absolutely clear. Second, much of modern science has striking visual dimensions, and the inclusion of, for example, a computer generated fractal image or an infrared satellite photograph of the Earth can do much to enhance a reader's appreciation of the artistic side of science. Third, illustrations can simply be used for a break, provided they are not used gratuitously. A cartoon or a lighthearted line drawing clearly related to the text can encourage readers onward to more difficult passages.

The best science popularizers must be skilled in the use of some distinct communications techniques. Still, popular science books are not, generally, beach reading; the material is usually difficult, sometimes very much so. The real trick is to use these techniques to make difficult technical content comprehensible to lay audiences. The next section looks more closely at the content of science popularizations, what happens to it when it is interpreted for general audiences, and how knowledge of these processes can aid librarians in developing science literacy collections.

THE CONTENT OF POPULAR SCIENCE WORKS

Keeping up with the literature is a major, ongoing responsibility of any scientist. Probably the most essential task involves regularly examining the primary journals that the scientist has identified as central to his/her research interests. Through these core journals groundbreaking research is presented for review and, thus, knowledge is furthered. The articles published in these journals are by and for knowledge producers and are written in the technical language they share. All other audiences are extraneous, because they are not expected to be part of the scholarly dialogue at this level.

In different ways, aficionados of popular science also seek to keep up with the literature. For them, however, the objective is not so much to participate in the production and qualitative review of new knowledge (although, in a democracy, they can influence research agendas through citizen activities); rather, they wish to keep abreast of issues, trends, controversies, and frontiers. Their motives vary from simple curiosity to activist passion, but in any case, they wish to use this information for different purposes than the scientists. Style and language are not the only things that change when scientific information is popularized; content does too.

Jeanne Fahnestock has identified several ways that popular writings "accommodate" science.[20] Perhaps the most significant and potentially insidious is a tendency for popular science works to gloss over or omit altogether situations where a scientist hedges or qualifies research conclusions. The following examples are enlightening. In the concluding statement of the original article, published in *Science* in 1980, the authors stated (italics added):

We favor the hypothesis that sex differences in achievement in and attitude toward mathematics result from superior male mathematical ability, which may in turn be related to greater male mathematical ability in spatial tasks. This male superiority is *probably* an expression of a combination of both endogenous and exogenous variables. *We recognize, however, that our data are consistent with numerous alternative hypotheses.* (Benbow & Stanley *Science* 210:1262–64)

In three popular versions of this article, the accounts were as follows:

The authors' conclusion: "Sex differences in achievement in and attitude toward mathematics result from superior male mathematical ability." ("Do Males Have a Math Gene?" *Newsweek* Dec. 15, 1980: 73)

According to its authors . . . males inherently have more mathematical ability than females. ("The Gender Factor in Math" *Time* Dec. 15, 1980: 57)

Two psychologists said yesterday that boys are better than girls in mathematical reasoning, and they urged educators to accept the possibility that something more than social factors may be responsible. ("Are Boys Better at Math?" *New York Times* Dec. 7, 1980: 107)

In the original research, the authors' conclusions were far less certain and definitive than those presented in the popular accounts. This, Fahnestock found, is common, and she suggests that it happens because lay audiences do not expect ambiguity from science; they expect facts. The danger, however, is that it dilutes science literacy and unwittingly creates misconceptions about how science works.

Another study confirmed Fahnestock's conclusions and made some new findings. In his book *Writing Biology: Texts in the Social Construction of Scientific Knowledge*, Greg Myers analyzed in great depth what he called "The Cnemidophorus File."[21] In 1980 two biologists published an article entitled "Sexual Behavior in Parthogenetic Lizards (Cnemidophorus)" in the *Proceedings of the National Academy of Science* (January 1980: 500). In it, the researchers examined certain physiological activities of a lizard species in which reproduction is conducted through female eggs, without males. When this article was later reported for the general public in *Time* magazine (February 18, 1981: 18), however, the title had become "Leapin' Lizards: Lesbian Reptiles Act Like Males." The tone of the work was changed drastically. Myers also found that these two articles exhibited an interesting paradox: scientific articles published in technical journals tend to emphasize methodology and technique, while popular science articles reporting the same content focus instead on the phenomenon being observed, sometimes to the point of portraying the scientists as passive observers and recorders of natural events, almost as though the scientists did nothing but stand back and watch.

Thus, Fahnestock found that some science popularizations misrepresent the conclusions of science by making them seem absolute and definitive, while Myers found that they also misrepresent the work of scientists by deemphasizing the rigors of experimental technique. Both found evidence of sensationalism in reporting. These findings show why poorly conceived popular science can impede rather than promote science literacy.

Why this trivialization of science occurs is debatable. Some argue that the intermediaries who are guilty of this are simply giving people what they want. That assumption, however, does not explain the existence of quality science popularizations and the substantial market they command. It might be, rather, that superficial works of popular science are seeking a different audience—the generally nonattentive public. In her book *Selling Science: How the Press Covers Science and Technology*, Dorothy Nelkin analyzes the editorial and economic pressures that science popularizers face when writing for the mass market.[22] She found that research results are sometimes overblown to make better headlines; for example, "breakthroughs" and "magic bullets" might be prematurely promised.

Likewise, research findings are sometimes spiced up or sensationalized. In this vein, Melvin Konner reflects on what happened after his paper, "Paleolithic Nutrition: A Consideration of Its Nature and Current Implications," was published in *New England Journal of Medicine*:

Days before we even saw the published paper, Eaton (the co-author) and I began to receive telephone calls from an array of newspaper and broadcast journalists ranging from science reporters to food editors. Several reporters adopted the phrase "Caveman Diet" and went on to use it despite our insistence that it was not only misleading but it was also insulting to contemporary hunting-and-gathering peoples. A few representative headlines: "Cavemen Cooked Up a Healthy Diet" (*U.S.A. Today*), "Cave Man Takes a

Healthy Bite Out of Today's Civilized Diet'' (*Atlanta Journal*), and "Check Ads for Specials on Saber-toothed Tigers" (*Atlanta Constitution*).[23]

Although he got a few laughs and some notoriety out of the experience, Konner saw the press's role as being ultimately misleading.

None of this is meant to impugn the work of science reporters. Many superb journalists—John McPhee, Tracy Kidder, John Noble Wilford, James Gleick, Richard Rhodes, Robert Kanigel, and others—have written extraordinary popular science expositions. Work of writers of this caliber is essential to furthering public science literacy. Still, the distinction between scientist and nonscientist popularizers is real and important. Historically, there have been tensions between scientists and the press; at issue is how authoritatively information is communicated to the public. Along with calls for increased popularization, however, scientists are realizing that such efforts require a "healthier symbiosis"[24] with the media. Toward that end, the Media Resources Service of the Scientists' Institute for Public Information provides a referral service for reporters seeking a specialist to comment on specific news issues. In 1993, ProfNet, a computer network, was established to perform similar services. The National Association of Science Writers, the National Science Foundation, the National Academy of Sciences, and the AAAS Committee on the Public Understanding of Science and Technology also provide liaisons between scientists, journalists, publishers, and the general public.

The problem, then, is one of authority. It presents a Catch-22. Invariably, the more technically accurate the information, the more difficult it will be for the general public to comprehend. If the information is diluted too much, however, it might make for interesting reading, but it might also be entirely misleading. Several in-between possibilities also exist.

Figure 3.1 shows the Poles of Technical Communication Model. At its extremes are the Research Level and the Tabloid Level; in between is found a "zone of authority." Material written at the research level is by and for peers— it is the stuff of scholarly journals. Material written at the tabloid level is for an audience seeking pure entertainment, at the expense of information accuracy—it is the stuff of "space alien" reports. "Low or technical content" materials are addressed to audiences of motivated general readers with no prior background in the subject. "Higher technical content" materials are for those without a technical background, but with some basic knowledge of the subject. Ideally, science literacy collections should contain materials representative of various levels within the zone of authority.

All popular science library materials should meet a minimum level of factual and conceptual accuracy. Anything less is unacceptable and potentially harmful if the goal is to promote science literacy. Although, unfortunately, no line can be drawn in the sand between what is and is not acceptable, appropriate works have several readily recognizable characteristics.

First, and probably most important, a science popularization written at a "no

Figure 3.1
The Poles of Technical Communication Model

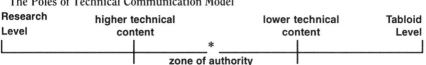

"prior knowledge" level should give the reader the background necessary to understand the information being presented and its significance. Doing so establishes the knowledge context of the information being reported, much as a literature review does in a professional article. Every scientific discovery is like a puzzle piece, and it is vital for the reader to have a sense of what the puzzle looks like to understand how the new piece fits into the whole. Thus, an overview of the state of the science at the time of the discovery is needed. Usually this requires a historical summary, which may stretch back to the very beginnings of science. That an overview of the field is needed may seem obvious, but too often it is omitted, which can leave the reader with the impression that this new discovery came out of nowhere; that, in turn, creates sensationalism.

Second, look for evidence throughout that the progress of science and discovery has been based on sound scientific methods. The reader should be suspicious of anything presented as fact without an explanation of *why* it is believed to be fact—how it was discovered and if it has been experimentally verified, what questions it answers, how it stands up to scientific scrutiny, and why it describes reality better than any alternative theories. To use the puzzle analogy, the author of a popular science work should explain why this particular piece fits better than any other piece, how it contributes to a more complete picture, and what new pieces might be added to it. Still, scientific progress is not always linear, and many different pieces might be tested before the one that fits best is discovered, and even then that piece might later be discarded when an even better one is found. In the primary literature, scientists review and debate the merits of new information. They do not always reach consensus, but knowledge is furthered by the process of debate. In popular science, when a new idea is presented, there should likewise be discussion of the prevailing scholarly opinions about that theory. Is it generally accepted as correct? Is it the subject of hot debate? It is on the fringe but still theoretically possible? Popular science must show some of the nuts and bolts of peer review and scientific methodology so that lay audiences can have confidence in its authority.

Finally, any popular science work must also meet minimum standards of readability. Again, there is no clearly demarcated line, but any librarian should be able to recognize when a book is written at or near too technical a level. If it reads like a technical communication or a laboratory notebook, it is probably not appropriate for a science literacy collection. As discussed in the previous section, the literary style of popular science must be carefully crafted. Look at sentence length and structure; look for effective rhetorical devices; look for wit, irony, cleverness, and perhaps even a sense of humor. Anecdotes about scientists and how they made their discoveries can enliven a story and make its technical content seem more human. Some writers might wax philosophical and speculative, and that, too, is entirely legitimate, provided they do so within the bounds of plausibility. Popular science readers expect and deserve some clues as to the frontiers and wonders of science.

Effective popular science accomplishes all of these things. It is not necessary to distort scientific information to make it comprehensible to a layperson. Popular science might not contain the nuances and esoteric details that are important in primary science, but it can and should be just as conceptually rich, perhaps more so. While some of the details are lost in popular science writing, so too are some of the human elements left out of primary science writing. Science consists of both.

TOWARD EFFECTIVE EVALUATIONS OF POPULAR SCIENCE MATERIALS

The principles described in this chapter can be applied to the very practical task of evaluating new works of popular science and developing science literacy collections in libraries. While many considerations go into these collection development decisions, evaluating popular science materials need not be a daunting or time-consuming task. Building on the principles set forth here, the next chapter examines some things to look for when evaluating popular science materials in several formats.

NOTES

1. C. P. Snow, *Two Cultures* and *A Second Look* (New York: Cambridge University Press, 1964).

2. Robert A. Day, *How to Write and Publish a Scientific Paper*, 2nd ed. (Philadelphia: ISI Press, 1983), xi.

3. Bob Coleman, "Scientific Writing and Scientific Discovery," *New York Times Book Review* (September 27, 1987): 1.

4. Leo Steinberg, "Art and Science: Need They Be Yoked?" *Daedalus* 115 (Summer 1986): 1–16; Mark Kipperman, "The Rhetorical Case Against a Theory of Literature and Science," *Philosophy and Literature* 10 (1986): 76–77.

5. David Locke, *Science as Writing* (New Haven, CT: Yale University Press, 1992).

6. Charles Bazerman, *Shaping Written Knowledge: The Genre Activity of the Experimental Article in Science* (Madison, WI: University of Wisconsin Press, 1988).

7. Richard Whitley, "Knowledge Producers and Knowledge Acquirers," in Terry Shinn and Richard Whitley, eds., *Expository Science: Forms and Functions of Popularization* (Boston: Reidel, 1985), 3–28.

8. Ibid.

9. Melvin Konner, *Why the Reckless Survive* (New York: Viking, 1990), vii.

10. Stephen Hawking, *A Brief History of Time* (New York: Bantam, 1988), vi.

11. Michael Shortland, and Jane Gregory, *Communicating Science: A Handbook* (Lon-

don: Longmans, 1991), 50–52; John A. Barry, *Technobabble* (Cambridge, MA: MIT Press, 1991).

12. Jeanne Fahnestock, "Accommodating Science: The Rhetorical Life of Scientific Facts," in Murdo William McRae, *The Literature of Science: Perspectives on Popular Scientific Writing* (Athens: University of Georgia Press, 1993), 17–36.

13. Shortland and Gregory, *Communicating Science*, 51.

14. Ibid., 75.

15. George Gopen, and Judith Swan, "The Science of Science Writing," *American Scientist* 78 (November/December 1990): 550–558.

16. J.B.S. Haldane, "How to Write a Popular Scientific Article," in *On Being the Right Size and Other Essays* (Oxford: Oxford University Press, 1985), 154–160.

17. Stephen Weinberg, *Dreams of a Final Theory* (New York: Pantheon, 1993), 78–81.

18. A. Zee, *An Old Man's Toy: Gravity at Work and Play in Einstein's Universe* (New York: Macmillan, 1989), 37–38.

19. John A. Barrow, *Theories of Everything* (Oxford: Oxford University Press, 1991), 37.

20. Fahnestock, "Accommodating Science."

21. Greg Myers, *Writing Biology: Texts in the Social Construction of Scientific Knowledge* (Madison: University of Wisconsin Press, 1990), 141–152.

22. Dorothy Nelkin, *Selling Science: How the Press Covers Science and Technology* (New York: Freeman, 1987).

23. Konner, *Why the Reckless Survive*, 43.

24. Marcell C. LaFollette, "Scientists and the Media," *The Scientist* (July 9, 1990): 13–14; see also Conrad Storad, "Newspaper Science Writers Evaluate Their Field," *Editor and Publisher* 117 (June 9, 1984): 100–101; Sharon Dunwoody and Michael Ryan, "Scientific Barriers to the Popularization of Science in the Mass Media," *Journal of Communication* (Winter 1985): 26–42; Sharon Friedman et al., eds., *Scientists and Journalists* (New York: Free Press, 1986).

Chapter Four

Evaluating Popular Science
Information Resources

We have discussed the importance of popular science in fostering science education and science literacy in the United States, outlined a brief history of scientific communications, and surveyed unique issues pertinent to the theory, style, and substance of this genre of literature. While these principles have been described for written works of popular science, the same basic issues and concerns remain relevant regardless of format. When evaluating any work of popular science for collection development, keep these general principles in mind and ask such questions as the following:

- How are jargon and mathematics handled?
- Are the pacing and structure of the work appropriate?
- At what technical level is this information presented?
- Are the scholarly context and significance of this information made clear?

When evaluating an item for possible inclusion in a library collection, the librarian engages in a process similar to that used by a reviewer. While the two processes are complementary, they are not coterminous. The evaluating process centers around deciding whether to purchase an item for a collection. The librarian assesses a book's relevance for a local collection, what gaps it might fill, and what its projected demand would be. Reviewing materials involves assessing their quality. Published reviews of the thumbs-up or thumbs-down variety are the result. When evaluating an information resource it is desirable to consult a number of reviews, then to use the information from them, combined with knowledge of the local collection and patrons, to make a purchasing decision. Librarians are particularly encouraged to consult reviews in professional and specialized journals. See Part II of this book for science periodicals that review books for general readers.

Here we will further examine some suggested procedures for evaluating works of popular science for library collections. Materials are discussed primarily by format, each format having its own distinct advantages and disadvantages as a vehicle for communicating science.

EVALUATING BOOKS

Books are perhaps the most common and arguably the most influential medium for communicating science to the general public. Their chief advantages are that (1) the written language lends itself to scrutiny, and thus to deeper understanding; and (2) because of their length and the degree of detail they permit, books allow for a more deliberate, progressive exposition of ideas than other formats. Reading a book requires a commitment of time and intellect. So, too, does the active cultivation of science literacy. While it is important to update and supplement book knowledge with that obtainable from other media, it is highly unlikely that a person can become and remain scientifically literate without investing some time in structured, self-directed reading.[1]

A word here regarding the obsolescence of science books: librarians know that science changes rapidly; in certain disciplines, the citation half-life of research articles averages around five years. Popular science materials, although subject to comparably rapid obsolescence, do not generally become outdated as quickly as original research. Before a scientific theory is popularized, it usually has gone through at least initial peer review, and scientists have reached some degree of consensus about its validity. Even books that are ten or more years old should accurately represent what was known or considered possible at the time they were written. Thus, they should not contain any ''wrong'' information, just information that needs to be supplemented. While timeliness is certainly preferred in science literacy collections, the useful shelf life of many books is longer than some librarians might expect. A few books, including several cited in the following chapters, are considered classics.

Numerous factors should go into a collection development decision when evaluating a new popular science book. Some apply to any book being considered, but even these obvious factors have different significance in the realm of popular science. Factors to examine include the book's authorship, publisher, and genre or format.

Authorship

Determine the author's credentials. Who wrote the book is important, but not necessarily for the same reasons that the authorship of a novel is important. A librarian might know, for example, that any novel by certain bestselling authors will be in high demand and thus, if for no other reason, merit purchase. Where popular science books are concerned, however, authorship reflects on the issue of authority; you should ask, For what purpose was the book written? Is the author qualified to write a book on this subject?

A writer's credentials are important. Look especially for authors who have a track record of successful popular science writing. Authors whose past work has been recognized in a "best books" list, such as *Library Journal*'s annual "Best SciTech Books for General Readers" feature (published annually in March), are usually trustworthy sources. Additionally, several professional groups and institutions, such as the American Institute for Physics, the McDonald Observatory, and the British Council for the Public Understanding of Science, sponsor prizes for science writing. Such recognition is another good measure of a writer's competence.

Some authors are also scientists. They fall in two general categories: (1) practitioners in the field who are popularizing general knowledge from their discipline, and (2) prominent scientists—often Nobelists or scientists known to the public—who are popularizing their own original work. Notable scientist/authors in the first category include Isaac Asimov, Jeremy Bernstein, John Gribbin, Alan Lightman, James Trefil, Barry Parker, and several others. While they enjoy solid reputations as scientists and have made their own original contributions, they are not superstars. They have respectable yet unspectacular academic credentials. They write their popular works while working on their research projects or between projects, and appear to see no rigid distinction between doing science and writing about it. They are not afflicted by the highbrow disdain for popular science common among many scientists. Not coincidentally, writers of this ilk often possess superior communications skills and manifest enthusiasm for their subjects and sincere empathy with their readers.

In the second category of scientist/authors are those who have made exceptional contributions to their fields and, as a result, have the "name appeal" that attracts attention and sells books. Steven Weinberg, Richard Feynman, Edward O. Wilson, Jane Goodall, Stephen Hawking, Roger Penrose, and others are examples. These are the best and the brightest in their fields, although popularizing science is not what they have chosen to do first, and sometimes not what they do best. Do not misunderstand that statement. These writers are among the most influential science popularizers and their books have contributed greatly to public understanding of their fields. In the best cases, these scientists popularize their work out of an earnest desire to share ideas with a broader audience. At worst, books by famous scientists get published because they sell, pure and simple. Still, even with the best of motives, these authors are, by trade and training, scientists first, communicators second. While their books might sell fabulously, other books available on the same subject may do a better job of teaching and explaining.

Popular science books may also be produced by professional writers or journalists. Here again, a distinction can be made between journalists who specialize in science writing and generalists who cover all fields.

Science writers, especially those affiliated with universities or reputable national newspapers or magazines, are uniquely trained professionals who often have some academic background in science, enjoy many contacts within scientific communities, and know the science beat thoroughly. (There are even specialists among specialists.) Though they are not scientists, they nonetheless qualify as science insiders. This is often reflected positively in the quality of their work.

The works of generalist writers should be evaluated carefully. Look for evidence that they have done their homework. Though they might have the noblest intentions, they often lack adequate resources to do work of the same caliber as the specialists. This is not to say that you must avoid books by these authors. Each book stands on its own merits. However, research indicates that popular science books by writers without professional connections to scientific communities tend to distort and sensationalize ideas more frequently than others. Some, in fact, appear to have been written for those very purposes.

Other authorial possibilities exist. In recent years, collaborations have become more common. Collaborations between scientists and journalists, scientists from different fields, or scientists and humanists can be especially effective. In true collaborations, the literary dynamics are livelier, the perspectives broader, and, ideally, each author contributes personal strengths and skills that add depth to the work. However, books sometimes appear ''by'' a noted scientist ''as told to'' an intermediary. The danger here is that the intermediary might choose to write about only those of the primary author's opinions or ideas that are likely to sell books. Miriam Pollet cautions, ''As long as a fast buck can be made on the human need for explanation, mystery, and fantasy—a lay fascination with the 'brave new world' that can be hustled by the media—librarians had better be wary.''[2]

The author's identity and credentials can reflect on the authority and accuracy of the information in the book. When the author is a superstar scientist, for example, you are probably safe assuming a high degree of technical accuracy, but at the same time you might wish to take a close look at how well that information is communicated for a lay audience. At the other extreme, if the author is a generalist with no particular history of popular science writing, inspect the work carefully for evidence that its content is well established by scientific consensus. Books that circumvent the peer review process are to be avoided.

Publishers

Determine the nature or institutional interests of the publisher. The publishing industry in the United States is enormous and heterogeneous. Huge, New York based commercial publishers dominate certain aspects of the industry, but myriad specialty and small presses also thrive within their own niches. Additionally, various professional, educational, and political groups publish as a secondary activity, in support of some larger objective. Publishers who sponsor books on popular scientific subjects are likewise very diverse. Librarians evaluating these books need to be aware of the types of science publishers and the kinds of books they produce.

In terms of numbers of good books *trade publishers* are probably the most prolific. This is also the most eclectic category. Major New York publishing houses are bottom-line, megabusinesses with vast resources. They conduct extensive market surveys before entering into a publishing enterprise. Typically, their catalogues list titles by known authors on current and topical subjects. Thus, the superstar scientists often publish with trade companies. For the same reason, trade publishers often list titles on scientific subjects that are hot or in the news. Topics extensively popularized in recent years include AIDS, the Human Genome Project, virtual reality, various environmental issues, and scientific funding, fraud, and mismanagement. In these cases, the librarian might have several titles from which to choose. (Here's a tip: the first book to appear on a hot topic is seldom the best.)

Numerically, *university presses* constitute the second largest category of popular science publishers. Some of the best science popularizations available come from publishing houses affiliated with academic institutions. Although they compete with trade publishers to some degree, the university presses generally operate under a clearer sense of mission. Like their parent institutions, university presses gain legitimacy by furthering knowledge and learning. Quality is important.

The authors of university press books are often scientists or scholars. These books seldom charge to the top of the bestseller lists, but they do reach a sizable audience, and may have a longer shelf life than trade books. University presses also produce books on subjects eschewed by the trade publishers. Sometimes, though, university press books approach or straddle the boundary between primary literature and technically inclined popularization. University presses tend to publish books appropriate for somewhat more knowledgeable readers. Less often, they publish books explicitly for readers with no prior knowledge of the subject.

The third category comprises *specialty publishers*. Although individual houses differ in their publishing orientation, all exist to fill a specific niche, and often to promote some particular cause. Some specialty publishers are offshoots of professional scientific societies. The American Institute for Physics, the American Chemical Society, and the Mathematical Association of America, to name just a few, have active publishing interests, which frequently include lines of popular works. Professional associations often produce works in series, and these can be identified and evaluated as parts of their wholes.

Another type of specialty publisher is represented by nonacademic societies that are organized around a philosophical or political mission. For example, a number of environmental alliances publish books on topics of interest to members of their organizations. When evaluating any book by a specialty publisher, it is important to recognize that the publisher has a vested interest in furthering a particular point of view on the subject. Do not expect a book published by an environmental organization to support development of nuclear energy resources or for a professional scientific society to bring forth a book that blows the whistle on fraud within the profession. Although these books might be excellent, they have inherent biases, suggesting to the librarian that other viewpoints need to be represented in the collection for the sake of balance.

Type of Book

Determine the book's genre or format. Several identifiable types of books can be found in the literature of popular science. Each offers something unique to the general reader seeking scientific information, and all might be included in a representative popular science collection.

The most common type is the *exposition*, written to "expound, explain, or appraise analytically a given subject." In practice, this basic function is accomplished in many different ways and for varying purposes. For example, an exposition might take the form of a general history, a remedial or refresher volume, a semi-scholarly treatise, a mass-market trade publication, an exposé or political discourse, a journalistic report of some new discovery, a personal or philosophical account based on the author's experiences, or, indeed, any combination of the aforementioned. What they have in common is that they tell the reader *about* something that has happened or is happening in the world of science. The most important factors to consider when evaluating an expository work relate to its style and substance, as discussed in Chapter Three.

Several subtypes of expositions have already been noted. A few rather distinct kinds, not mentioned previously, merit brief attention. The first is the *scientific essay*. The word is derived from the French *essayer* (to try); accordingly, a notable strength of the genre is the amount of literary freedom it accords the author. The purpose of the essay is to expound a point of view or to persuade, and the author's success in that endeavor depends on the eloquence of the words as much as the force of the intellectual argument. This kind of writing stands in stark contrast to cookbook science writing. For that reason, many scientists with unfulfilled literary aspirations have turned to the essay as an alternative vehicle of expression. Loren Eiseley, scientist and author of a popular history of evolution, *The Immense Journey*, wrote:

That the self and its minute adventures may be interesting every essayist from Montaigne to Emerson has intimated, but only if one is utterly, nakedly honest and does not pontificate. In a silence [on] which nothing could impinge, I shifted away from the article as originally intended. A personal anecdote introduced it, . . . and yet I had done no harm to the scientific data.[3]

Popular science theorists suggest that the essay can be an important bridge between technical and public scientific communications.[4] There is no doubt that scientific essays are popular among general readers.

A second expository subtype that deserves specific mention here is the *scientific biography*. Through the lives of great scientists, the skillful biographer can animate an entire era of science and society. A well-written biography (or autobiography) explores the entirety of the individual's thoughts, experiences, and social influences. The distinct value of a biography is that not only is it about an individual, but in a sense it is also about that individual's entire personal universe. Thus, through the telling of an individual's own story, a broad social, cultural, and intellectual portrait emerges. Biographies are great eye-openers for those who regard science writing as narrow and impersonal. Some biographies are written at a more popular level than others, but even a comprehensive scholarly biography, although difficult going in parts, should make for compelling general reading.

Another common expository format is the *anthology*, usually under the general editorship of a single person or an editorial board. Anthologies can be useful vehicles, particularly when they gather the work of diverse writers—for example, scientists, civil servants, knowledgeable laypersons—who lend breadth of opinion to the subject. Look for uniform quality, regardless of point of view, and thematic coherence.

Another subtype, the *textbook*, is often disdained in collection development handbooks, but should be considered for science literacy collections. Standard wisdom holds that textbooks are inappropriate additions to library collections because (1) they are written for classroom use exclusively, (2) they present no new information, and (3) they are rapidly superseded by new editions.[5] Many library collection development policies explicitly (and perhaps peremptorily) state that textbooks are not purchased.

Nevertheless, textbooks can fill gaps in a basic science literacy collection that cannot be filled by any other kinds of library materials. Suppose you are an adult who, several years out of college, wishes to brush up on calculus, long fallen into disuse. Probably you would find a good introductory text most useful. (You are not likely to find the mechanics of calculus described in expository form, for example.) The value of textbooks is that they are written to teach. If you are prompted by general intellectual curiosity, texts will be of little service, but if you are seeking an applied overview of a broad technical field, a standard text can provide just the right kind of treatment. The core concepts of some areas of science—calculus is a good example, as are inorganic chemistry and electrical engineering—are best assimilated when learned from a textbook. These subjects simply do not lend themselves to the literary treatment found in expositions. Further, with textbook knowledge mastered, the reader may be prepared to advance to works of a more technical nature.

The *coffee table book* is yet another subtype sometimes maligned in collection development handbooks. These books are often viewed as vanities, to be purchased for show and displayed, unread, for others to see. Still, they have a place in science literacy collections. In 1992 Niles Eldridge's oversize book *Fossils* was nominated for the prestigious Council for the Public Understanding of Science's annual award for the best science popularization.[6] Some reviewers declared it unworthy of the award because it was a coffee table book. This prompted Eldridge's colleague, Stephen Jay Gould, to defend the book by citing the author's paleontological credentials and pointing out that, glossy illustrations aside, the book's text was intelligible, accurate, and informative.

Gould's point was well taken. If you look beyond the flashy format of many such books, they can be discovered to contain much valuable information. In the best of these books, the narratives are intelligent and the photographs are instructional; if they are intrinsically appealing, so much the better.

Certainly, a core of *reference books* is essential to a science literacy collection. Minimally, the library should have a comprehensive science dictionary, a general or semitechnical encyclopedia with broad disciplinary coverage, a chronology or timeline of science, a biographical source (if not covered by the encyclopedia), and selected handbooks, as appropriate. Discipline-specific resources might be added to suit the needs of local patrons.

In recent years, publishers have brought forth several new reference books that have been quite consciously developed to serve science literacy information needs. These are especially useful in that, while replete with facts and figures, they keep technical content to a minimum and, through such devices as sidebars and introductory section overviews, attempt to give a "big picture" of the information being presented.

In sum, books are and should remain the core components of a science literacy collection. Still, the book format is limited in certain ways.

First is the time factor. Even a brand new science book might contain some obsolete information. Further, books can bring readers up to date but cannot *keep* them up to date. Second, not everyone has the time or willingness to read a whole book on science. Many prefer to keep up with science by seeking information in other media. For example, many find the most brilliantly written book less engaging than a video presentation. Some people learn better from what they see and hear than from what they read. For those who view the sciences as passive and impersonal, videos can be used to bring them to life.

These two limitations of books can be compensated for by supplementing the science literacy collection with appropriate periodicals for general readers, and by developing representative videotape collections. These formats will be addressed briefly.

EVALUATING PERIODICALS

Periodicals are vital for science literacy collections for essentially the same reason that journals are vital for science research collections.[7] The periodical format originated to report news of interest to readers. What constitutes "news" to the scholar would probably be recent research and conference reports, while to the layperson it might be the mission of the latest Space Shuttle flight or who won the Nobel Prize. In either case, this information is likely to be sought in a periodical resource.

The Poles of Technical Communication model (see Figure 3.1) is a useful yardstick for evaluating the level, scope, and purpose of scientific periodicals. There are several basic types of periodicals for general readers, and their defining characteristics are fairly consistent, depending on where they fall in the "zone of authority." Those nearer the upper end, and thus closer to the primary literature, are more technical and specialized, aiming for a strongly motivated audience with some prior knowledge of the subject area. The information in these sources will usually be accurate and authoritative, but difficult to assimilate. Periodicals in the middle of the zone target a larger audience, which, while intellectually motivated to follow science in general, may or may not know anything about a given subject. Those closer to the Tabloid Level are written for all readers, and the science features they publish compete for attention with whatever else appears within their pages, from political scandals to baseball scores. Here, the absolute accuracy of the information may be suspect, but it easily assimilated by the reader.

Upper Zone of Authority Periodicals

Popular periodical titles near the upper end of the zone of authority include *American Scientist, American Mathematical Monthly, BioScience, Byte, Natural History, Scientific American, Sky and Telescope*, and many others (see "Periodicals for General Readers" sections in following chapters). They report on breaking research and trends in science in general, or in specific disciplines. Although they do not publish original research, they follow what happens to new research after peer review has lent further insight and, most important, summarize and synthesize that information so that the reader can deduce how it all fits together. The contributors tend to be scientists and practitioners. These periodicals are published by professional societies, special associations, or commercial houses with strong interests in science. Many of these periodicals belong in any science literacy collection, but the librarian should recognize that they are not for dilettantes; their readers are amateurs or hobbyists, general readers with ongoing special interests, and even scientists who use them to keep up with the goings-on in other disciplines.

Both general and specialized science periodicals at this level do something else for those aspiring to become or remain science literate: they expose various "subcultures" of science. If you take out all of the articles, columns, reviews, and pictures, you are left with letters, announcements, and advertisements placed in the magazine on the assumption that they will attract the attention of its readers. In essence, these extraneous bits and pieces represent the interests of those who belong to the subculture. In a popular astronomy magazine, for example, there might be an ad for a new telescope model, an announcement of an upcoming "star party," the address of a regional association seeking members, and a listing of new books available through a special book club. This kind of information is invaluable to anybody who wants to learn more about a field of science.

Middle Zone of Authority Periodicals

Middle zone science periodicals, among which might be counted *Science News, Discover, Omni,* and *Popular Science,* are for those who are consciously attentive to science, but in a rather discursive way. They generally seek information that is broad, not deep. They might start reading a technical article, but, upon realizing that it is not making sense, abandon it for something else. Doing so is not a sign of intellectual capriciousness; rather, it simply shows that the information is not "right" for them. The contributors to these periodicals are primarily science journalists and writers who know their science, but whose greatest skills are in communications. Typically, these titles are produced by commercial publishers and include glossy photographs, various regular columns, and sometimes even a little comic relief.

The middle of the zone of authority is actually pretty wide and heterogeneous. Some titles tend more toward the technical side, while others, in the quest for more subscribers, might creep, over time, toward the opposite end of the continuum. Complicating matters even more, other popular magazines, which might not have a strong emphasis in science per se, still occasionally publish features that are grounded in some scientific issue. Discussions of environmental topics, for example, can be found in outdoors magazines, business publications, and even travel periodicals. Some of the writing is very good; some is atrocious. Evaluating these eclectic resources thus requires careful scrutiny, an informed assessment of a title's audience, and monitoring on an ongoing basis.

Lower Zone of Authority Periodicals

Finally, those periodicals at the low end of the zone of authority are general interest magazines and newspapers. Communicating science is not their primary purpose, and they do so only when science is considered newsworthy. The quality of science writing might be uneven within an issue. For example, a feature on a major scientific event, written by one of the magazine's best reporters, may win honors for its accuracy and objectivity. Buried deeper within the same issue, however, one might find a sensationalized account of a new, yet to be published theory on the extinction of the dinosaurs. These titles should probably not be evaluated from the standpoint of developing collections for science literacy. If they belong in a library collection for other reasons, then purchase them, but do not do so for their science content.

EVALUATING VIDEO RESOURCES

Video resources offer some potent advantages over print formats as tools for science literacy. They engage and stimulate people in a more dynamic way, and thus can serve as touchstones for different types of intellectual activities. The efficacy of videos in classroom teaching at all levels is well studied.[8] Likewise, groups advocating adult science literacy have explicitly called for more and higher-quality media-based popularization. Potentially, video science can reach a larger audience than book science, especially when it is produced as part of regular network programming, such as PBS's *Nova*, various offerings of the Discovery Channel, or even occasional major network specials. Many people are likely, by preference, to seek information from video media. The importance of video science resources in library collections is more than just a function of their popular appeal, however; they fill a genuine patron need.

Video science resources add immediacy to information. They create a sensory experience that cannot be duplicated by print resources. Thus, their chief advantage is that they can be used to bring science to life by investing it with the qualities of motion and sound. Experiments can be demonstrated. Sophisticated computer imagery makes it possible to show things on video that until recently only scientists could have visualized. Video is also a humanizing medium: even scientists with the most formidable intellects can be shown as human beings who occasionally laugh or stumble over words. The most effective video is that which makes science real, something that viewers can experience.

When evaluating a video media resource, begin by determining what stylistic category it belongs to. Three main script categories are commonly recognized: (1) narrative documentary, (2) drama documentary, and (3) dramatic.[9]

Narrative Documentary

In a nutshell, narrative documentary is the model used in the most "serious" productions. It tends to be content-loaded and aims to deliver information in a very direct way. It often resembles a lecture from a talking head. In these productions, the script is all-important, for it is the words that will be remembered. Visual or sound bites are used infrequently, and may actually seem distracting or gratuitous.

Drama Documentary

Drama documentaries are also essentially designed like lectures and must be carefully scripted, but they usually depart from the discourse at several key points to show real-life events related to the script. In popular science videos, these segments might take viewers into a science lab, to the brink of a glacier, or into outer space. Because drama documentaries are both script-based and action-oriented, they must be balanced, and a production that is weighted too heavily toward lecture might become tedious, while if the imbalance is toward action it can seem superficial.

Dramatic

The predominantly dramatic production is sometimes divided into two sub-categories: reality drama and fantasy drama. An effective use of reality drama is the recreation of an event—say a scientific discovery—usually by actors in a staged environment, though actual footage can be used with good results. The litmus test for these productions is how successfully they communicate their informational content and how honestly they depict the events and personalities involved. Fantasy drama, on the other hand, tours the unknown but possible, and it takes considerably more imaginative license. In popular science, this style is used for exploring the frontiers—space in the next century, the future course of evolution, or the effects of progressive environmental degradation. Clearly, the pitfall of fantasy drama is that it becomes just that, fantasy. Thus it is important to establish the scientific credibility of the ideas being presented.

The issue of levels of relevance, which is very important for evaluators developing book collections, is somewhat less crucial where video formats are concerned. Highly technical information does not readily lend itself to video explication. Straight documentaries tend to be more probing. Fantasy dramas, even when thoroughly grounded in fact, tend toward artistic expression, sometimes even playfully so. Still, except for those videos whose sole purpose is demonstration of some technical procedure (these are generally used only for advanced pedagogical purposes), videos, by their nature, are created to appeal to broad audiences. Some subjects lend themselves more readily to documentary treatment, others to dramatic, but in neither case should the information that they contain be more than moderately technical.

As with books, accuracy and authority in information content remain extremely important considerations in evaluating video resources. Just as it is wise not to believe everything you read, what you see should be questioned as well. In written formats, there are literary conventions and techniques by which an author can demonstrate that the information is contextually and substantively accurate. These include citing references, presented information sequentially so as to establish its historical and scholarly context, and answering rival theories and interpretations. Additionally, the author's credentials and the publisher's reputation give indications of authority.

Video is quite different, however, and although analogous factors can be considered in assessing authority, the format does not as easily lend itself to evaluative scrutiny. Most informational videos last no more than one hour; if written, their scripts would probably amount to twenty to thirty double-spaced pages. Obviously, this is not an expansive format, and the result is that details are often sacrificed. Even so, at the very minimum expect the script to address these essential questions:

- What do scientists know about the subject?
- How and by whom was this knowledge obtained?
- What remains to be known, and how could it be discovered?
- What are the consequences of this information for all of us?

These questions address issues related to (1) the context of the information, (2) the scientific processes, (3) topics under current investigation, and (4) the practical effects of science. Look for specifics. Testimonials from credentialed scientists are usually trustworthy. There should be some linkage between the information and its sources in primary literature. You should also expect an indication of where you might find additional information.

The nearest thing to an "author" or "publisher" of a video recording is usually its producer. Video producers are difficult to classify, especially with respect to their popular science outputs, but some general types include commercial and film companies, specialized professional associations, television stations and cable companies, educational bodies, and many, many independent outfits. Few generalities apply across the board, but here are some things to consider:

- Was the video produced by a for-profit entity?
- Does the producer have an educational mission?
- Does the producer (or funding agency) represent a special interest?
- Was the videotape originally shown in some other medium, such as film or TV?

The answers to these questions give some indication of the production's purpose and intended audience, which can, in turn, reflect on its accuracy and objectivity.

Many people find science concepts more comprehensible and easier to assimilate when presented in video format. Indeed, many people who would probably never invest the time to read a book on, say, relativity theory, might eagerly borrow a video on the subject. The need for video materials in a science literacy collection is obvious. Still, they can do little more than introduce a person to a subject area. Library patrons who are attentive to science will likely seek further information from resources that, by their nature, offer more detailed, rigorous treatment.

EMERGING SCIENCE INFORMATION RESOURCES

The traditional resources discussed above comprise the core of science literacy library collections, and will probably continue to do so for years to come. Even so, automated technologies with powerful library applications are beginning to impact science education practice. For example, CD-ROM databases are now common in library reference departments. New technologies have opened frontiers of information access, retrieval, and visualization. As tools for promoting science literacy, automated resources have great promise, for they uniquely use technology to teach technology. The scope of this work permits only cursory mention of these resources. Still, as practice and research suggest new applications, they will become increasingly important as library resources in general, and as science literacy resources in particular.

Interactive videodiscs, which, like audio CDs, are read by laser players, are durable media for storing and displaying video information.[10] They are particularly useful in illustrating what happens in dynamic events, like chemical reactions, medical procedures, or meteorological phenomena. Further, they can be "repurposed," which means that whole new programs can be designed from the contents of an existing videodisc and used to highlight or recombine certain segments. They can be used in classroom tutorials, to create visual databases, or for simulations.

When videodisc technology is controlled by hypermedia programs, it allows for the creation of so-called multimedia resources. Hypermedia programs allow persons to organize and explore information in associative, nonlinear ways. Users can make their own choices about how to explore a subject and retrieve, store, and present that information. Programs might allow for seamless access to text, audio, and graphics. For example, a patron might choose, usually by clicking an icon, to investigate the workings of a tornado—a dynamic image of a tornado being formed will appear on the screen, along with window graphics giving detailed information about its force, accompanied by authentic audio effects. Hypermedia can be used in the classroom, as a database manager, to create computer "slide shows," or just for browsing.

The last topic in this brief sketch on new library technologies for science literacy is the universe of Internet accessible resources. Numerous full text and informational databases are available through the Net. For example, users with network access can connect remotely via "telnet" (a command that enables logon) to databases such as NASA's Spacelink (the address is space-link.msfc.nasa.gov), an informational database for students and teachers. Hundreds of such databases exist covering topics including AIDS, the environment, the Human Genome Project, astronomical events, and current weather information. When these databases allow for anonymous logon and use of file transfer protocols, information can be retrieved digitally, then loaded into local systems for later use.

"Gophers," which are developed on a client/server model by colleges, computer centers, businesses, and government agencies, further facilitate access to these resources. Hierarchical gopher menus allow you to see databases listed by categories, choose from that listing, and make a direct connection to that source without even knowing its electronic address. In that sense, gophers "point" at each other and allow users to "tunnel" from one logon site to another. A good source for information on various gophers and the subject resources to which they connect is "Gopher Jewels," developed at the University of Southern California; the gopher can be reached at cwis.usc.edu.

Finally, a promising and rapidly developing Internet resource called World

Wide Web (also called WWW or W³) is similar to gopher, but uses hypertext as its basic organizational design. Every piece of information in the web provides "pointers" or "links" that can be used to find the same or related subjects. Some striking graphic images can be retrieved from WWW resources.

Again, these and other cutting-edge technologies provide information that, although accessed through nontraditional means, can supplement and expand library resources dramatically. Of paramount interest here is that they make available current, authoritative information on a variety of scientific subjects of general public interest. Electronic formats offer the additional advantages of allowing for file transfer and downloading. Any library patron with a modicum of computer skills will most likely find interactive, online searching versatile, challenging, and enjoyable.

This concludes discussion on science literacy, scientific communication, and library resources. The basic principles covered should provide librarians and media specialists with a functional understanding of popular science and related collection development issues. The resources listed in the annotated bibliographies in Part II of this book were selected and compiled accordingly.

NOTES

1. Self-directedness in learning is a core concept in the field of adult education that can be fruitfully applied to studies in science literacy. Information seeking is, almost by definition, a self-directed behavior. For more information, see Malcolm Knowles, *Self Directed Learning: A Guide for Learners and Teachers* (Chicago: Association Press, 1975); Stephen Brookfield, "Self Directed Adult Learning: A Critical Paradigm," *Adult Education Quarterly* 35 (1984): 59–71; Brenda Dervin and Michael Nalin, "Information Needs and Users" in Martha Williams, ed., *Annual Review of Information Science and Technology*, vol. 21 (Washington, DC: American Society for Information Science, 1986): 3–33.

2. Miriam Pollet, "Criteria for Science Book Selection in Academic Libraries," *Collection Building* 4 (Fall 1984): 42–47; see also Beth Clewis, "Selecting Science Books for General Readers," *Collection Building* 10 (1989): 12–15.

3. See Andrew Angyal, "Loren Eiseley's *Immense Journey*: The Making of a Literary Naturalist," in William Murdo McRae, ed., *The Literature of Science* (Athens: University of Georgia Press, 1993), 54–72.

4. Ben Agger, *Reading Science: A Literary, Political and Sociological Analysis* (New York: General Hall, 1989), 210–229.

5. Katz lists textbooks among a category meriting "Instant Rejection." William Katz, *Collection Development: The Selection of Library Materials* (New York: Holt, Rinehart and Winston, 1980), 98.

6. Michael Kenward, "Evolution on the Bookshelf," *New Scientist* 134 (May 23, 1992), 39.

7. Lydia F. Knight, "SciTech Magazines for Non-specialists," *American Libraries* (November 1982): 628–632 (an annotated bibliography, somewhat dated).

8. Bernard Dixon, ''Books and Films: Powerful Media for Science Popularization,'' *Impact of Science on Society* 36 (1986): 379–385. See also Kathryn Wolff, et al., *The Best Science Films, Filmstrips, and Videocassettes for Children* (Washington, DC, 1982) (now somewhat dated); Leroy Dubeck, Suzanne Moshier, and Judith Boss, *Science in Cinema* (New York: Teacher's College Press, 1988); Ortrum Zuber-Skerrit, *Video in Higher Education* (New York: Nichols, 1984); James A. Brown, *Television Critical Viewing Skills* (Hillsdale, NY: Erlbaum, 1991).

9. Kenneth O'Bryan, *Writing for Instructional Television* (Washington, DC: Corporation for Public Broadcasting, 1981).

10. Angus Reynolds, and Ronald Anderson, *Selecting and Developing Media for Instruction*, 3rd ed. (New York: Van Nostrand Reinhold, 1992); Ann Barron and Gary Orwig, *New Technologies for Education: A Beginner's Guide* (Littleton, CO: Libraries Unlimited, 1993).

Part II

Subject Guides to Popular Science Information Resources

The following nine chapters contain annotated subject bibliographies for popular information resources in the major disciplines of modern science and some of the specific issues and concepts within them. The goal is to recommend a core of popular science resources that could represent these fields—their foundations and frontiers—in general library collections.

Entries are qualitatively selective, having been chosen with consideration to the evaluative criteria discussed in Chapters Three and Four. By necessity, entries are also quantitatively selective. Not all subjects are equally represented in the literature of popular science. While there is a glut of current books and other resources on certain topics, other topics have been almost totally ignored. Some whole disciplines, such as chemistry and mathematics, are comparatively underrepresented in popular literature. Thus, while the numbers of citations listed by general disciplines and special topics within them are roughly indicative of the volume of publishing output for each, there is also an attempt to include other subjects that are not particularly hot, but which are nonetheless basic to their fields, whenever such resources are available.

Within each chapter, the specific subjects listed are those that are suggested for representation in general library collections. For each subject, one or more core titles are cited, along with others that are listed under the heading "Also Recommended." The purpose of this organization is, first, to focus on the relevant subjects that fill unique needs in science literacy collections and, second, to look comparatively at some of the resources by which they can be represented. Cross-references between subjects are also given, where appropriate. Nonbook resources are listed by form—CD-ROMs, audiovisual materials, and so on—in sections separate from the subject categories.

In total, the subjects listed in each chapter represent the body of current, significant popular literature for an entire discipline. Maximum breadth of coverage has been sought for each discipline to the degree that appropriate popular resources exist for various specialties and distinct topics.

For core entries listed, the complete bibliographic citation includes author, title, publisher, date, content notes, and descriptive annotations (generally 25–75 words). For entries listed as "Also Recommended," citations do not include content notes, and annotations are briefer. Up to three book reviews are cited for every title. Sources of book reviews are abbreviated (for example, *Book Review Digest* is abbreviated as BRD); see the Appendix, "Book Review Source Abbreviations," for a key to these.

In annotations, consideration is given to such matters as a resource's level of relevance, its scope and comprehensibility, and its uniqueness. For obvious reasons, currency is favored (over 70 percent of materials cited have been published since 1990), but enduring classics are given their due. Frequently, comparisons are made between resources, especially the ways in which they complement or contrast with one another. Such information can be useful not only for collection development purposes, but also for making reference recommendations or designing programmatic learning activities. The substance of the annotations reflects the concept that cultivating science literacy is an ongoing, progressive activity—a matter of lifelong learning.

Chapter Five

Science—General

One thing I have learned in a long life: that all of our science, measured against reality, is primitive and childlike—and yet it is the most precious thing we have.

Albert Einstein

. . . he that increaseth knowledge, increaseth sorrow.

Ecclesiastes 1:18

It seems to me that when it's time to die, and that time will come to all of us, there'll be a certain pleasure in thinking that you had utilized your life well, that you had learned as much as you could, gathered in as much as possible of the universe, and enjoyed it.

Isaac Asimov

We must know. We must know.

David Hilbert (inscription on his tomb)

The word "science" means many things to many people. To some, it suggests inquisitiveness and objectivity—a "discipline of curiosity." To others, it conveys images of the laboratory, where exacting observations and experiments are conducted using rigorous methods. Some see it as bold and visionary, while others see it as mechanistic and dogmatic. We have pure science, the realm of theory; experimental science, where theories are tested; and applied science, where knowledge is put to work. Today, science can also be viewed in terms of the political and economic factors under which it is done: "Big Science," large-scale, institutionalized research, in contrast to "Little Science," the work of individuals and small teams. Thus, the concept of "science," like others of comparable scope within human experience—"art" and "religion," for example—is all of these things, but is also bigger than the sum of its combined parts.

Central to all the sciences is something that has been called "the scientific method." In practice, there are many scientific methods, but the objective of them all is to achieve *results*. Observations are made in order to draw conclusions. Experiments are conducted to test hypotheses. Products are designed to work. Though the results of scientific undertakings might be quite different, they are obtained by means of basic standardized procedures that have been proven to yield reliable results in the past. To verify that appropriate procedures have been employed, science relies on a process of peer review, in which fellow scientists assess the work of their colleagues. To be sure, this simplified model is variously interpreted and, like any other work of humans, sometimes fails to meet the ideal. Still, a hallmark of science is that knowledge must be *validated*. No idea that has not been subject to this manner of scrutiny can claim to be truly scientific.

Three broad subject categories of materials are cited in this chapter. The first deals with science as a belief system and how it works. These materials explain scientific methodologies and the values they embody, often with reference to historical or contemporary case studies. The second category examines the social roles of science. Science is a product of society, and it thus reflects the whole cultural, social, economic, and political fabric of that society. Indeed, some sociologists have put forth global theories regarding the differential acceptance of scientific principles among and between cultures. This subject, in particular, has been relatively underrepresented in popular literature, and it can be hoped that, in an age where people look to science to solve many of the world's problems, more accessible books and articles will be produced exploring the complex interplay of science and society. Finally, included in this chapter are substantive works that cover science at a general level. In addition to reference works, these tend to be either primers for nonspecialists, or popular presentations of newsworthy topics and discoveries.

Materials covering general topics in science can be gateways to other science information resources. The irony is that the greatest volume of publishing activity and popular interest is generally focused on specific issues, like the Human Genome Project or artificial intelligence. Library patrons who start with these specialized materials, however, frequently return to the more general works for background and reference, or for perspective. Perhaps a better approach to guided or self-directed science literacy explorations would be to start with one or more of the general works cited here, then advance to topical works. In either case, however, a core of general science materials and reference works is essential to a science literacy collection.

GENERAL WORKS

Reference

Academic Press Dictionary of Science and Technology. Christopher Morris, ed. Academic Press, 1992. 2432 p., illus. 0–12–200400–0.

A massive work, optimally organized for showing the interconnectedness of the sciences. Over 124 broad fields are covered, from acoustical engineering to zoology, and throughout each of the nearly 125,000 terms that are defined are linked to at least one of these fields. Each of the major fields is introduced by a "window," a boxed essay on the general subject written by a noted expert. Most definitions are under 100 words. (A Lib v24 My 1993 p388; Choice v30 F 1993 p935; Nature v360 D 10, 1992 p546)

Brennan, Richard. *Dictionary of Science Literacy.* Wiley, 1991. 334 p., illus. 0–471–53214–2.

Brennan wrote and compiled this dictionary to meet the American Association for the Advancement of Science's guidelines for science literacy. Some 650 terms were selected on the basis of their core importance to modern science. Each entry contains a concise definition, followed by an explanation of the concept's relevance to daily life. Thoroughly cross-referenced. (BL v89 F 1, 1992 p1053; LJ v116 D 1991 p132; SBF v28 Mr 1992 p43)

Bunch, Bryan. *Henry Holt Handbook of Current Science and Technology.* Holt, 1992. 689 p., index. 0–8050–1829–8.

Subtitled "A Sourcebook of Facts and Analysis Covering the Most Important Events in Science and Technology," this impressive tome is arguably the best one-volume source for current, substantive information on recent and future scientific developments. Authoritative articles summarize the state of the discipline for all major fields and give information on breaking discoveries. Throughout, there are tables, charts, chronologies, and various other sources useful for ready reference. (A Lib v24 My 1993 p388; ARBA v24 1993 p614; Choice v30 F 1993 p981)

Chambers Concise Dictionary of Scientists. David Millar et al., eds. Cambridge University Press, 1989. 461 p., illus. 1852963549.

Most science dictionaries and encyclopedias provide rather superficial biographical coverage. While the biographies in this book are brief, it nonetheless fills a gap and is much more affordable than the multivolume *Dictionary of Scientific Biography,* listed elsewhere in this section. (RSSEMA p43; TES S 29, 1989 p31; TLS Ag 4, 1989 p845)

General Science Index. H. W. Wilson, 1978– . m. 0162–1963.

This distinguished reference source indexes the contents of over 100 core journals in all fields of the sciences. This is the best gateway to periodical literature for those seeking general scientific information.

Hellemans, Alexander, and Bryan Bunch. *The Timetables of Science*. Touch-stone/Simon and Schuster, 1991. 660 p., index. 0–671–73328–1.

The drama of science can best be understood within a broad historical con-text. For that reason, every library should have a basic chronology of science. This one is recommended because it juxtaposes major fields of science within a time continuum, so that at a glance you can see the major historical accom-plishments in several fields at approximately the same time. (SA v266 Ja 1992 p148)

Malinowsky, H. Robert. *Reference Sources in Science, Engineering, Medicine, and Agriculture*. Oryx Press, 1994. 355 p., index. 0–89774–3.

Listing over 2,400 current resources, this is the best literature guide to the sciences available. Totally up to date. Annotations are succinct but very helpful.

Van Nostrand's Scientific Encyclopedia. 7th ed. Van Nostrand, 1989. 2 vols., index. 0–442–21750–1.

This workhorse encyclopedia offers general overviews of major fields and more concise reviews of specific topics in the sciences. Explanations are direct and minimally technical. For libraries unable to afford a more expensive mul-tivolume encyclopedia, this will serve in probably 80 percent of the cases. (RSSEMA p35)

Also Recommended

Almanac of Science and Technology. Richard Golob and Eric Brus, eds. Har-court Brace Jovanovich, 1990. Broad, readable overviews on key ideas and events within all disciplines. (KR v58 Ja 15, 1990 p100; LJ v115 Mr 1, 1990 p86)

Dictionary of Scientific Biography. Charles Coulston Gillispie, ed. Scribner's, 1980. The most thorough biographical reference for information on scientists, this multivolume set issues regular supplements. Articles are probing and semi-scholarly. (NYTBR v85 Ag 31, 1980 p6; SA v243 Ag 1980 p35)

Historical Science Experiments on File. Facts on File, 1994. Gives details on seventy-five famous experiments. Loose-leaf format. Has great pedagogical value.

McGraw-Hill Encyclopedia of Science and Technology. 7th ed. Sybil Parker, ed. McGraw-Hill, 1992. This standard 20-volume encyclopedia is an expensive but valuable investment. Covers all areas at an appropriate technical level for informed generalists. The parent title has spawned several specialized volumes. Also available on CD-ROM. A brand new 1994 abridged edition is available. (RSSEMA p35)

Science and Technology Desk Reference. Carnegie Library of Pittsburgh, Sci-ence and Technology Department, eds. Gale, 1993. Ready answers to technical

questions compiled by the staff of the Science and Technology Department at
the Carnegie Library of Pittsburgh. (A Lib v24 My 1993 p420; Choice v30 Je
1993 p1619; LJ v188 Mr 15, 1993 p74)

History of Science

Album of Science. I. Bernard Cohen, ed. Macmillan, 1980–1989. 5 vols., biblio.,
index. 0–684–19074–5.
   Volumes in this well-received set cover antiquity to the Middle Ages, 1450–
1800, the nineteenth century, and biological and physical sciences in the twentieth century. Very well illustrated; could also serve as a guide to the history of
scientific illustration. (RSSEMA p44)

Boorstin, Daniel. *The Discoverers: A History of Man's Search to Know His
World and Himself.* Random House, 1983. 717 p., biblio., index. 0394402294.
   A Pulitzer Prize winning historian and former Librarian of Congress, Boorstin
portrays the history of Western science as an epic of discovery that has been
pushed forward by an insatiable human need to understand the world. The style
is readable and the scope is grand. It is at once the story of entire civilizations
and of the careers and personalities of the individual makers of science. Covers
science from ancient times through the early twentieth century. (BL v80 O 1,
1983 p185; Choice v21 Ap 1984 p1183; LJ v108 N 15, 1983 p2156)

Cohen, I. Bernard. *Revolution in Science.* Belknap Press/Harvard University
Press, 1985. 711 p., refs., index. 0–674–76777.
   Intellectual revolutions irreversibly change not only the content of belief and
practice within a discipline, but entire social outlooks as well. Cohen begins by
establishing, at some length, the "revolutionary" theory in the history of science, then from that foundation tells the sweeping history of the last four centuries of scientific inquiry. Stresses the historical evidence for the occurrences
of true revolutions. Scholarly but fascinating. (KR v53 Mr 1985 p213; LJ v110
Je 1, 1985 p134; Nature v315 My 30, 1985 p433)

Companion to the History of Modern Science. R. C. Olby et al., eds. Routledge,
1990. 1081 p., index. 0–415–01988–5.
   This ambitious anthology gathers articles from leading scholars who explore
diverse historical and contemporary philosophical problems, turning points, topics and interpretations, and themes related to science. This would best be used
as a guide and handbook for those with a more than passing interest in the
subject. Could open doors to a lifetime's worth of reading. (SBF v27 Ja 1991
p5)

Lindberg, David C. *The Beginnings of Western Science: The European Scientific
Tradition in Philosophical, Religious and Institutional Context, 600 B.C. to*

1450. University of Chicago Press, 1992. 455 p., illus., maps, index. 0–226–48230–8.

The revolution in Western science that began with Copernicus in some ways represented a major break from intellectual traditions of the past, but in many other ways also continued them. This book documents the essential fabric of that tradition, from the dawn of Greek science through medieval times. (LJ v117 Jl 1992 p117; NYTBR v97 S 20, 1992 p49)

On the Shoulders of Giants (series). Facts on File, 1993– . 5 vols., biblio., gloss., index.

For high school students and beginners, the five volumes in this series comprise the best available general summary of the history of science. Each volume is brief (under 200 pages), limits itself to core concepts, and provides lists for further reading. (BL v90 N 1, 1993 p514; SBF v29 N 1993 p234; SLJ v39 D 1993 p146)

Also Recommended

Needham, Joseph, and Colin Ronan. *The Shorter Science and Civilization in China*. Cambridge University Press, 1985. An abridgment of Needham's multivolume landmark work *Science and Civilization in China*. (TES Ja 24, 1986 p34)

Schwartz, Joseph. *The Creative Moment: How Science Made Itself Alien to Modern Culture*. HarperCollins, 1991. By examining several key moments in the history of Western science, the author shows how its core concepts and institutions have retreated farther and farther from common experience. (BL v88 My 1, 1992 p1572; KR v60 Mr 15, 1992 p381; LJ v117 Ap 15, 1993 p118)

(*Note*: Almost all general histories of science are written from a Western and European perspective. More balanced world histories of science should appear in coming years.)

General and Miscellaneous

Asimov, Isaac. *Frontiers*. Truman Talley: Dutton, 1991. 390 p. 0–525–24662–2; Asimov, Isaac, and Janet Asimov. *Frontiers II: More Recent Discoveries about Life, Earth, Space and the Universe*. Truman Talley: Dutton, 1993. 369 p., index. 0–525–93631–9.

These companion books, the second of which was published after Isaac Asimov's death in 1992, summarize in approximately 1,000-word essays some of the very latest scientific discoveries and near-future research horizons. Good for anybody seeking a quick, painless way to catch up. (BL v89 Jl 1993 p1930; LJ v118 Jl 1993 p111)

Gardner, Martin. *The New Ambidextrous Universe*. 3rd ed. Freeman, 1990. 392 p., illus., index. 0–7167–2092–2.

The first edition of this tour de force by *Scientific American*'s mathematics wizard was published in 1964, and over the years it has entertained and enlightened thousands of readers. Gardner writes about symmetries and asymmetries in nature, some of which are part of our everyday experience, such as the reversed image we see in a mirror, and others which are more exotic, like those found in the structures of molecules. The lively brainteasers throughout make it well worthwhile to spend some time with this book. (LJ v116 Mr 1, 1991 p63)

Greenwood, Addison et al. *Science at the Frontiers*. National Academy Press, 1991. 254 p., index. 0–309–04480–4.

The National Academy of Science's own report, which was produced as the result of a series of "Frontiers" symposia. Covers ten subjects in the biological and physical sciences as well as engineering. Articles are written at a level for informed and motivated laypersons. (SBF v29 Ag 1993 p167)

Rutherford, F. James. *Science for All Americans*. Oxford University Press, 1990. 246 p., index. 0–19–506771–1.

This book states the recommendations and substance of the AAAS's Project 2061, which is aimed at the elimination of science illiteracy in America. In its fifteen chapters, readers will get a crash course in the basic concepts of the major scientific fields. The final chapters are for teachers and suggest practical reforms that can improve science education. (LATBR Mr 10, 1991 p10)

Trefil, James, and Robert Hazen. *Science Matters: Achieving Science Literacy*. Doubleday, 1991. 294 p., index. 0–385–24796–6.

The double entendre in the title describes both the content and the importance of this book. The authors, two scientists and popularizers, take the view that certain basic, unifying principles can be used to explain the rudiments of science in a way that generalists can grasp and also apply in their daily lives. They thus explain these bedrock concepts and how they relate to all fields. A good primer for anybody aspiring to science literacy. (LJ v117 Mr 1, 1992 p47; Kliatt v26 Ap 1992 p42; PW v238 D 6, 1991 p69)

The World Treasury of Physics, Astronomy and Mathematics. Timothy Ferris, ed. Little, Brown, 1991. 859 p., index. 0–316–28129–8.

A rich and highly readable omnibus, this collection gathers the work of over seventy writers, including many contemporary and historical giants of science, as well as nonscientists with relevant insights. There are four general, topical categories: "The Realm of the Atom," "The Wider Universe," "The Cosmos of Numbers," and "The Ways of Science." Selections show not only science's technical foundations, but also its expressive and philosophical sides. (KR v58 D 1, 1990 p1666; New Sci v131 Ag 24, 1991 p49; SBF v27 Je 1991 p136)

Also Recommended

Asimov, Isaac. *Asimov's New Guide to Science*. Rev. ed. Basic Books, 1984. Unfortunately dated, but this book was until recently the best single source for comprehensive, general information in all fields of science. Widely found in libraries. (KR v52 S 1, 1984 p834)

Hall, Stephen. *Mapping the Next Millennium*. Random House, 1991. Vistas of science rendered visual through the use of high-tech cartography. (LJ v116 D 1991 p187; NYTBR v97 Mr 15, 1992 p20; PW v238 O 18, 1991 p46)

Masters of Science (series). John Brockman, general ed. HarperCollins, 1994– . This is a very promising new series of some dozen monographs, all under 200 pages, written by leading science popularizers on topics of current relevance. Contributors include John Barrow, George Smoot, Marvin Minsky, Stephen Jay Gould, Jared Diamond, Richard Leakey, and others.

The New York Times Book of Science Literacy. Richard Flaste, ed. Times Books, 1991. A somewhat desultory question-and-answer format, strongest in areas of human and cognitive sciences. (BL v87 Ja 15, 1991 p987; KR v58 D 15, 1990 p1733; LJ v116 F 15, 1991 p218)

On the Frontiers of Science: How Scientists See Our Future. Nigel Calder and John Newell, eds. Facts on File, 1989. A photographic tour of new developments in the fields of "Space," "Earth," "Body," "Mind," and "Humanity." High school age and up. (BRpt v9 My 1990 p62)

Royston, Robert. *Serendipity: Accidental Discoveries in Science*. Wiley, 1989. These vignettes show why Pasteur said, "Chance favors the prepared mind."

Trefil, James. *A Scientist in the City*. Doubleday, 1994. 266 p., index. 0–385–24797–4. From various urban vantage points, Trefil divines basic scientific principles and shows how they apply to experience. (BL v90 Ja 1, 1994 p795; LJ v118 N 15, 1993 p96; Nature v368 Mr 24, 1994 p367)

SPECIAL TOPICS

Essays

Asimov, Isaac. *The Relativity of Wrong*. Doubleday, 1988. 225 p. 0–385–24473–8; *Out of Everywhere*. Doubleday, 1990. 0–385–26201–9; *The Secret of the Universe*. Doubleday, 1991. 256 p. 0–385–41693–8.

Of all his writings, Asimov's essays were the favorites of many. The selections from these later works were culled from his regular column in the *Magazine of Fantasy and Science Fiction*. (BL v86 My 15, 1990 p1764; KR v58 Ap 15, 1990 p543; PW v238 Ja 25, 1991 p42)

Bernstein, Jeremy. *Experiencing Science*. Basic, 1978. 275 p. 0–465–02185–9; *Quarks, Cranks and the Cosmos*. Basic, 1993. 0–465–08897–x.

Bernstein is a physicist and a renowned popularizer whose approach is to profile a person or event in science. Most of these pieces were originally published in the *New Yorker.* (Choice v30 Jl 1993 p1789; KR v60 D 15, 1992 p1543)

Doing Science. John Brockman, ed. Prentice-Hall, 1991. 296 p. 0–13–795097–7; *Ways of Knowing.* John Brockman, ed. Prentice-Hall, 1991. 284 p. 0–13–517236–5.

Essays from members of the Reality Club, a select group of scientists and scholars chartered to "arrive at the edge of the world's knowledge, seek out the most complex and sophisticated minds, put them in a room together, and [have] them [ask] each other the questions they are asking themselves." (BL v87 D 1, 1990 p695; PW v237 D 7, 1990 p79)

Dyson, Freeman. *Infinite in All Directions.* Harper, 1988. 319 p. 0–06–039081–6; *From Eros to Gaia.* Pantheon, 1992. 372 p. 0–679–41307–3.

By training Dyson is a physicist, but as these essays demonstrate, he possesses a voracious intellect and is not afraid to venture controversial, sometimes chimerical opinions. (Aud v94 S 1992 p116; Choice v26 N 1988 p509; LJ v114 Mr 1, 1989 p44)

Great Essays in Science. Martin Gardner, ed. Prometheus, 1994. 427 p. 0–97975–853–8.

Gardner, a leading figure in popular science, selected the thirty-four essays in this collection because they "have something to say about science and say it forcibly and well." Has a historical context. (SBF v30 Ap 1994 p70)

Konner, Melvin. *Why the Reckless Survive.* Viking, 1990. 306 p. 0–670–82936–6.

Thoughts and opinions from psychology, anthropology, neuroscience, and medicine. (LJ v116 Mr 1, 1991 p63; NYTBR v96 14, 1991 p32)

Medawar, Peter. *The Threat and the Glory.* HarperCollins, 1990. 1991 p., index. 0–06–039112–x.

Twenty-three essays from a Nobel laureate explore the theme that, in science, the threat and the glory are two sides of the same coin. (LJ v116 Mr 1, 1991 p64; Nature v352 Ag 22, 1991 p676; TLS Ja 11, 1991 p21)

Morowitz, Harold. *The Thermodynamics of Pizza.* Rutgers University Press, 1991. 245 p. 0–8135–1635–8.

The author's works have appeared in numerous national publications. In this and other collections, he revels in the oddities and idiosyncrasies of science with a questioning mind and sharp wit. (Choice v29 O 1991 p299; LJ v116 Ap 1, 1991 p146)

Mysteries of Life and the Universe. William Shore, ed. Harcourt Brace Jovanovich, 1993. 0–15–163972–8.

Thirty of the very finest contemporary science essayists contributed to this collection, the proceeds of which were donated to a hunger relief organization.

An innovative compilation and an intriguing science sampler. (BL v89 N 1, 1992 p474; LJ v117 N 1, 1992 p111; PW v239 O 5, 1992 p62)

Perutz, Max F. *Is Science Necessary?* Dutton, 1989. 285 p. 0–525–24673–8.

These general essays reflect the author's belief that "true science thrives best in glass houses." The title essay should be required reading for science literacy. (BL v85 D 15, 1988 p670; KR v56 D 15, 1988 p1796; Nature v339 My 18, 1989 p190)

Raymo, Chet. *The Virgin and the Mousetrap.* Viking, 1991. 199 p. 0–670–83315–0.

In his essays, Raymo searches for the "soul" of science; his writings imbue his subjects with mystery and wonder. (BL v87 Ag 1991 p2085; KR v59 Je 1, 1991 p718; SBF v27 N 1991 p228)

Sagan, Carl. *Broca's Brain: Reflections on the Romance of Science.* Random House, 1979. 347 p. 0–394–50169–1.

Some of the facts are dated, but the inquisitive zeal that Sagan conveys is timeless. Essays cover topics of "human concern" and others. (Choice v16 D 1979 p1328; NYTBR N 25, 1979 p74; SBF v15 Mr 1980 p190)

(*Note*: The preceding represent just a handful of the fine anthologies available in the tradition of science essays. Elsewhere throughout this book, collections that are topic-specific are listed where appropriate. Since essays are perhaps the most common form of popular science writing, any science literacy collection should contain a core of titles, although what constitutes that core can be widely and quite differently interpreted.)

Methods and the Nature of Scientific Work

Collins, Harry, and Trevor Pinch. *The Golem: What Everyone Should Know about Science.* Cambridge University Press, 1993. 164 p., index. 0–521–35601–6.

"This book is for the general reader who wants to know how science really works and how much authority to grant to experts." The authors address this audience by examining several case studies in the history of science. In doing so, they debunk certain myths of the scientific process and give due credit— and *dis*credit—to human factors. (New Sci v138 My 29, 1993 p41)

Grinnell, Frederick. *The Scientific Attitude.* 2nd ed. Guilford Press, 1992. 180p. 0–89862–018–x.

"What do scientists do when they do science?" The author poses this question and discusses the challenges that it presents. Covers making observations, formulating hypotheses, designing experiments, and interpreting data. Also elaborates on the social and political factors that affect how science is done. Some in the field believe this is the book for any student thinking about becoming a scientist. (SBF v28 Je 1992 p135; SciTech v16 My 1992 p3)

Harre, Rom. *Great Scientific Experiments: Twenty Experiments that Changed Our View of the World.* Oxford University Press, 1981. 0–19–5204–36.

Three categories of famous experiments: (1) those that exemplify formal methodologies; (2) those emphasizing content and theory; and (3) a few that illustrate techniques. (Choice v21 Mr 1984 p1001; LJ v108 D 15, 1983 p2339)

Rothman, Milton A. *The Science Gap: Dispelling the Myths and Understanding the Reality of Science.* Prometheus, 1992. 254 p., index. 0–87975–710–8.

For the educated lay reader, this book examines sixteen myths about science, such as "Scientists don't have any imagination" and "All scientists are objective," and exposes their clichés, inaccuracies, and stereotypes. The aim is to eliminate popular misconceptions. (Astron v20 S 1992 p94; SBF v28 Je 1992 p141)

Also Recommended

Carr, Joseph. *The Art of Science: A Practical Guide to Experiments, Observations, and Handling Data.* HighText, 1992. A handy student guidebook. (SBF v29 Mr 1993 p44)

The Nobel Prize

McGrayne-Bertsch, Sharon. *Nobel Prize Women in Science.* Birch Lane Press, 1992. 419 p., index. 1–55972–146–4.

Asking why so few women have achieved the honor of a Nobel Prize in science, McGrayne-Bertsch makes the answer quite clear in these inspiring biographies. The women featured here possessed the unique combination of drive, intelligence, and passion for inquiry that enabled them to prevail in a male-dominated system. (BL v89 D 15, 1992 p705; Choice v31 O 1993 p311; LJ v118 F 1, 1993 p105)

Taubes, Gary. *Nobel Dreams: Power, Deceit, and the Ultimate Experiment.* Random House, 1986. 503 p., index. 0–394–54503–6.

In 1984 Carlo Rubbia won the Nobel Prize for his work in experimental physics. This is the story of how he made his discoveries. Instead of giving a romanticized account of a scientific breakthrough, Taubes depicts the world of cutting-edge science as being highly competitive and driven by the desire for glory. (BRD 1987 p1832; LJ v112 Mr 1, 1987 p84; NYTBR Ja 25, 1987 p11)

Philosophy of Science

Harre, Rom. *The Philosophies of Science: An Introductory Survey.* New ed. Oxford University Press, 1984. 203 p., index. 0–19–289201–0.

In this, a standard text for introductory courses in the philosophy of science, the author shows how views about the essential nature of science are related to various schools of philosophy. Arguments look at concrete historical episodes.

Kuhn, Thomas S. *The Structure of Scientific Revolutions.* 2nd ed. University of Chicago Press, 1970. 210 p., biblio.

Kuhn's study of the paradigms of scientific progress is a watershed work in the history and philosophy of science and remains essential reading for anybody with a serious interest in the field. Appropriate for large to mid-size public libraries, and core for college libraries. (See *World Changes: Thomas Kuhn and the Nature of Science*, Paul Horwich, ed. MIT Press, 1993.)

Strahler, Arthur. *Understanding Science: An Introduction to Concepts and Issues.* Prometheus, 1992. 409 p., index. 0–87975–724–8.

This book was written to serve as an introduction to the philosophy of science for students and educated laypersons. Part One presents basic concepts and issues central to science. Part Two, perhaps the most enlightening, examines science and its interrelations with other fields of knowledge. The wealth of ideas contained here compensates for the rather dry style in which the book is written. (PW v239 O 5, 1992 p62; SBF v29 My 1993 p101)

Wolpert, Lewis. *The Unnatural Nature of Science: Why Science Does Not Make (Common) Sense.* Harvard University Press, 1993. 191 p., notes, index. 0–674–92980–2.

Acknowledging that science is often misunderstood by the public, Wolpert asks why. In a nutshell, he argues that science forces you to think "unnatural thoughts," which require a leap of intellectual faith, but which, once made, can be infinitely rewarding. (Choice v31 O 1993 p312; LJ v118 Mr 1, 1993 p105; SBF v29 O 1993 p198)

Also Recommended

Bauer, Henry H. *Scientific Literacy and the Myth of the Scientific Method.* University of Illinois Press, 1992. A very useful book for anybody interested in the social phenomenon of science literacy. (Choice v30 O1992 p319; SBF v28 O 1992 p197; Sci v256 My 15, 1992 p1304)

Cromer, Alan. *Uncommon Sense: The Heretical Nature of Science.* Oxford University Press, 1993. 240 p., index. 0–19–508213–3. A study of the Greek origins of scientific thought with emphasis on how and why it developed. The central premise is that scientific thinking is, in a sense, "unnatural." A good companion to Wolpert's work, cited above. (BL v90 S 1, 1993 p17; KR v61 Jl 15, 1993 p907)

Snow, C. P. *The Two Cultures* and *A Second Look.* Cambridge University Press, 1990. 107 p. The two cultures of which Snow wrote are those of science and humanism. In this very important critical analysis, he argued that communications between the cultures have become strained, almost to the point of breaking down completely. Originally published in 1964. (BRD 1964 p1101; Sci Am v210 Je 1964 p134)

(*Note*: See also titles listed in other chapters under the subject "Philosophical Topics.")

Pseudosciences

Gardner, Martin. *On the Wild Side*. Prometheus, 1992. 257 p., index. 0–87975–713–2; *Science: Good, Bad and Bogus*. Prometheus, 1989. 412 p. 0–87975–573–3.

Prolific and occasionally acerbic science writer Martin Gardner offers his views on the trivial, fraudulent, and sometimes just silly information that is passed off as science in various public arenas. (SBF v28 O 1992 p203)

Also Recommended

Stegner, Victor. *Physics and Psychics*. Prometheus, 1990. A scientist and skeptic debunks paranormal phenomena. (LJ v116 Mr 1, 1991 p63)

Social Aspects of Science—General Topics

Foerstel, Herbert. *Secret Science: Federal Control of American Science and Technology*. Praeger, 1993. 227 p., index. 0–275–94447–6.

In science, access to information is critical to free inquiry. The government, however, is often reluctant to release information that it regards as sensitive. The author, a librarian, argues that scientific secrecy is counterproductive and dangerous. (BL v89 Ap 1, 1993 p1391; Choice v30 Je 1993 p1648; KR v61 Mr 1, 1993 p274)

Nelkin, Dorothy. *Selling Science: How the Press Covers Science and Technology*. Freeman, 1987. 224 p., index. 14413514.

The author alleges that many in the press treat news as just a vehicle for selling newspapers. This has negative consequences for the nation's science literacy and public support of science. (Choice v25 S 1987 p156; LJ v112 My 15, 1987 p90; Sci v236 My 22, 1987 p872)

Price, Derek J. de Solla. *Little Science, Big Science*. Columbia University Press, 1986. 301 p., illus. 0–231–04956–0.

The science of such geniuses as Galileo, Newton, and Einstein was largely that of individuals—hence, it was, in Price's terms, "Little Science." Conversely, a "Big Science" approach to research usually involves whole departments of scientists working at or for large bureaucratic institutions. (LJ v112 Mr 1, 1987 p35; SciTech v10 N 1986 p3)

Restivo, Sal. *Science, Society, and Values: Toward a Sociology of Objectivity*. Lehigh University Press, 1994. 270 p., notes, index. 0–934223–21–1.

Although semi-scholarly, this could serve as an introduction to the sociology of science for undergraduates and general readers. The author, a working social historian of science, describes intellectual models for the sociology of science from the nineteenth century to the present. For motivated readers, this work invokes a broad context for the field and raises stimulating questions.

Smith, Bruce L. R. *The Advisors: Scientists in the Policy Process*. Brookings Institution, 1992. 238 p., index. 0–8157–7990–9.

Science advisors are scientists and technical experts who counsel political leaders on policy issues. Smith focuses on their significance in influencing presidential decisions through their activities within the Department of Defense, the Environmental Protection Agency, and other entities. He shows the system in action—how it works and, when it fails, why. (SBF v28 N 1992 p230)

Smith, Bruce L. R. *American Science Policy since World War II*. Brookings Institution, 1990. 230 p., refs., index. 0–8157–7997–6.

American "science policy" has not always been science, or always policy. What were commonly shared beliefs between scientists and political leaders at the beginning of the post–World War II era became increasingly fragmented and often contentious as the nation advanced into the Vietnam War years. This book examines those tumultuous years in public science, the effects of which are still profoundly felt today. (Sci v251 Ja 11, 1991 p210)

Also Recommended

Big Science: The Growth of Large Scale Research. Peter Galison and Bruce Hevly, eds. Stanford University Press, 1993. Examines the social and institutional forces behind scientific bureaucracies. (Choice v30 D 1992 p655; New TB v77 Jl 1992 p836)

The Discipline of Curiosity: Science in the World. Janny Groen, ed. Elsevier, 1990. Experts from many disciplines and professions reflect on roles for science in society within a global context. (New TB v76 Jl 1991 p904)

Primack, Joel, and Frank von Hippel. *Advice and Dissent: Scientists in the Public Arena*. Basic, 1974. The role of technical experts in society and politics. Considers mistakes and successes. Still cited in the literature. (BRD 1975 p1020; LJ v99 N 15, 1974 p2972; NYTBR Je 29, 1975 p4)

Reingold, Nathan. *Science, American Style*. Rutgers University Press, 1991. Readable, intelligent essays on the practice of science in America. (Choice v29 My 1992 p1416; New TB v77 Mr 1992 p296)

Social Aspects of Science—Economic Issues

Mukerji, Chandra. *A Fragile Power: Scientists and the State*. Princeton University Press, 1989. 252 p., biblio., index. 0–691–08538–2.

The author writes, "Americans have forgotten why they funded science to begin with, and have not paid attention to the forces that have distorted their memories.'' In examining this thesis he conducts a cultural analysis of the relationship between scientific power and laboratory practice, with special emphasis on the role of the government as cultural authority. (Nature v344 Ap 26, 1990 p888; NYTBR v95 Ap 8, 1990 p18; Sci v248 My 18, 1990 p872)

Teitelman, Robert. *Profits of Science: The American Marriage of Business and Technology.* Basic, 1994. 258 p., notes, index. 0–465–03983–9.

Far from the ivory towers of academic research, much—if not most—technological innovation today is generated within the R&D departments of corporations. Market factors and corporate cultures thus dictate the kinds of research done. Teitelman argues for an ''ecology of technology,'' which would enhance our understanding of technological change and its relation to economic productivity. (Nature v368 Mr 24, 1994 p372; NYTBR v99 My 13, 1994 p9; PW v241 Ja 24, 1994 p49)

Social Aspects of Science—Ethics and Fraud

Bell, Robert. *Impure Science: Fraud, Compromise, and Political Influence in Scientific Research.* Wiley, 1992. 301 p., index. 0–471–42913–3.

Fundamental to good science are experimental neutrality, empirical objectivity, and a peer review system of checks and balances. Unfortunately, when huge sums of money are at stake, conflicts of interest and political pressures can sometimes subvert the ideal. Bell looks at several recent celebrated cases where external factors led to the production of questionable science. (BL v88 Je 15, 1992 p1793; KR v60 Ap 15, 1992 p506; LJ v117 My 15, 1992 p114)

Blum, Deborah. *The Monkey Wars.* Oxford University Press, 1994. 306 p., notes, index. 0–19–509412–3.

The use of primates in biomedical research is a controversial and polarizing issue. Based upon her Pulitzer Prize winning articles on the subject, Blum's book introduces key figures, conveys their values and convictions, and seeks a viable middle ground that addresses the needs of research and the ethical imperatives for human treatment of animals. (Note: No reviews are available for this book.)

Ethical Issues in Scientific Research: An Anthology. Edward Erwin et al., eds. Garland, 1994. 413 p., index. 0–8163–0641–5.

Essays on relevant topics in chapters entitled ''Science and Values,'' ''Fraud and Deception in Scientific Research,'' ''Experimentation on Humans,'' ''Animal Research,'' ''Genetic Research,'' and ''Controversial Research Topics.''

Sarasohn, Judy. *Science on Trial: The Whistle-Blower, the Accused, and the Nobel Laureate.* St. Martin's, 1993. 304 p., index. 0–312–09247–4.

Nobel winner David Baltimore was one of the most respected figures in science. When a subordinate began pointing out inconsistencies in a landmark article that he had coauthored, however, Baltimore stonewalled investigation.

Although the misdeeds were found to have been committed by a colleague of
Baltimore's, the work was tainted, and the resulting examination of the case
raised many issues on ethics, politics, and science. (KR v61 Ag 15, 1993 p1059;
NYTBR v98 O 17, 1993 p12)

Taubes, Gary. *Bad Science: The Short Life and Weird Time of Cold Fusion.*
Random House, 1993. 503 p., index. 0–394–58456–2.
   Cold fusion, which was announced with much fanfare in 1988, never existed.
Today the whole incident stands as a case study in how science ought not to
be done. Taubes gives an exhaustive and objective account. (BL v88 My 15,
1993 p1662; KR v61 Ap 15, 1993 p518; Nature v364 Ag 5, 1993 p497)

   Also Recommended

LaFollette, Marcel C. *Stealing into Print: Fraud, Plagiarism, and Misconduct
in Scientific Publishing.* University of California Press, 1992. An indictment of
certain practices in the world of scientific publishing. (LATBR S 27, 1992 p4;
PW v239 Jl 13, 1992 p40)

Research Fraud in the Behavioral and Biomedical Sciences. David Miller and
Michel Hersen, eds. Wiley, 1992. Examines cases and factors contributing to
scientific fraud. (SciTech v16 Ag 1992 p3)

Social Aspects of Science—Gender and Race Issues

Gornick, Vivian. *Women in Science: 100 Journeys into the Territory.* Revised
and updated ed. Touchstone, 1990. 0–671–41738–x.
   In over 100 vignettes on the lives and careers of spirited women scientists,
Gornick covers a lot of territory in describing the obstacles faced by women
scientists, as well as the rewards that come from transcending them. (Kliatt v19
Spring, 1985 p46; NYTBR v90 Ja 27, 1985 p36)

Schiebinger, Londa. *The Mind Has No Sex? Women in the Origins of Modern
Science.* Harvard University Press, 1989. 355 p., biblio., index. 0–67457623–
3.
   The fundamental question Schiebinger raises is, Why have women scientists
been so historically rare? In attempting to answer this question, she examines
several social and institutional factors in the practice of science in Europe during
the seventeenth and eighteenth centuries that led to the exclusion of women
from scientific endeavor. Mostly for college and university libraries. (KR v57
Ag 1, 1989 p1145)

Shipman, Pat. *The Evolution of Racism: Human Differences and the Use and
Abuse of Science.* Simon and Schuster, 1994. 318 p. biblio., index. 0671754602.
   As society today becomes more sensitive to racial and cultural human diversity, so has science made great advances in understanding the genetic basis of
that diversity. In the past, eugenics was used as a supposedly scientific instru-

ment for denying basic human rights. Could new discoveries be used in the same way? Shipman looks at the history and current issues of this extremely delicate topic. (KR v62 My 1, 1994 p618; LATBR J 17, 1994 p4; PW v241 Je 13, 1994 p56).

Shepherd, Linda Jean. *Lifting the Veil: The Feminine Face of Science.* Shambhala, 1993. 329 p., index. 0–87773–656–1.

In this "personal journey," the author writes of the gradual insights that led her to an understanding of what it means to be a woman in science. She claims that there are certain values inherent in good science that are, by their nature, feminine. (BL v89 My 15, 1993 p1660; LJ v118 Ap 15, 1993 p123)

Also Recommended

Keller, Evelyn Fox. *Reflections on Gender in Science.* Yale University Press, 1985. Keller's goal is "the reclamation, from within science, of science as a human instead of a masculine project." She offers a thoughtful feminist critique of the ways by which science became associated with masculine values. (Choice v22 Je 1985 p1516; LJ v110 My 1, 1985 p63; NYTBR v90 Ap 21, 1985 p36)

The Outer Circle: Women in the Scientific Community. Harriet Zuckerman, Jonathan R. Cole, and John T. Bruer, eds. Norton, 1991. Essays on issues related to women in science, and interviews with noted women scientists. (KR v59 Ag 1991 p1003; LJ v116 Ag 1991 p138; PW v238 Jl 5, 1991 p52)

The "Racial Economy" of Science. Sandra Harding, ed. Indiana University Press, 1993. Essays on diversity in the history and practice of science. Some pieces are specialized, but others are of broad relevance. (LJ v118 O 1, 1993 p124; SciTech v18 Ja 1994 p5)

Social Aspects of Science—Science and Religion

Davies, Paul. *The Mind of God: Scientific Basis for a Rational World.* Simon and Schuster, 1992. 254 p., index. 0–671–68787–5.

Ever since Galileo, science has progressively ventured—some would say trespassed—into the realm of religion. Today, with modern physics and cosmology posing questions about the ultimate origin and fate of the universe, it is fair to ask if science might not supply better answers than religion. Unlike other books that have dealt with science and spiritualism, this is respectful to both and avoids New Age superficiality. (BL v88 F 15, 1992 p8; Nature v354 N 28, 1991 p632; SBF v28 Je 1992 p133)

Tipler, Frank. *The Physics of Immortality: Modern Cosmology, God and the Resurrection of the Dead.* Doubleday, 1994. 517 p. 0–385–46798–2.

Tipler writes: "Physics has now absorbed theology; the divorce between science and religion, between reason and emotion, is over." In defense of that claim, he makes sophisticated arguments for the compatibility of Judeo/Christian/Islamic cosmology and physical law. He also accepts the doctrine of resurrection of the dead, albeit in electronic form. Brazen, but thoughtful speculations. (KR v62 J 15, 1994 p970; PW v241 J 25, 1994 p38)

Also Recommended

Clements, Tad. *Science vs. Religion.* Prometheus, 1990. A defense of the humanistic, scientific worldview. (SBF v27 Mr 1991 p37)

(*Note*: See also entries listed in various sections under "Philosophy of" and "Philosophical Topics.")

Technical Writing

Rubens, Philip. *Scientific and Technical Writing: A Manual of Style.* Holt, 1992. 513 p., index. 0–8050–1831–x.
   A comprehensive guide to style and formatting for scientists and science students. Gives techniques for communications, grammar, audience assessment, notation handling, and layout. Includes electronic document preparation techniques. (ARBA v24 1993 p408; BL v89 O 15, 1992 p458)

Also Recommended

Day, Robert A. *Scientific English.* Oryx Press, 1988. A grammar and syntax handbook.

OTHER RESOURCES

Periodicals for General Readers

American Scientist Sigma Xi—Scientific Research Society, bi-m., revs. 0003–0996.
   Published by a national scientific honors society, this attractive and widely read magazine features topical articles by scientists and has an excellent book review section.

Discover. Walt Disney Publishing Group, m.   0319–8480.
   Glossy news articles designed to catch the eye. Features are written for maximum readability and public appeal, so they tend to emphasize the more fantastic conjectures and breaking frontiers.

Issues in Science and Technology. National Academy of Sciences, q. 0748–5492.
   Ideas and public policy issues presented with objectivity and authority, as befitting the highly respected academy.

Nature. Macmillan, w., revs. 0028–0836.

A tremendously respected British journal of new research in all of the sciences. Refereed. Often considered the status equivalent of the AAAS's major journal, *Science*.

New Scientist. IPC Magazines w., revs. 0077–8923.

A colorful British news magazine for all fields of science. Especially strong on the commentaries and editorials. Features the works of many leading British science writers. Widely read in the States also.

Popular Science. Times-Mirror, m. 0161–7370.

Popular articles of the ''what's new'' in technological innovations and equipment variety. Includes blueprints and how-to advice.

Science. American Association for the Advancement of Science, w., revs. 0036–8075.

The leading primary journal that covers professional American science comprehensively. The research is accessible only to specialists, but general readers can get a lot from the news and reviews.

Science Books and Films. American Association for the Advancement of Science, 5/yr., revs. 0098–342x.

A marvelous review source for information on books and audiovisual materials for lifetime science learners, from elementary students to general adults.

Science News. Science Service, w., revs. 0036–8423.

The leading American newsmagazine of the sciences. Reading this magazine cover to cover (most issues are less than twenty pages) every week is highly recommended as a way to keep up with science.

Scientific American. Scientific American, m., revs. 0036–8733.

The well-established and equally well-respected journal covers all areas of the sciences with intelligent articles written by specialists.

Audiovisual Resources

Powers of Ten. Pyramid Film and Video, 1978 (videocassette, 9 minutes, color).

Deals with scientific scales on the order of forty powers of ten, from the realm of the atom to cosmic distances.

A Private Universe. Pyramid Film and Video, 1989 (videocassette, 22 minutes, color).

This video has been widely used at conferences and teachers' meetings where science literacy is discussed. Interviews Harvard graduates and ninth grade students about their understanding of elementary astronomy concepts, and discovers that both populations have misperceptions that hinder science literacy.

Scientific Methods and Values. Hawkhill Associates, 1990 (videocassette, 30 minutes, color).

An overview of scientific methodologies with emphasis on the personalities and human factors that go into discovery. Includes passages on pseudosciences and values in science. (SBF v27 Je 1991 p148)

CD-ROMs

General Science Index. H. W. Wilson, q.
The equivalent of the well-known print index of the same name.

McGraw-Hill Science and Technology Reference Set (release 2.0). McGraw-Hill, 1991.
CD contains contents of print *Concise Encyclopedia of Science and Technology* and *Dictionary of Scientific and Technical Terms.*

SciTech Reference Plus. R. R. Bowker. 3598402279.
Disc provides information from *American Men and Women of Science*, Directory of American Research and Technology database, and the scientific and technical *Books in Print* volume.

Time Tables of Science and Innovations. Xiphias Co.
An original, interactive, media-based product that has programmed over 6,000 stories of scientific work and discoveries, from the beginning of time to beyond the future.

Astronomy and Space Sciences

Socrates: Shall we set down astronomy among the subjects of study?

Glaucon: I think so, to know something about the seasons, the months, and the years is of use for military purposes, as well as for agriculture and navigation.

Socrates: It amuses me to see how afraid you are, lest the common herd of people should accuse you of recommending useless studies.

Plato, The Republic

From our home on the Earth, we look out into the distances and strive to imagine the sort of world into which we were born. Today, we have reached far into space. Our immediate neighborhood we know rather intimately. But with increasing distance our knowledge fades. . . . The search will continue. The urge is older than history. It is not satisfied and will not be suppressed.

Edwin Hubble

In a sense, astronomy is the oldest science. Although its prescientific origins are shrouded in ritual and mythology, some of the same ancient cultures that sought omens from the stars and created fantastic stories of celestial deities also developed accurate techniques for predicting the positions of the planets, the dates of equinoxes and solstices, and, in the case of the Greek astronomer Thales, a solar eclipse. Today, where city lights overwhelm starlight for many, it might be difficult to comprehend the recondite meanings the stars had for ancient and medieval civilizations. Still, judging by the volume of astronomy popularizations produced every year, many curious people still gaze at the night sky and see in its lights possible answers to basic human questions: Who are we? Where did the universe come from? What other worlds exist?

Until about the time of Isaac Newton, astronomy was principally descriptive. The Greek astronomer Ptolemy designed a system explaining planetary motion that survived, albeit with major modifications, until Copernicus and Kepler. Aristotle's physics remained intact until Galileo's telescopic observations challenged its authority. It was Newton's laws of motion, however, that inextricably wed physics and astronomy, ushering in a new scientific era for the study of the cosmos. The emphasis changed from *describing* the sun, stars, and planets to *understanding* their composition and dynamics. In the twentieth century, the union is so strong that, ironically, scientists study the largest objects in the universe, stars and galaxies (and even the universe itself), by investigating their smallest components—basic particles like quarks, neutrinos, and X particles. Other disciplines, such as geology, meterology, and chemistry, have also contributed to astronomical research.

When Galileo first fixed his telescope on the sun and moon, he changed forever how we think of the universe. Today, however, a professional astronomer needs much more than a simple telescope to do research. Astronomers gather data from the entire electromagnetic spectrum and must perform exacting analyses in order to match theory with observation. These data are collected in enormous amounts and minute detail by complex instruments, including many satellite high technologies like the Cosmic Background Explorer and the recently repaired Hubble Space Telescope. Supercomputers are often employed to organize and analyze these raw data. Astronomy is thus an extremely technology dependent science. Even so, theorists may still do much of their work with just a pencil, paper, and a good deal of insight, while amateurs play an important role by keeping their eyes trained on the visual universe.

The popular information resources of the field reflect its eclecticism. Descriptive information, such as star charts or observers' guides, has longevity. Most good historical works written during the last twenty years remain authoritative up to the time of their publication. In the last decade, though, expansive new frontiers, both technological and theoretical, have opened in astronomy. This is particularly true in cosmology, which has developed a rich body of popular literature. Several new titles in cosmology and related subjects appear every year, and many of these become obsolete in five to ten years. Look for forthcoming books on such topics as dark matter, gravity waves, galaxy formation, and the mapping of the universe. Meanwhile, solar system astronomy is also making great advances. In the field of space sciences, politics affects progress as much as any other factor in terms of funding for research. Publication numbers reflect the boom or bust pattern.

ASTRONOMY-GENERAL

Reference

Cambridge Encyclopedia of Space. Michael Rycroft, ed. Cambridge University Press, 1990. 386 p., illus., photos. 0–521–36426–4.

The expository essays in this cross-referenced collection are organized in four sections: "From Dream to Reality," "Going into Space," "Exploring the Universe," and "Living in Space." The format allows for service as a reference book, but it also can be browsed or read for edification. (BRD 1991 p286; BL v87 Mr 1 1991 p1420; Choice v28 My 1991 p1454)

Moore, Diane. *The HarperCollins Dictionary: Astronomy and Space Science.* HarperPerennial, 1992. 338 p. 0–06–271542–9.

Over 2,300 entries cover the entire scope of astronomy and space sciences, from Abenezra (a lunar crater) to Zurich Number (a measure of sunspot activity). Included are technical terms as well as colloquialisms. Most terms are defined rather concisely, but lengthier treatments, often with diagrams or timelines, also appear. (SBF v29 Mr 1993 p44; S&T v84 S 1992 p289; SciTech v16 S 1992 p15)

Also Recommended

The Astronomy and Astrophysics Encyclopedia. Stephen P. Maran, ed. Van Nostrand, 1992. A definitive, moderately technical reference with over 500 entries written by specialists. (BL v88 Mr 15, 1992 p1399; Choice v29 Mr 1992 p1037; SciTech v16 F 1992 p8)

Extraterrestrial Encyclopedia. Joseph Angelo, Jr., ed. Facts on File, 1991. Space technologies and concepts related to the Search for Extraterrestrial Intelligence Program. (RSSEMA p48)

Autobiography and Biography

Drake, Stillman. *Galileo: Pioneer Scientist.* University of Toronto Press, 1994. 262 p. 0–8020–2725–3.

The trial of Galileo has become a symbol for the necessity of intellectual freedom in scientific discourse. Today, Galileo's remains one of the most recognizable names in the history of science. Drake is probably the leading English-speaking Galileo scholar alive; he lends considerable authority to this popularized biography.

Koestler, Arthur. *The Watershed: A Biography of Johannes Kepler.* University Press of America, 1985. 280 p. (originally published in 1960). 0–819–14339–1.

Johannes Kepler, who calculated the laws governing planetary orbital motions, is one of the most enigmatic figures in the history of science. Koestler, an acclaimed novelist, was fascinated by Kepler, and thus wrote of his torments and triumphs with unique empathy. This is also a story of a time, around the turn of the seventeenth century, when science and superstition coexisted.

Lightman, Alan, and Roberta Brawer. *Origins: The Lives and Times of Modern Cosmologists.* Harvard University Press, 1990. 561 p., notes, gloss. 0–674–64470–0.

In interviews with twenty-seven leading contemporary cosmologists, Lightman and Brawer explore their individual and collective genius, and their ruminations on the future of this most speculative of sciences. The book's introduction contains one of the most succinct and readable surveys of the salient issues in modern cosmology available, and the glossary is also very helpful. (Atl v266 S 1990 p113; LJ v115 Je 1, 1990 p168; NYTBR v95 S 16, 1990 p23)

Overbye, Dennis. *Lonely Hearts of the Cosmos: The Story of the Scientific Quest for the Secret of the Universe.* HarperCollins, 1991. 438 p., index. 0–06–015964–2.

Due to the indefatigable efforts of a motley collection of scientists and scholars, astronomy and cosmology have made tremendous advances in the last forty years. Overbye tells the story of those years of discovery by examining the personalities of several of the key players in the quest. The result is an intimate look at these individuals, some of whom are as eccentric as they are ingenious, and the scientific subculture to which they belong. (BL v87 Ja 15, 1991 p987; LJ v116 Ja 1991 p142; PW v237 N 9, 1990 p49)

Also Recommended

Gamow, George. *My World Line: An Informal Autobiography.* Viking, 1970. Offerings from astronomy's mischievous genius.

(*Note*: See also entries in Chapter Twelve under "Autobiography and Biography.")

History of Astronomy

Ferris, Timothy. *Coming of Age in the Milky Way.* Morrow, 1988. 495 p., illus., index. 0–688–05889–0.

Probably the most widely read history of astronomy available for general readers. It is much more than just a straight history, though, for it lends considerable insight into the processes of science and the personalities of important astronomers. Excellent illustrations and diagrams aid comprehension. (BL v84 Jl 1988 p1768; KR v56 Je 15, 1988 p872; NYTBR v93 Jl 17, 1988 p1)

Lightman, Alan. *Ancient Light*. Harvard University Press, 1991. 170 p., gloss., index. 0–674–03362–0.

The history of cosmology is complex, but Lightman's concise overview introduces the major figures and developing ideas necessary for anybody who wishes to be literate in the subject. A good first book on cosmology. (NYTBR v96 S 29, 1991 p22; PW v238 Ag 16, 1991 p44)

Also Recommended

Hoskin, Michael, gen. ed., *The General History of Astronomy*. Cambridge University Press, 1984. A comprehensive study, 4 vols. For colleges and larger public libraries. (S&T v69 Je 1985 p515; Sci v226 N 30, 1984 p1067)

(*Note*: Many books cited in this chapter provide historical overviews of their subjects.)

General Topics

Comins, Neil F. *What If the Moon Didn't Exist? Voyages to Worlds the Earth Might Have Been*. HarperCollins, 1993. 304 p., index. 0–06–016864–1.

The title sounds like (and in fact was inspired by) a child's innocent question, but in reality it represents a perfectly plausible astronomical situation. In answer to that question: if not for the moon's gravitational force, the Earth would rotate much faster, such that a day would be approximately eight hours long, and torrential winds would blow almost constantly. This intriguing and delightful book explores several possible planetary scenarios. (Astron v22 F 1994 p94; BL v90 N 15, 1993 p588; KR v61 O 15, 1993 p1302)

Goldsmith, Donald. *The Astronomers*. St. Martin's, 1992. 332 p., photos, gloss., index. 0–312–09245–8.

A chapter in this book is entitled "We Are All Astronomers," by which the author means that anybody who has ever gazed in wonder at the night sky possesses the same intuitive curiosity that makes an astronomer. Throughout this book, he maintains a populist approach to science in which he documents the technical discoveries and questions, but also conveys a sense of wonderment at the same time. Mostly covers the origin of the universe and star formation. (B Rpt v10 N 1991 p51; WLB v65 Je 1991 p162)

Lightman, Alan. *Time for the Stars: Astronomy in the 1990's*. Viking, 1992. 124 p., index. 0–670–83976–0.

In the early 1990s the National Academy of Science's Astronomy and Astrophysics Survey Committee issued a report assessing the current state of the science and outlining future initiatives. Lightman's summary of that report for the general reader is an accessible overview of major trends in astronomy and can provide a gateway to other, more specialized books. (BL v88 D 1, 1991 p666; Choice v30 S 1992 p145 S&T v83 My 1992 p529)

Malin, David. *A View of the Universe*. Cambridge University Press, 1993. 266 p., photos, index. 0–521–44477–2.

Oversize in format, with lavish color photographs, this book would look good on any coffee table, but it contains a wealth of information in addition to its visual splendor. Malin, one of the world's foremost astrophotographers, discusses how sophisticated techniques of his field not only reveal the beauty of the universe, but also lead to new scientific discoveries. (New Sci v141 F 5, 1994 p42)

Also Recommended

Abell, George, David Morrison, and Sidney Wolff. *Realm of the Universe*. 5th ed. Harcourt Brace Jovanovich, 1992. A popular textbook.

Asimov, Isaac. *The Measure of the Universe*. Harper, 1983. Cosmological distances made comprehensible. (BL v79 Je 15, 1983 p1309; KR v51 Ap 1, 1983 p415; LJ v108 Je 1, 1983 p1146)

Field, G. B. *The Invisible Universe*. Birkhauser, 1985. Astronomy in nonvisual segments of the spectrum. (BL v81 F 1, 1985 p745; Choice v22 Ap 1985 p1185; LJ v110 F 1, 1985 p104)

Jastrow, Robert. *Red Giants and White Dwarves*. 3rd ed. Norton, 1990. The latest edition of a classic. Covers origins and evolution of the universe. (LJ v116 Mr 1, 1991 p60)

Moore, Patrick. *Fireside Astronomy*. Wiley, 1991. Essays on various topics by a master popularizer. (LJ v117 Je 1, 1992 p166; New Sci v135 Ag 8, 1992 p42)

ASTRONOMY—SPECIAL TOPICS

Atlases and Star Guides

Ridpath, Ian, and Wil Tirion. *The Monthly Sky Guide*. New ed. Cambridge University Press, 1991. 63 p., index. 0–521–38665–9.

A handy beginner's guide to stargazing, this manual employs a failsafe method for locating objects using the naked eye, binoculars, or small telescopes. The narratives describe the unique features of several of the most visually compelling celestial objects. For Northern Hemisphere observers. (Astron v19 Ja 1991 p96)

Tirion, Wil, and Patrick Moore. *Sky Atlas 2000.0*. Cambridge University Press, 1991. 74 p., illus., maps. 0–521–26322–0.

This atlas is designed to serve the advancing needs of those who have found Ridpath and Tirion's previously cited work useful, but now need something more sophisticated. There are detailed monthly star maps, plus six all-sky maps showing nebulae, clusters, and other objects. Primarily for amateur astronomers who have access to small or medium-sized telescopes. (Astron v20 Ap 1992 p94; Choice v30 S 1992 p92; New Sci v133 F 22, 1992 p45)

(*Note*: Numerous atlases and star guides are available, any number of which might also be recommended. A core science literacy collection should contain at least one beginner's guide. Atlases for more serious amateurs should be added when merited. NASA publishes many very specific and detailed atlases.)

Comets

Levy, Stephen. *The Quest for Comets*. Plenum, 1994. 280 p., photos, index. 0–306–44651–0.
Levy, a foremost amateur astronomer, has discovered nineteen comets, most of them while working in his backyard, and is well qualified to write this general history of comets and comet hunting.

Yeomans, Donald K. *Comets: A Chronological History of Observation, Science, Myth, and Folklore*. Wiley, 1991. 484 p., illus., index. 0–471–61011–9.
To this day, the appearance of a comet will cause more people to gaze sky-ward than almost any other celestial phenomenon. The history of humankind's experiences with comets is part myth, part science, and this expansive reference captures the breadth of that experience. The chronologies are also useful. (Astron v19 My 1991 p97; BL v87 F 1, 1991 p1102; LJ v117 Mr 1, 1992 p44)

Cosmology—General Topics

Darling, David. *Deep Time*. Delacorte, 1989. 192 p. 0–385–29757–2.
Darling surveys modern cosmology by tracing the journey of a single suba-tomic particle from the Big Bang, to the formation of the solar system, to the very end of the universe. A readable and entertaining approach to a complex subject. (BL v85 Ag 1989 p1931; KR v57 Je 15, 1989 p888; LJ v114 My 15, 1989 p86)

Halpern, Paul. *Cosmic Wormholes: The Search for Interstellar Shortcuts*. Dut-ton, 1992. 236 p., index. 0–525–93477–4.
This fanciful but feasible journey into the world of exotic matter—black holes, white holes, wormholes—is for science fiction fans who are interested in science fact. The subject captures the imagination; the writing is clever and rich in illustrative metaphors. (KR v60 Ag 1, 1992 p964; PW v239 Ag 3, 1992 p50)

Hawking, Stephen. *A Brief History of Time: From the Big Bang to Black Holes*. Intro. by Carl Sagan. Bantam, 1988. 198 p., illus., index. 0–553–05430–x.
Influenced chiefly by widespread public fascination with Hawking and his triumph over a debilitating disease, this history of cosmological thought became the all-time bestselling work of popular science. It is not the most comprehen-

sible book on the subject for the interested layperson, but it is certainly the most popular. (BRD 1988 p760; LJ v113 Ap 15, 1988 p87; NYTBR Ap 3, 1988 p10)

Morris, Richard. *Cosmic Questions* Wiley, 1993. 200 p. 0–471–59521–7.
Writing in a straightforward question and answer format, Morris successfully anticipates and addresses the salient issues of modern cosmology that most intrigue lay readers. He concludes with a philosophical summary of the place of humankind in the universe. (Astron v22 My 1994 p125; LJ v118 N 1, 1993 p144)

Smoot, George, and Keay Davidson. *Wrinkles in Time*. Morrow, 1994. 388 p., photos, biblio., index. 0–688–12330–9.
In 1992 Smoot and his team from Berkeley announced the discovery of the largest and most distant objects in the universe—"wrinkles in time," remnants of the Big Bang which fill in a missing piece in the theory. The sudden discovery, however, was the culmination of twenty years of sustained research. The story of Smoot's labors provides a dramatic illustration of how "Big Science" works. (BL v90 N 15, 1993 p589; KR v61 O 1, 1993 p1258; LJ v118 D 1993 p166)

Thorne, Kip. *Black Holes and Time Warps: Einstein's Outrageous Legacy*. Norton, 1994. c. 600 p., illus., biblio., index. 0–393–03505–0.
Black holes, so far undetected but widely believed to exist, are among the catalogue of exotic astronomical objects predicted by relativity theory. Thorne, a leading researcher, expounds the theory and how it relates to practice and observation in astronomy. A good book for readers who have finished Halpern's and want a more rigorous treatment. (LJ v119 Ap 15, 1994 p106)

Zee, A. *An Old Man's Toy: Gravity at Work and Play in Einstein's Universe*. Macmillan, 1989. 272 p., illus., index. 0–02–633440–2.
Gravity is the weakest of the primary physical forces, but across the distances of space it exerts the most influence on massive objects. The fate of the entire universe may be determined by gravitational forces. Zee writes, "Physics began with gravity, but it may also end with gravity." (KR v57 Ja 15, 1989 p16; LJ v114 Mr 1, 1989 p86; NYTBR v94 Jl 30, 1989 p3)

Also Recommended

Bartusiak, Marcia. *Thursday's Universe*. Times Books, 1986. Each chapter deals with a separate facet of modern cosmology. (BL v83 O 1, 1986 p175; PW v230 O 17, 1986 p51)

Cohen, Nathan. *Gravity's Lens*. Wiley, 1988. A general, nontechnical overview of new ideas in cosmology. (BL v84 Ag 1988 p1874; PW v233 Je 24, 1988 p101)

Kaku, Michio. *Hyperspace: A Scientific Odyssey Through Parallel Universes,*

Time Warps, and the Tenth Dimension. Oxford University Press, 1994. Discovery and frontiers of cosmological astronomy.

Cosmology—The Big Bang

Barrow, John. *The Origin of the Universe.* Basic, 1994. 176 p. 0–465–05354–8; Davies, Paul. *The Last Three Minutes: Conjecture about the Ultimate Fate of the Universe.* Basic, 1994 c. 176 p. 0–465–04892–7.

These were the first two titles released under the publisher's excellent Science Masters series, which aims to provide succinct introductions to important issues in contemporary science. These books on the beginning and ending of the universe make a nice companion pair.

Gribbin, John. *In Search of the Big Bang.* Bantam, 1986, 413 p., illus., index. 0–553–4258–4.

The structure and evolution of the universe may well have been determined within the first few seconds after the Big Bang. Gribbin describes the subatomic particles and basic forces that science has identified in the universe today and, working backward, demonstrates how the laws they obey created the universe we experience today. Well written and witty, though not an easy read. (BL v82 Je 1, 1986 p1423; Nature v322 Ag 7, 1986 p505)

Parker, Barry. *Creation: The Story of the Origin and Evolution of the Universe.* Plenum, 1988. 295 p., illus., gloss., index. 0–306–42952–7.

As good an introduction to the complex science of the Big Bang theory and its empirical study as can be found. The personal anecdotes of the lives of the key scientists in the search make for somewhat lighter reading, and the illustrations and line drawings aid comprehension. (BL v85 S 1, 1988 p16; KR v56 Ag 1, 1988 p1133; PW v234 S 16, 1988 p71)

Also Recommended

Barrow, John D., and Joseph Silk. *The Left Hand of Creation.* Basic, 1986. Two scientists write on the origin of the universe, stars, nebulae, and galaxies. (Choice v23 Jl 1986 p1692; Nature v320 Mr 27, 1986 p315; NYTBR v91 F 16, 1986 p20)

Lerner, Eric. *The Big Bang Never Happened.* Random House, 1992. An important counterpoint to the dominant theory of the universe. (Astron v20 Ap 1992 p112; Choice v29 D 1991 p614; LJ v117 Mr 1, 1992 p43)

Trefil, James. *The Moment of Creation.* Scribner's, 1983. Although a bit behind the times, this book can be found in many libraries. (BL v80 O 1, 1983 p214; KR v51 Jl 15, 1983 p818; LJ v108 O 1, 1983 p1884)

Weinberg, Steven. *The First Three Minutes.* Basic, 1977. Classic, but now somewhat dated. (BL v73 Ap 15, 1977 p1226; Choice v14 Jl 1977 p711; KR v45 F 1, 1977 p152)

Cosmology—Dark Matter

Bartusiak, Marcia. *Through a Universe Darkly*. HarperCollins, 1992. 383 p., refs., index. 0–06–018310–1.

In order to understand the dynamics of the universe, such as how galaxies are born, as well as the ultimate fate of the universe, science must know its composition. In the past, it was believed that space was permeated with a substance called either. Today, scientists search for so-called dark matter, which some estimate may make up 90 percent of the cosmos. Bartusiak traces the history of thought on this subject. (BL v89 Je 1, 1993 p1752; NYTBR v98 Ag 15, 1993 p9; PW v240 Jl 12, 1993 p66)

Also Recommended

Riordan, Michael. *The Shadows of Creation*. Freeman, 1991. More on dark matter; a bit more technical but accessible. (KR v59 F 15, 1991 237; Nature v352 Ag 22, 1991 p677; PW v238 F 15, 1991 p83)

Trefil, James. *The Dark Side of the Universe*. Scribner's, 1988. Particle physics, cosmology, and "missing matter." (BL v85 S 1, 1988 p16; KR v56 Jl 1, 1988; NYTBR v93 O 16, 1988 p14)

Cosmology—Philosophical Topics

Barrow, John D., and Frank N. Tipler. *The Anthropic Cosmological Principle*. Oxford University Press, 1986. 706 p. 0–198–51949–4.

Anthropic cosmology observes that the universe in which we live is the only universe in which we *could* live. The physical laws within it seem tailor-made to our existence. Human centeredness is key to this philosophy. The intellectual consequences of this view are deep, and this is the most complete source available on the subject. A moderately difficult read. (Choice v23 Jl 1986 p1692; NYTBR v91 F 16, 1986 p20; Sci v232 My 23, 1986 p1036)

Gribbin, John, and Martin Rees. *Cosmic Coincidences*. Bantam, 1989. 302 p., index. 0–553–34740.

Had any of a number of the physical properties of the universe been even slightly different, life may never have evolved. Rees and Gribbin explore several of the physical laws that have made the universe and the intelligence within it possible. An intriguing perspective that would prepare a reader for Barrow's more advanced book on the Anthropic Principle. (BL v85 Ag 1989 p1932; KR v57 Je 15, 1989 p893; PW v236 Jl 14, 1989 p70)

Also Recommended

Davies, P.C.W. *Superstrings: A Theory of Everything?* Cambridge University Press, 1988. 234 p., index. A discussion of proposed objects important to many theories of cosmology. (Sci v242 D 16, 1988 p1584; SciTech v12 O 1988 p12)

(*Note*: See also entries in Chapter Twelve under "Philosophical Aspects" and "Grand Unified Theories.")

Galaxies

Hodge, Paul. *Galaxies.* Harvard University Press, 1986. 174 p., illus., index. 0–674–34065–5.
Fifty years ago, astrophysicist Hallow Shapley wrote a book with the same title. Hodge undertook the formidable task of bringing it up to date with half a century's new discoveries. This fine introduction to galactic study lives up to its lineage. (Choice v24 S 1986 p155; SciTech v10 Ap 1986 p9)

Parker, Barry. *Colliding Galaxies: The Universe in Turmoil.* Plenum, 1990. 298 p., illus., index. 0–306–43566–7.
As viewed from Earth, distant galaxies seem stable and majestic. Even so, they are the centers of incredible cosmic violence, some of the most extreme of which occurs when two galaxies interact. Starting with a history of the discovery and classification of galaxies, Parker brings readers to the frontiers of galactic research. (KR v58 S 1, 1990 p1234; LJ v116 Mr 1, 1991 p60; SciTech v14 D 1990 p9)

Also Recommended

Ferris, Timothy. *Galaxies.* Harrison House, 1987. Informational and magnificently illustrated.

Hubble, Edwin. *Realm of the Nebulae.* Dover, 1958. An influential popularization of an earlier era, still worth a read.

Handbooks for Amateurs

Harrington, Philip S. *Touring the Universe Through Binoculars: A Complete Astronomer's Guidebook.* Wiley, 1990. 294 p., illus., biblio., index. 0–471–51337–7.
A wealth of astronomical objects can be viewed and appreciated with the aid of a pair of binoculars. This is precisely the way that many laypersons have become interested in descriptive astronomy. This practical book gives information on where to look to see over 1,000 objects. (Astron v19 Mr 1991 p100; BL N 1, 1990 p488; LJ v116 Mr 1, 1991 p60)

Newton, Jack. *The Guide to Amateur Astronomy.* 2nd ed. Cambridge University Press, 1994. 0–521–34028–4.

A useful handbook for beginning to intermediate hobbyists. A distinct strength of this book is that it contains contributions (especially photographic) from other amateurs, so it speaks to a kindred audience. Contains practical advice on selecting and using a telescope, as well as information on basic astronomical principles.

Also Recommended

Schaaf, Fred. *The Starry Room: Naked Eye Astronomy in the Intimate Universe.* Wiley, 1988. An ode to stargazing. (BL v85 D 15, 1988 p671; LJ v113 D 1988 p121; S&T v78 O 1989 p375)

(*Note*: Numerous naked-eye field guides and star finders are available. A core collection in any small library should have at least one.)

Planetary Astronomy

Hartman, William K. *Moons and Planets.* 3rd ed. Wadsworth, 1993. 510 p., illus., 0–534–18894–x.

Now in its third edition, this book has become a standard text in undergraduate courses, but it can be a gold mine for readers of any age. The physical principles of all planetary bodies and many of their satellites are discussed in separate, integrated chapters—for example, atmospheres, interiors, orbital characteristics, and so on.

Littmann, Mark. *Planets Beyond: Discovering the Outer Solar System.* Wiley, 1990. 318 p., gloss., index. 0–933–34655–7.

The outer planets from Jupiter onward are worlds quite unlike the nearer, terrestrial planets. This American Institute of Physics Award winning book was fully updated to incorporate information gathered by Voyager 2. (Astron v18 Mr 1990 p91; New Tech B v74 Mr 1989 p601)

Also Recommended

The New Solar System. 3rd ed. J. Kelly Beatty et al., eds. Sky Publishing, 1990. Articles by experts cover all planets and various topics.

(*Note*: Some fine books have been written about almost all of the individual planets and their moons. These may be added as appropriate to the benefit of any core collection.)

Prehistoric Astronomy

Calvin, William. *How the Shaman Stole the Moon: In Search of Ancient Prophet-Scientists from Stonehenge to the Grand Canyon.* Bantam, 1991. 223 p., illus., index. 0–553–07740–6.

Calvin, a neurobiologist, tours such intriguing sites as Delphi, Stonehenge, and Mesa Verde in expounding the remarkable degree of sophistication that ancient civilizations possessed in making accurate astronomical observations and predictions. (KR v59 O 1, 1991 p1254)

Also Recommended

Aveni, Anthony. *Conversing with the Planets: How Science and Myth Invented the Cosmos.* Times Books, 1992. A somewhat more analytical discussion of astronomical influences on culture. (KR v60 Jl 1, 1992 p819; PW v239 Jl 13, 1992 p40)

Krupp, Edwin C. *Echoes of Ancient Skies.* New American Library, 1984. A good introduction to archaeoastronomy.

SETI (Search for Extraterrestrial Intelligence)

Drake, Frank, and Dava Sobel. *Is Anyone Out There? The Search for Extraterrestrial Intelligence.* Delacorte, 1992. 272 p., photos, index. 0–385–30532–x.

In 1960, when Frank Drake first began using the National Radio Astronomy Observatory telescope to search for intelligent signals from deep sky sources, many thought he was deluded. Today, however, he is president of the government-funded SETI Institute. He confidently asserts that proof of extraterrestrial intelligence will be discovered by the year 2000. Here is the story of the development of a discipline from fringe science to the cutting edge. (Astron v20 O 1992 p104; BW v22 S 13, 1992 p13; KR v60 Ag 15, 1992 p1032)

Trefil, James, and Robert T. Rood. *Are We Alone?* Scribner's, 1981. 262 p., illus., index. 0–684–16826–x.

This book has a twofold purpose: (1) to discuss aspects of science that bear on the question of extraterrestrial life, and (2) to advance the authors' own (somewhat skeptical) conclusions. In all, Trefil and Rood cover the gamut of scientific and philosophical issues related to SETI, as well as the technological means by which such beings might be contacted. (BL v77 Jl 1, 1981 p1386; KR v49 Ap 15, 1981 p559; LJ v106 Ap 15, 1981 p892)

Also Recommended

Horowitz, Norman. *To Utopia and Back: The Search for Life in the Solar System.* Freeman, 1986. A former chief of bioscience at Jet Propulsion Laboratory recounts search efforts for life in the solar system. (Choice v24 N 1986 p505; Nature O 2, 1986 p402)

Sagan, Carl, and I. S. Shklovskii. *Intelligent Life in the Universe.* Holden-Day, 1978. A classic; one of the first serious works to describe the possibilities of life in space. (TES My 26, 1978 p53)

Space Travel and Exploration

Burrows, William. *Exploring Space: Voyages in the Solar System and Beyond.*
Random House, 1990. 502 p., photos, biblio., index. 0–393–56983–0.

The American space program was a product of Cold War politics as much as
of emerging technology. *Exploring Space* presents the history of NASA and the
program from its inception to the present day in epic fashion, with due consideration to the full range of social, economic, political, and technical factors that
affected it. Burrows also offers cautionary observations on NASA's future. (BL
v87 D 15, 1990 p799; Choice v28 Je 1991 p1658; S&T v82 Jl 1991 p43)

Chaikin, Andrew. *A Man on the Moon: The Voyages of the Apollo Astronauts.*
Viking, 1994. 670 p., photos, index. 0–670–81446–6.

The author envisioned this work as the true and candid story the twenty-four
men who have walked on the moon would tell if they were to gather to tell it.
Eight years in preparation, based on hundreds of hours of interviews, this book
is well documented and poignantly honest.

Gump, David. *Space Enterprise: Beyond NASA.* Praeger, 1990. 220 p. 0–275–
93314–8.

The industrialization of space and the emergence of private enterprise there
may be the wave of the economic future. Some plausible enterprises include
mining asteroids, collecting antimatter for energy, and cleaning the polluted
atmosphere from space-based operations. A bully, entrepreneurial perspective.
(Choice v27 Mr 1990 p1182)

Moore, Patrick. *Mission to the Planets: The Illustrated Story of Man's Exploration of the Solar System.* Norton, 1990.

Space probes have visited and returned data from the moon and all of the
planets save Pluto. Moore, a prolific writer, devotes at least one chapter to each,
and also gives information on other bodies within the solar system, notably
Halley's comet. The appendices list characteristics of planets and the various
interplanetary vessels—both Soviet and U.S.—that have visited them. (Choice
v28 Ja 1991 p803; S&T v81 Ap 1991 p390)

Shipman, Harry L. *Humans in Space: Twenty-First Century Frontiers.* Plenum,
1989. 351 p. 0–306–43171–8.

Possible scenarios for the next 100 years in space include the establishment
of colonies on Mars, generational travel beyond the solar system, and the preservation of our own ''spaceship Earth.'' Shipman, an astronomer, asks fascinating questions and speculates as to their answers. (BRD 1989 p1523; LJ v114
My 15, 1989 p86; NYTBR Ap 9, 1989 p32)

Walter, William J. *Space Age.* Random House, 1992. Photos, biblio., index. 0–
679–40295–0.

This companion to the PBS series of the same name has a broad scope and serves as a fine introduction to the past, present, and future of space sciences. The emphasis is on how technologies designed for space exploration have affected human culture in real, practical ways. (BL v89 S 1, 1992 p4; KR v60 Ag 1, 1992 p980; LJ v117 Ag 1992 p143)

Wilford, John Nobel. *Mars Beckons.* Knopf, 1990. 244 p., photos, index. 0–394–58359–0.
Mars is the planet most like the Earth and, throughout history, has commanded a special place in the human imagination. In this book, the Pulitzer Prize winning author describes a possible voyage to Mars, which, he argues, would be our next great space adventure. (BL v86 My 15, 1990 p1755; KR v58 My 15, 1990 p721; LJ v115 Je 15, 1990 p1230)

Also Recommended

Aldrin, Buzz, and Malcolm McConnel. *Men from Earth.* Bantam, 1989. An insider's account of Apollo and the moon landing. (BL v85 My 15, 1990 p721; KR v57 My 15, 1989 p738; LJ v114 Je 15, 1989 p75)

Lewis, Richard S. *Space in the Twenty-First Century.* Columbia University Press, 1990. A journalistic account of NASA's possible future endeavors.

Lovell, Jim, and Jeffrey Kluger. *Lost Moon: The Perilous Voyage of Apollo 13.* Houghton, 1994. An inside account of the heroic efforts undertaken to save the crew of a severely damaged spacecraft. (Note: No reviews are available for this book.)

Ordway, Frederick I., and Randy Liebermann, eds. *Blueprint for Space: From Science Fiction to Science Fact.* Smithsonian, 1991. Popular perceptions of space from the past and present. (BL v88 F 1, 1992 p999; Choice v29 Jl 1992 p1703; LJ v117 F 15, 1992 p193)

Ordway, Frederick I., and Mitchell R. Sharpe. *The Rocket Team.* MIT Press, 1982. Werner von Braun and early rocketry.

Shepard, Alan, and Deke Slayton (with Jay Barbree and Howard Benedict). *Moon Shot: The Inside Story of America's Race to the Moon.* Turner, 1994. A firsthand account of the early years of the space program by two veteran astronauts and two space reporters. (LJ v119 Ap 15, 1994 p106)

Wolfe, Tom. *The Right Stuff.* Farrar, 1979. A very popular account of the first American astronauts. (BL v76 O 15, 1979 p318; LJ v104 O 15, 1979 p2228; Time v114 S 24, 1979 p81)

(*Note*: See also entries in Chapter Thirteen under "Aviation.")

The Sun and Stars

Cohen, Martin. *In Darkness Born: The Story of Star Formation.* Cambridge University Press, 1988.

In a historical sketch, a researcher in the field describes how observations made in all regions of the spectrum have led to our understanding of how stars are born, evolve, and die. A high school level background in physics is helpful. (BRpt v7 S 1988 p52; Choice v26 O 1988 p341)

Noyes, Robert W. *The Sun, Our Star.* Cambridge University Press, 1982. 263 p. illus., photos, index. 0–674–85435–7.

Part of the Harvard University Books on Astronomy series, this book is still the best general introduction to the physics of the sun and the history of solar research. (KR v50 S 1, 1982 p1048)

Also Recommended

Gribbin, John. *Blinded by the Light: The Secret Life of the Sun.* Harmony, 1991. The interior structure of the sun. (New Sci v130 Je 8, 1991 p49; Nature v356 Ap 2, 1992 p396; SBF v28 Mr 1992 p39)

Marschall, Laurence A. *The Supernova Story.* Plenum, 1988. An account of massive stellar explosions. (BL v85 S 1, 1988 p16; PW v234 S 9, 1988 p116)

Moore, Patrick. *Astronomer's Stars.* Norton, 1989. Famous stars and what astronomers know about them. (BL v85 Jl 1989 p1856; LJ v114 Jl 1989 p103)

Telescopes

Asimov, Isaac. *Eyes on the Universe: A History of the Telescope.* Houghton Mifflin, 1975.

Almost twenty years after its publication, Asimov's remains a lucid, first-rate history of telescopes. Indeed, Asimov, something of a visionary, successfully anticipated development of new telescope technologies, which has extended this book's shelf life beyond the norm for popular astronomy books. (BL v72 O 15, 1975 p268; KR v43 Jl 15, 1975 p808; LJ v100 S 15, 1975 p1643)

Tucker, W., and K. Tucker. *The Cosmic Inquirers: Modern Telescopes and Their Makers.* Harvard University Press, 1986. 221 p. 0–674–17435–6.

Modern telescopes, such as the Very Large Array Telescope, the Infrared Satellite, and those in the Gamma Ray Observatory, do much more than just magnify starlight—they search the "invisible universe" for unseen scientific clues in the electromagnetic spectrum. This book is a good companion to Asimov's, because it picks up close to where that one ends. Although pre-launch to the Hubble, it also gives some information on that most ambitious of telescope projects. (Choice v24 S 1986 p156; LJ v111 Ap 1, 1986 p156; S&T v72 Ag 1986 p136)

Also Recommended

Chaisson, Eric J. *The Hubble Wars*. HarperCollins, 1994. The technology and politics of the Hubble Space Telescope from an insider's perspective.

Field, George, and Donald Goldsmith. *The Space Telescope: Eyes above the Atmosphere*. Contemporary Books, 1989. This book got a second life when the Hubble was successfully repaired. (Astron v18 S 1990 p100; BL v86 N 1, 1989 p508; LJ v114 D 1989 p160)

OTHER RESOURCES

Periodicals for General Readers

Astronomy Magazine. Kalmbach Publishing, m., revs. 0091–6358.

A glossy commercial publication featuring news and monthly sky guides for amateur astronomers.

Mercury Magazine. Astronomical Society of the Pacific, bi-m. 0047–6773.

This is the official publication of the Society, the mission of which is "to increase the public understanding and appreciation of astronomy."

Sky and Telescope. Sky Publishing, m., revs. 0037–6604.

The magazine with the largest circulation of any for amateur astronomy. Includes popular accounts of new scientific ventures.

Yearbook of Astronomy. John Guy and Patrick Moore, eds. Norton, annual. 0084–3660.

Reviews and summaries of each year's significant astronomical events and discoveries.

Audiovisual Materials

The Astronomers (series). PBS Videos/KCET TV-Los Angeles, 1991 (six videocassettes, 360 minutes total).

Titles in this series include "Where Is the Rest of the Universe?," "Searching for Black Holes," "A Window to Creation," "Waves of the Future," "Stardust," and "Prospecting for Planets." Collectively, they cover the gamut of history, contemporary knowledge, and future horizons in astronomy. (LJ v116 N 1, 1991 p140; S&T v82 O 1991 p375)

Origin and Evolution of the Universe (series). Films for the Humanities and Sciences, 1990. (videocassettes, c. 60 minutes each).

Describes the physical laws and processes by which the universe and objects in it have evolved. Titles include "The Grand Design," "The Origin of Quasars," "The Origin of Galaxies," "The Birth of Stars and the Great Cosmic Cycle," and "Origin of the Universe." (SBF v30 Mr 1993 p34)

Space Age (series). Public Media Video/WQED Television, Pittsburgh, 1992 (six videocassettes, c. 58 minutes each).

   Some intriguing future possibilities are introduced in these videos. The images are striking. The six titles are "Celestial Sentinels," "Mission to Planet Earth," "Quest for the Planet Mars," "The Unexpected Universe," "What's Heaven For?," and "To the Moon and Beyond." (Variety v348 O 12, 1992 p189)

Also Recommended

Cosmos (series). Carl Sagan Films/PBS, 1980 (videocassettes). This hugely successful early eighties production can still be used with good results.

Flying by the Planets. NASA Jet Propulsion Laboratory/Astronomical Society of the Pacific, 1991 (videocassette, 35 minutes; includes pamphlet and reading list). In eight segments, shows computer enhanced images of Venus, Earth, Mars, Jupiter, and Saturn. (SBF v30 Mr 1993 p34)

Living and Working in Space: The Countdown Has Begun. FASE Productions; distributed by PBS (videocassette, 60 minutes). Daily life on a space station. (SBF v30 Mr 1994 p34)

CD-ROMs

Recommended

Amazing Universe. Hopkins Technology/NASA Jet Propulsion Lab, 1992 (one CD). Detailed images from spacecraft. Includes solar system and deep sky objects.

A Brief History of Time. Freeman, 1994. An interactive CD "adventure" that borrows its title and themes from the popular Stephen Hawking book. Explores from the subatomic landscape to "Relativity Street," and offers a ride to the brink of a black hole.

Hubble Space Telescope Guide Star Catalog. Astronomical Society of the Pacific (two CDs). Extensive data on almost 20,000 stars. For colleges or serious amateurs.

NASA Space Encyclopedia. NASA Jet Propulsion Lab. A fully illustrated aerospace encyclopedia with text, images, and audio.

Space Time and Art. Wayzata Technology/Updata Publications. Over 300 space images, some creatively reworked, from Voyager and other spacecraft. Includes a database of astronomical organizations and relevant institutions. Emphasizes interrelatedness of space sciences and art.

Voyages to the Planets. NASA, Jet Propulsion Lab (seven CDs). A huge collection of high resolution images from Voyager and Viking spacecraft. Covers objects from the solar system. Several abridged versions are also available.

Chapter Seven

Biological Sciences

If you press a piece of underwear soiled with sweat together with some wheat in an open mouth jar, after about twenty-one days the odor changes and the ferment, coming out of the underwear and penetrating through the husks of the wheat, changes the wheat into mice. But what is more remarkable is that mice of both sexes emerge (from the wheat) and these mice successfully reproduce with mice born naturally from parents.
Jan Baptiste Von Helmont, 1667

The smallest living cell probably contains over a quarter of a million protein molecules engaged in the multitudinous coordinated activities which make up the phenomenon of life. At the instant of death, whether a man or microbe, that ordered, incredible spinning passes away in an almost furious haste. . . . I do not think, if someone finally twists the key successfully in the tiniest and most humble house of life, that many of these questions will be answered, or that the dark forces which create lights in the deep sea and living batteries in the waters of tropical swamps, or the dread cycles of parasites, or the most noble workings of the human brain, will be much if at all revealed.
Loren Eiseley, 1946

And now I see with eye serene
The very pulse of the machine.

William Wordsworth

In fact, it's quite easy to create life.
Thomas Ray, University of Delaware, 1992

The scientific study of life, at least in any modern sense, is a relatively recent enterprise. While humans have always speculated as to the origin of life on Earth, the differentiation of living things, and the mechanisms of heredity, the tools and principles employed by biologists today are distinctly contemporary. As in the physical sciences, Aristotelian empiricism was predominant in fashioning an intellectual framework for the life sciences until the Middle Ages and beyond. Aristotle's thought provided the points of departure for successors such as the Roman anatomist Galen, the great Renaissance physician Vesalius, and William Harvey, who discovered the circulation of blood. While serviceable theories of human and animal physiology were developed from this Aristotelian foundation, notions about the origin of life and the transmission of heredity were at best conjectural, and sometimes, as in the passage cited by Von Helmont, wildly fanciful.

The biological sciences began to mature in the late seventeenth century. The Dutch scientist Anton van Leeuwenhoek published his observations of the microscopic world, thus stimulating the development of the field of microbiology and the formation of cell theory. Increasingly, the fields of biology and chemistry became wed. As the eighteenth century Enlightenment opened new vistas of inquiry, Linnaeus and others systematized the science of taxonomy, and this heightened understanding of the organization of species gave rise to the kinds of speculation that led to Darwin's and Wallace's theories of evolution by natural selection. When combined with advances in the understanding of cell physiology and genetics, evolutionary theory provided a comprehensive, unifying context for all of the life sciences.

In the twentieth century, biologists have revolutionized the discipline through their examinations of the basic dynamics of life. The impact on the biological sciences of Watson and Crick's 1953 discovery of the structure of DNA, the molecule of life, has been likened to that of relativity theory on physics. The analogy goes farther; some suggest that the revolution in biology now upon us will have at least as great an impact on society as that in physics did earlier in this century. As scientists learn more about the workings of living things, they also learn how to *control* those processes, and that knowledge, in turn, raises tremendous possibilities as well as deep concerns. These issues and technologies have been the subject of many recent popularizations, and will doubtless be followed by many others. Ironically, however, as much as scientists know about living things, the central question—"What is life?"—remains puzzling.

Biology is a broad, inclusive discipline with sometimes fuzzy boundaries. The general focus of this chapter is on what has been called the "hard" biological sciences, that is, the laboratory sciences, such as embryology, microbiology, molecular biology, physiology, the neurosciences, and so on. Also included are botany and zoology. In regard to the latter, popular zoological literature is rich and voluminous. The few, but representative books cited here were selected, first, because they describe species about which there is broad general interest or concern and, second, because in doing so they refer to or draw upon current scientific research, or at least reflect scientific ways of thinking in their observations. The subjects covered in this chapter overlap significantly with many discussed elsewhere in this book, most notably in Chapter Eleven, "Natural History," and Chapter Ten, "Medicine and Health Sciences." Cross-references are provided as appropriate.

BIOLOGICAL SCIENCES—GENERAL

Reference

The Cambridge Encyclopedia of the Life Sciences. Adrian Friday and David S. Ingram, eds. Cambridge University Press, 1985. 432 p., 0–521–25696–8.

This encyclopedia covers a hierarchy of biological interactions in three general parts: "Processes and Organization," "Environments," and "Evolution and the Fossil Record." Presents biology as a science linked with broad areas of intellectual understanding. (Nature v314 Ap 25, 1985 p700; TES My 17, 1985 p50)

A Concise Dictionary of Biology. 2nd ed. Oxford University Press, 1990. 266 p., illus., 0–19–286109–3.

Entries are related to biology and biochemistry, along with other natural science areas that apply. There are approximately 1,500 definitions of 75–150 words each (Sch Lib v34 Je 1986 p197; Tes My 23, 1986 p28)

Also Recommended

Encyclopedia of Human Biology. Renato Dulbeco, ed. 8 vols. Academic Press, 1991. A work of serious scholarship that can be used for broad reference purposes in biological and medical sciences. (Choice v29 O 1, 1991 p256; LJ v116 Ap 15, 1991 p82)

Facts on File Dictionary of Biology. Rev. ed. Elizabeth Tootill, ed. Facts on File, 1988. A good general dictionary. (RSSEMA p54)

King, Robert, and William Stansfield. *Dictionary of Genetics.* 4th ed. Oxford University Press, 1990. Comprehensive; covers terms used in genetic research. (RSSEMA p54)

Lawrence, Eleanor. *Henderson's Dictionary of Biological Terms.* 10th ed. Wiley, 1989. 637 p. Quite comprehensive. Brief definitions. Includes a chemical structures appendix.

(*Note*: See also reference sources listed under specific categories in this chapter, as well in Chapters Eight, Nine, and Eleven.)

Autobiography and Biography

Carlson, Elof Axel. *Genes, Radiation and Society: The Life and Work of H. J. Muller.* Cornell University Press, 1981. 0801414044.

Muller won a Nobel Prize (along with Thomas Hunt Morgan) for his work on chromosome theory and mutagenesis. He was also one of the first to study the biological effects of radiation. Still, he is probably most remembered for his views on eugenics and his Communist Party activities in the first half of this century. (Choice v19 My 1982 p1268; LJ v107 F 1, 1982 p253; Sci v214 D 11, 1981 p1232)

Clark, Ronald W. *J.B.S.: The Life and Work of J.B.S. Haldane.* Rev. ed. Oxford University Press, 1986. 326 p., photos, index. 0–19–281430–3.

Haldane was something of a scientific Renaissance man. Educated in the classics, he made his greatest contributions to the field of genetics, but he is probably best remembered for his popular essays and the socialist undertones they frequently conveyed. He was a gadfly, an absolute iconoclast, and an ardent social activist. (KR v36 N 1, 1968 p1256; LJ v93 O 1, 1968 p3549; PW v194 Jl 15, 1968 p52)

Kanigel, Robert. *Apprentice to Genius: The Making of a Scientific Dynasty.* Macmillan, 1986. 271 p. 0025606506.

Kanigel looks at the social phenomenon of science mentoring through the relationships of four neuroscientists, all of whom were nominated for the Nobel Prize. The focus is on Steve Brodie, a pioneer in the study of metabolism. (BL v83 O 1, 1986 p178; KR v54 S 15, 1986 p1427; NYTBR v91 O 5, 1986 p55)

Keller, Evelyn Fox. *A Feeling for the Organism: The Life and Work of Barbara McClintock.* Freeman, 1983. 235 p. 0–7617–1433–7.

McClintock won the Nobel Prize for her discovery of transpositional, or so-called jumping genes. This biography by a leading scholar of feminist issues in science gives insight into the mind of a reclusive scientist, but also shows some of the biases and difficulties facing women in the scientific community. (LJ v108 Je 1, 1983 p1146; Nature v304 Jl 28, 1983 p377; PW v223 Ap 8, 1983 p48)

Nisbett, Alec. *Konrad Lorenz.* Harcourt Brace Jovanovich, 1977. 240 p., photos, index. 0–15–147286–6.

Lorenz allowed himself to be observed and interviewed by the author for this book. The image of Lorenz, who studied how following behavior is imprinted in newborn goslings, being shadowed in lockstep by young birds is familiar to many people, and that very personal association comes through in this book. (BL v73 Ja 15, 1977 p32; KR v44 D 1, 1976 p1293)

Wilson, Edward O. *Naturalist.* Island Press, 1994. 380 p., photos, index. 1–55963–288–7.

Through his compassionate and insightful popular writings, Wilson has become a preeminent voice in modern scientific environmentalism. As this work shows, Wilson came upon these commitments honestly. For all the respect he is accorded, however, he is no stranger to controversy, and some of his theories have sparked harsh criticism. (Note: No reviews are available for this book.)

Also Recommended

Fischer, Ernst Peter, and Carol Lipson. *Thinking about Science: Max Delbruck and the Origins of Molecular Biology.* Norton, 1988. Delbruck brought his knowledge as a physicist into his work in evolution and genetics. (BL v85 S 1, 1988 p15; KR v56 Jl 1988 p1026)

Scheffer, Victor B. *Adventures of a Zoologist.* Scribner's, 1980. The author recounts his fifty years as a wildlife biologist, with a specialty in marine biology. (HB v57 F 1981 p90; LJ v106 My 1, 1981 p940; SBF v16 Ja 1981 p141)

History of Biological Sciences

Borell, Merriley. *The Biological Sciences in the Twentieth Century.* Scribner's, 1989. 306 p. 0684164833.
Like other works in the Album of Science series to which this book belongs, Borell's is authoritative and well illustrated. (BL v86 Ja 1, 1990 p942; Choice v27 Ja 1990 p824; SBF v25 Ja 1990 p132)

Moore, John A. *Science as a Way of Knowing: The Foundations of Modern Biology.* Harvard University Press, 1993. 530 p., index. 0–674–79480–x.
Science is not just a matter of observation and experiment—it is also founded in worldviews and philosophical means of understanding. The author expounds this theme in four areas: "Understanding Nature," "The Growth of Evolutionary Thought," "Classical Genetics," and "Enigmas of Development." (Choice v31 O 1993 p315; LJ v118 Ap 15, 1993 p123; SBF v29 Ag 1993 p169)

Serafini, Anthony. *The Epic History of Biology.* Plenum, 1993. 350 p., index. 0–306–44511–5.
This up-to-date and easy-to-read history covers the biological sciences in their entirety, from prehistory to modern times. The early chapters tend to focus on philosophical and ideological issues, while the later ones favor medicine, molecular biology, and biotechnological methods. (SBF v30 Ap 1994 p71; SciTech v17 D 1993 p17)

Also Recommended

Mayr, Ernst. *The Growth of Biological Thought.* Harvard University Press, 1982. A magnum opus on philosophical paradigm shifts in biology. A classic frequently cited in professional literature. (KR v50 F 15, 1982 p258; LJ v107 Ag 1982 p1471; Sci v216 My 21, 1982 p842)

General Topics

Cairns-Smith, A. G. *Seven Clues to the Origin of Life: A Scientific Detective Story.* Cambridge University Press, 1990. 131 p. 0–521–39828–2.

The author examines contemporary clues regarding conditions on the early planet Earth in search of how life originated. (Choice v23 Ap 1986 p1230; LJ v111 Mr 1, 1986 p45; Nature v318 N 14, 1985 p119)

de Duve, Christian. *Vital Dust: Life as a Cosmological Imperative.* Harper-Collins, 1995. 0–465–09044–3.

Fascinating speculations on life on Earth in seven planetary epochs, from the "Age of Chemistry," the prebiotic era, to "The Age of the Unknown," an educated guess at what the future holds for living things.

From Gaia to Selfish Genes: Selected Writings in the Life Sciences. Connie Barlow, ed. MIT Press, 1991. 272 p., index. 0–262–02323–7.

A selective anthology of writings from past and current leaders in various fields of the biological sciences, including several with histories of writing for popular audiences. Editorial sidebars help establish connecting themes. (New Sci v131 S 14, 1991 p53; SBF v27 N 1991 p231)

Man and Beast Revisited. Michael Robinson and Lionel Tiger, eds. Smithsonian, 1991. 373 p., index. 0–87474–775–9.

Sequel to *Man and Beast: Comparative Social Behavior* (Eisenberg and Dillon, 1971), this book contains twenty-one articles on subjects related to comparative animal behavior. Topics include growth and development, consciousness, social life, and mating, among others. Comparisons with human beings are made throughout. (Nature v352 Jl 25, 1991 p296; SBF v27 Ag 1991 p168; Sci v253 S 27, 1991 p1564)

Margulis, Lynn, and Karlene V. Schwartz. *Five Kingdoms: An Illustrated Guide to the Phyla of Life on Earth.* Freeman, 1987. 376 p., illus., index. 0–7167–1885–5.

Taxonomy, cytology, and life cycles of the five kingdoms: prokaryotae, protoctists, fungi, animalia, and plantae. (BWatch v9 Mr 1988 p1)

Morowitz, Harold J. *Mayonnaise and the Origin of Life: Thoughts on Mind and Molecules.* Scribner's, 1986. 244 p. 0–684–18444–3.

Morowitz wanders the landscape of the biological sciences, pen in hand, with zest and curiosity. Essays in this collection are divided between zoology and natural histories, ruminations on the history of biology, and occasional odd voyages into literature and other fields. Several of the author's other collections could also be recommended. (Nature v318 N 14, 1985 p116; NW v107 Ja 20, 1986 p74; SBF v22 S 1986 p48)

Pool, Robert. *Eve's Rib: Searching for the Biological Roots of Sex Differences.* Crown, 1994. c. 320 p., index. 0–517–59298–3.

Chronicles scientific knowledge and interpretation—some of it controversial—regarding the chromosomal and hormonal sources of biological gender differences. Also ventures into the realms of psychology and society. (BL v90 My 1, 1994 p1572; KR v62 Mr 1, 1994 p284; LJ v119 My 1, 1994 p134)

Smith, John Maynard. *The Problems of Biology*. Oxford University Press, 1986. 134 p. 0–19–289198–7.
The issues on the front burner for biological researchers include, as a natural beginning, "What Is Life?," and continue through topics related to the origin of life, the biochemical stability and organization of organisms, the methods of evolution, the interactions of heredity and evolution, and others. This is a popular summary for general readers. (Nature v320 Ap 17, 1986 p640)

Wilson, Edward O. *The Diversity of Life*. Belknap Press/Harvard University Press, 1992. 424 p., illus., maps. 0–674–21298–3.
Cyclic episodes of extinction have been part of the natural order of organic history ever since the origin of life on Earth. Although they have resulted in the loss of countless species, they have also triggered new evolutionary events. Biodiversity has been extremely resilient on Earth. Still, according to Wilson, a Harvard entomologist and Pulitzer Prize winner, a new cycle is upon us, this one due to human activity, and biodiversity may never recover. An absolute must for all libraries. (NW v120 S 14, 1992 p70; NYTBR v97 O 4, 1992 p1)

Also Recommended

Biological Science. 5th ed. William Keeton and James L. Gould, eds. Norton, 1993. A voluminous and well-organized text, standard for beginning to intermediate college courses. (New Sci v139 S 25, 1993 p44)

Eiseley, Loren. *The Immense Journey*. Random House, 1957. A personal journey "to understand and enjoy the miracles of this world, both in and out of science." Classic. (BRD 1957 p284; BL v54 S 15, 1957 p40; KR v25 Je 15, 1957 p433)

Fausto Sterling, Anne. *Myths of Gender: Biological Theories about Men and Women*. Basic, 1986. The author combines a knowledge of genetics with a personal commitment to feminist principles. (Choice v23 Ap 1986 p1235; Nature v319 Ja 23, 1986 p271; LJ v111 F 1, 1986 p86)

Haldane, J.B.S. *On Being the Right Size and Other Essays*. Oxford University Press, 1985. Haldane, a brilliant British geneticist and social critic, was the consummate science essayist of the middle part of this century.

The Rhythms of Life. Edward S. Ayensu and Philip Whitfield, eds. Crown, 1982. "The rhythms of life," as depicted in this handsome book, include circadian, seasonal, lifespan, environmental, and so on. (BL v78 Jl 1982 p1406; LJ v107 Jl 1982 p1335)

BIOLOGICAL SCIENCE—SPECIAL TOPICS

Anatomy and Physiology

Ackerman, Diane. *A Natural History of the Senses*. Random House, 1990. 352 p. 0–674–77660–7.

"Synesthesia" is the experience where one sensory perception triggers another sensory or mental event. Ackerman, the author of a popular series of natural histories published in the *New Yorker*, explains how the senses work and tells of the unusual and sometimes obsessive ways that the desire to stimulate the senses has influenced whole cultures. (BL v86 My 1, 1990 p1675; LJ v115 My 1, 1990 p108; NYTBR v95 Jl 29, 1990 p8)

Gray's Anatomy. 37th ed. Churchill Livingston, 1989. 1598 p. 0443025886.

Every library should have a recent edition of this classic, first published in 1858. Covers all areas of anatomy and includes bibliographies.

Kittredge, Mary. *The Human Body: An Overview*. Chelsea House, 1990. 144 p. 0791000192.

As good a one-volume treatment of human anatomy and physiology as any available. The book's major organization is by body systems. The concluding chapter is intriguingly entitled "The Future of the Body." (SLJ v36 Mr 1990 p245)

National Geographic Association. *The Incredible Machine*. National Geographic, 1992. 384 p. 0870446207.

The body is compared to a machine, more efficient and complex than any made by human beings. The photographs, some of which make use of sophisticated technology, penetrate to the microscopic level. Also discussed are whole-body issues like aging, nutrition, and reproduction. (BL v82 Ag 1982 p1646)

Wolpert, Lewis. *The Triumph of the Embryo*. Oxford University Press, 1991. illus., index. 0–19–854243–7.

The miracle of embryology, as Wolpert describes it, is that from a single cell—a fertilized egg—there can arise all of the complexities and variations that we see in species. In this book, the subject of embryology is covered in general, with a special emphasis on human development. (BL v87 Jl 1991 p2020; LJ v116 O 1, 1991 p138)

Also Recommended

Crapo, Lawrence. *Hormones: The Messengers of Life*. Freeman, 1985. An overview of hormones and endocrinology.

Gonzalez-Crussi, F. *Three Forms of Sudden Death and Other Reflections on the Grandeur and Misery of the Human Body*. Harcourt Brace Jovanovich, 1986. Gonzalez-Crussi, a practicing pathologist, writes in an expansive literary style to describe selected oddities of human physiology. (BL v83 O 15, 1986 p314; NYTBR v91 O 15, 1986 p314)

Tortora, Gerard. *An Introduction to the Human Body: The Essentials of Anatomy and Physiology.* 3rd ed. HarperCollins, 1994. College level.

(*Note*: See also entries in Chapter Eight under "Biochemistry.")

Botany

Bernhardt, Peter. *Wiley Violets and Underground Orchids.* Morrow, 1989. 0–688–08350–0; *Natural Affairs: A Botanist Looks at the Attachments Between Plants and People.* Villard/Random House, 1993. 256 p., index. 0–679–41316–2.

In these two books, Bernhardt draws upon his experiences as a botanist in the field and in the laboratory to give general readers some insight into the plant kingdom. He delights in the unusual, such as the bizarre survival strategies of the Asian slipper orchid. He also dwells on how plants have shaped human culture. (Aud v95 My 1993 p138; LJ v118 Mr 1, 1993 p102; NYTBR v98 My 2, 1992 p9)

Friedman, Sara Ann. *Celebrating the Wild Mushroom.* Dodd, 1986. 256 p. 0396087558.

Fungi rarely stimulate as much passion as the author of this book exhibits. Friedman does not limit herself to dry, scientific facts, but also looks at mushroom myth and lore. Anecdotes and experience provided much of her background information. (LJ v111 My 15, 1986 p72)

Imes, Rick. *The Practical Botanist.* Fireside/Simon and Schuster, 1991. 151 p., illus., gloss., index. 0–671–69306–9.

All of the essentials of the taxonomy, physiology, morphology, and ecology of plants and plant communities presented in a way that encourages readers to pause and take a look at the botanical riches around them. (LJ v116 Je 15, 1991 p100; SBF v27 N 1991 p234)

Plotkin, Mark J. *Tales of a Shaman's Apprentice: An Ethnobotanist Searches for New Medicines in the Amazon Rain Forest.* Penguin, 1993. 319 p., gloss. 0–670–83137–9.

In the 1980s, when public awareness of rainforest environmental problems was first heightened, the author was actively researching the flora and fauna of Amazonia. He not only catalogued rare plants, but also documented native uses of them. An intriguing work of ethnobotany. (KR v61 Je 1, 1993 p704; LJ v118 Jl 1993 p113; SBF v29 D 1993 p265)

Also Recommended

Flora of North America North of Mexico. Flora of North America Editorial Committee, eds. Oxford University Press, 1993. A proposed fourteen-volume series in progress. For references in larger libraries. (SBF v30 Ap 1994 p71)

Huxley, Anthony. *Green Inheritance: The World Wildlife Fund Book of Plants.* Doubleday, 1985. Human/plant interactions and interdependence, especially in imperiled botanical systems. (BL v81 Ap 15, 1985 p1145; LJ v110 Ap 15, 1985 p78; Nature v314 Ap 25, 1985 p698)

Venning, Frank D. *Wildflowers of North America: A Guide to Field Identification.* Golden, 1984. Another in the popular field guide series.

Microbiology and Cell Biology

Burgess, Jeremy, Michael Martin, and Rosemary Taylor. *Under the Microscope: A Hidden World Revealed.* Cambridge University Press, 1990. 208 p., photos, index. 0–521–39940–8.

In nine chapters on subjects from "The Human Body" to "The Industrial World," the text and especially the photographs in this book show the living world that is too small to be seen. (Nature v345 Je 21, 1990 p677)

Dixon, Bernard. *Power Unseen: How Microbes Rule the World.* Freeman, 1994. 237 p., photos, biblio., index. 0–7167–4504–6.

Bacteria, viruses, and protozoa, and the astonishing degree to which they affect human life. Describes these interactions in seventy-five vignettes. (LJ v119 Mr 1, 1994 p114)

Postgate, J. R. *Microbes and Man.* 3rd ed. Cambridge University Press, 1992.

Postgate, a British scientist, gives an informative exposition of microbiology and microorganisms. Ever since the first edition appeared in 1969, this has remained a popular first source for curious general readers. (Nature v355 F 20, 1992 p689)

Sagan, Dorion, and Lynn Margulis. *Garden of Microbial Delights.* Harcourt Brace Jovanovich, 1988. 231 p.

Anton van Leeuwenhoek, the pioneer of microscopy, shocked his seventeenth century compatriots by declaring that there were more living creatures in his mouth than people in Holland. Sagan and Margulis give a glimpse into this unseen world that is part of our daily experience. (BL v85 S 15, 1988 p107)

Also Recommended

de Duve, Christian. *A Guided Tour of the Living Cell.* Scientific American, 1985. A two-volume boxed set on the cellular structure of living things. (BL v81 Ap 1, 1985 p1086; Choice v22 Je 1985 p1520; Nature v314 Ap 25, 1985 p695)

de Kruif, Paul. *The Microbe Hunters.* Harcourt, Brace, 1954. A classic that has stood the test of time.

Thomas, Lewis. *The Lives of a Cell.* Viking, 1974. Culled from Thomas's *New England Journal of Medicine* column "Notes of a Biology Watcher," the twenty-nine essays in this collection present his views on the biology and the practice of medicine. (BRD 1974 p1209; Choice v11 N 1974 p1339; Time v104 Jl 22, 1974 p84)

(*Note*: See also entries in Chapter Eight under "Biochemistry" and in Chapter Thirteen under "Biotechnology and Genetic Engineering."

Molecular Biology—General

Daudel, Raymond. *The Realm of Molecules.* Translated by Nicholas Hartman. McGraw-Hill, 1993. 132 p., index. 0–07–015642–5.

Quantum molecular science is a hybrid field that has encompassed research from physics, chemistry, and biological sciences. Within the context of the life sciences, researchers hope to apply theories toward better understandings of such things as what triggers the proliferation of cancer cells and how diseases propagate within the body. This is a short, nontechnical introduction to a very difficult field. (SBF v29 Ap 1993 p74; SciTech v17 F 1993 p14)

Gros, François. *The Gene Civilization.* McGraw-Hill, 1992. 136 p., index. 0–07–024963–6.

From McGraw-Hill's Horizons of Science series, this book begins with Darwinian theory in looking at the progress of genetics and the field of molecular biology. In the near future, the author contends, we will live in a society in which genetic technologies will affect us on a daily basis. Translated from the French. (BL v88 My 15, 1992 p1323; SBF v28 Ap 1992 p72; SciTech v16 Je 1992 p3)

Judson, Horace Freeland. *The Eighth Day of Creation: Makers of the Revolution in Biology.* Simon and Schuster, 1979. 686 p., illus., biblio., index. 0617225405.

Fifteen years after its publication, this book is still the best general history of the field of molecular biology. The text is divided into three parts: the history of DNA prior to the Watson-Crick model, a discussion of how DNA's genetic role was discovered and understood, and an examination of the role of the protein hemoglobin in respiration and physiology. (BRD 1979 p659; LJ v104 Ja 15, 1979 p178; NYTBR Ap 8, 1979 p1)

(*Note*: See also entries in Chapter Thirteen under "Biotechnology and Genetic Engineering.")

Molecular Biology—DNA

Pollack, Robert. *Signs of Life: The Language and Meaning of DNA.* Houghton Mifflin, 1994. 201 p., biblio. 0–395–64498–4.

In its four base chemicals, DNA contains the instructions for all living things. Pollack likens DNA's coding to a language, a metaphor he develops throughout this book. As with literary texts, scientists study DNA in search of meanings. (KR v61 D 1, 1993 p1508)

Watson, James. *The Double Helix: A Personal Account of the Discovery of the Structure of DNA*. Atheneum, 1968. 226 p., photos. 0–689–70602.

Watson's first person account of his and Francis Crick's discovery of the double helix structure of the DNA molecule is a classic of popular science and remains an intriguing case study of the processes of scientific investigation. The style is very candid and yields unique insights into the personalities of the author and many leading biochemists of the day. (BRD 1968 p1386; Choice v5 My 1968 p368; LJ v93 My 15, 1968 p2136)

Also Recommended

McCarty, Maclyn. *The Transforming Principle: Discovering that Genes Are Made of DNA*. Norton, 1985. An inside-the-lab, first person account. (Choice v23 N 1985 p468; LJ v110 My 1, 1985 p69; Nature v317 S 19, 1985 p209)

Molecular Biology—Genetics, General

Berg, Paul, and Maxine Singer. *Dealing with Genes: The Language of Heredity*. University Science Books, 1992. 269 p., gloss., index. 0–935702–69–5.

Basic cell chemistry and recombinant DNA technology, with emphasis on techniques. Also discusses current areas of high interest, like virology, immunology, and genomic evolution. The authorial collaboration of a Nobel laureate and the president of the Carnegie Institute is first rate. (Choice v30 Je 1993 p1792; SBF v20 Ap 1993 p73; SciTech v17 Ja 1993 p14)

Dawkins, Richard. *The Blind Watchmaker*. Norton, 1986. 332 p., illus. 0–393–02216–1.

Dawkins begins, "Natural selection is the blind watchmaker, blind because it does not see ahead, does not plan consequences ... yet the living results of natural selection overwhelmingly impress us with the appearance of design." This book attacks a fundamental creationist argument by demonstrating how evolution can proceed along a desultory path yet achieve complex, highly organized results. (KR v54 O 15, 1986 p1548; PW v230 O 24, 1986 p67; TLS S 26 1986 p1047)

Dawkins, Richard. *The Selfish Gene*. New ed. Oxford University Press, 1989. 352 p., biblio., index. 0–19–286092–5.

Dawkins, a zoologist, extends Darwin's theory of natural selection to the study of ethology, or animal behavior, to argue that many of the actions and rituals of animals are genetically programmed for the overriding purpose of reproduction; or, as he puts it, "An organism is just a gene's way of making another gene." This "gene's-eye view" of evolution was controversial when originally published in 1976, but now represents the scientific orthodoxy. (Choice v14 My 1977 p402; LJ v101 D 1, 1976 p2500; TES D 23, 1988 p9)

Jones, Steve. *The Language of Genes*. Anchor, 1994. 272 p., index. 0–385–47372–9.

Jones has a unique ability to communicate the esoteric workings of genetics and how they have functioned in our species in its evolutionary, cultural, and intellectual history. A good general overview to a fast-moving field. (KR v62 Je 1, 1994 p755; New Sci v140 N 20, 1993 p41)

Lee, Thomas F. *Gene Future: The Promise and Perils of the New Biology*. Plenum, 1993. 339 p., index. 0–306–44509–3.

The "new" biology is founded upon an ever-increasing knowledge of genetic processes and extremely sophisticated technologies for altering and enhancing those processes. This book looks at a full range of practical issues at a general to intermediate level. (Nature v366 N 25, 1993 p379; SBF v30 Ja 1994 p9)

Levine, Joseph, and David Suzuki. *The Secret of Life: Redesigning the Living World*. WGBH, 1993. 265 p., index. 0–9636881–0–3.

Companion to a television series of the same name, this book is a good basic introduction for generalists. In too many recent books, genetics is discussed from a heavily technological view; thus, the "unity of life" perspective, which relates ecological and environmental topics, is welcome. (PW v240 S 20, 1993 p56; SBF c29 N 1993 p232)

Also Recommended

Hall, Stephen. *Invisible Frontiers: The Race to Synthesize a Human Gene*. Atlantic Monthly, 1987. How biotechnology, with impetus from both academe and business, seeks to create better genes. (BL v83 Ag 1987 p1707; LJ v112 S 1, 1987 p190; Nature v329 O 1, 1987 p399)

(*Note*: See also entries in Chapter Thirteen under "Biotechnology and Genetic Engineering.")

Molecular Biology—Human Genetics

Bishop, Jerry, and Michael Waldholz. *Genome*. Simon and Schuster, 1990. 352 p., index. 0–671–57335–8.

This, an early popular account of the Human Genome Project, remains the first choice for general readers because of its journalistic style, its rapid pacing, and its unique emphasis throughout on social, economic, and ethical issues. The authors are *Wall Street Journal* science correspondents. (BL v86 Ag 1990 p2136; LJ v116 Mr 1, 1991 p61; WSJ v216 Ag 10, 1990 pA9)

Shapiro, Robert. *The Human Blueprint: The Race to Unlock the Secrets of Our Genetic Script*. St. Martin's, 1991. 412 p., photos, index. 0–312–05873–x.

At this time, the Human Genome Project—a massive international effort to map the entire human genetic code—is making rapid progress. This book gives a broad historical background to the study of genetics, a thorough, nontechnical overview of the project, and a thoughtful discussion of the ethical and philosophical questions it raises. Be advised, though, that many new discoveries are being made, and supplementary readings may be useful for updating the material presented here. (BL v88 S 1, 1991 p14; KR v59 Ag 1, 1991 p998; LJ v116 S 1, 1991 p226)

Also Recommended

The Code of Codes. Daniel J. Kevles and Leroy Hood, eds. Harvard University Press, 1992. A collection of essays on the Human Genome Project by specialists from various fields. A good supplemental reading for those who want more information than that provided in general works. (SBF v28 N 1992 p233)

Cook-Deegan, Robert. *The Gene Wars: Science, Politics, and the Human Genome.* Norton, 1994. An account of the political and scientific organization of the Human Genome Project. (LJ v119 Ap 1, 1994 p127)
(*Note*: See also entries in Chapter Ten under "Medical Genetics.")

Neurosciences and Psychology

Edelman, Gerald. *Bright Air, Brilliant Fire: On the Matter of the Mind.* Basic, 1992. 282 p., notes, index. 0–465–00764–3.
  "We are at the beginning of a neuroscientific revolution," the author, a Nobelist, writes in the preface. The ultimate result of this revolution could be an understanding of how the mind works and, by extension, how we come to know the world. The author likens the brain to an ecosystem, constantly changing and interacting with other systems. (BL v88 Ap 15, 1992 p1485; LJ v117 Ap 1, 1992 p142; New Sci v134 Je 13, 1992 p48)

Gazzaniga, Michael. *Nature's Mind: The Biological Roots of Thinking, Emotions, Sexuality, Language and Intelligence.* Basic, 1992. Index. 0–465–07649–1.
  The evolution of the structure of the brain has been studied fairly extensively. Not so the evolution of brain functions and the human behavioral attributes they control. The author speculates that nature has endowed us with certain ingrained psychological capabilities that we experience as thought, and which have aided us in evolution. (KR v60 S 1, 1992 p1103; PW v239 Ag 24, 1992 p66; SBF v29 D 1992 p5)

Flanagan, Owen J. *The Science of the Mind.* 2nd ed. MIT Press, 1991. 424 p., index. 0–262–06137–9.
  Psychology, the science of thinking, has always enjoyed a symbiotic relationship with philosophy, the discipline of thinking. Here is an intelligent look at how cognitive sciences have influenced philosophy and, by extension, culture. Discussions of Freud, Skinner, Piaget, William James, and others whose work straddles both fields are included. (SBF v27 N 1991 p228)

Johnson, George. *In the Palaces of Memory*. Knopf, 1991. 225 p. 0–394–58348–5.

In a sense, we are our memories. Johnson writes, "Whenever you read a book or have a conversation, the experience causes physical changes in your brain." This book is about those changes and how they affect or create personalities. Of various books on the subject, this is the first choice of many. (BL v87 F 15, 1991 p1164; LJ v116 F 1, 1991 p90; Nature v351 My 16, 1991 p200)

Kosslyn, Stephen M., and Olivier Koenig. *Wet Mind: The New Cognitive Neuroscience*. Free Press, 1992. 548 p., biblio., index. 0–02–917595–x.

"Wet mind" research capitalizes on the idea that "the mind is what the brain does"; that is, mental events are the results of neurological events in the brain. This view, which favors the nature side of the nature versus nurture debate, has opened new vistas in understanding a wide variety of human behaviors, and also has applications in the field of artificial intelligence. Case studies and examples drive home the point. (KR v60 F 15, 1992 p235; LJ v117 My 15, 1992 p114)

Pinker, Stephen. *The Language Instinct: How the Mind Creates Language*. Morrow, 1994. 494 p., notes, gloss., index. 0–688–12141–1.

A leading researcher in psycho-linguistics has written the best book available on the universal human phenomenon of language and its cognitive foundations in the species. General readers will delight in the wordplay; more informed audiences will be stimulated by his bold theories, which are based upon the latest genetic and neurological research. (Atl v272 Mr 1994 p130; BL v90 F 15, 1994 p1041; Nature v368 Mr 24, 1994 p360)

Wright, Robert. *The Moral Animal: Why We Are the Way We Are, the New Science of Evolutionary Psychology*. Pantheon, 1994. notes, biblio., index. 0–679–40773–1.

Evolutionary psychology, a relatively new discipline that integrates genetics and zoology with the social and behavioral sciences, speculates that what we do and think are byproducts of evolutionary attributes developed by natural selection. (KR v62 J 15, 1994 p971; PW v241 J 25, 1994 p38)

Also Recommended

Harth, Eric. *The Creative Loop: How the Brain Makes the Mind*. Addison-Wesley, 1993. The sometimes "messy" mental processes and associations that lead to creativity. (LJ v119 Mr 1, 1994 p56; PW v240 S 6, 1993 p79)

Osherson, Daniel. *An Invitation to Cognitive Sciences*. MIT Press, 1990. A three-volume collection that covers psychology in relation to vision, language, and thought. (Choice v28 S 1990 p226; SciTech v14 Ap 1990 p1)

(*Note*: See also entries in Chapter Ten under "Diseases—Neurological Disorders," in Chapter Eleven under "Evolution, Human," and in Chapter Thirteen under "Computer Sciences—Artificial Intelligence and Robotics.")

Philosophy of Biology

Ruse, Michael. *Philosophy of Biology Today*. State University of New York Press, 1988. 155 p., notes, index. 0–88706–910–x.

A summary handbook on topics including evolutionary theory, population genetics, molecular biology, systematics, human biology, and the philosophical implications of each. (Choice v30 Ap 1989 p1348)

Sociobiology

Badcock, C. R. *Evolution and Individual Behavior: An Introduction to Human Sociobiology*. Blackwell, 1991. 303 p. 0631174282.

The bottom line as to whether a species is successful is if it is able to reproduce. Thus, it is not surprising that species possess evolutionary traits that have been designed by nature to ensure that measure of success. As the author shows, this is just as true of human beings as of any other animal species. (Choice v29 Je 1992 p1617)

Goldsmith, Timothy. *The Biological Roots of Human Nature*. Oxford University Press, 1991. 161 p. 0–19–506288–4.

From its inception, sociobiological theory was subject to misconceptions and sensationalization. The author attempts to correct those errors in his discussions of evolution and human behavior. (Choice v29 Je 1992 p1568)

Schwartz, Barry. *The Battle for Human Nature*. Norton, 1986. 348 p. 0393023192.

The biological basis for human behavior is not well understood by scientists or the general public. Often, the view is that humans are, by their natures, ultimately self-centered. Schwartz considers ethical and philosophical issues from a sociobiological standpoint. (Choice v24 N 1986 p517; LJ v111 My 15, 1986 p69; NYTBR v91 O 16, 1986 p10)

Also Recommended

Durham, William H. *Coevolution: Genes, Culture, and Human Diversity*. Stanford University Press, 1991. Cultural change and human diversity in a biological context. (BioSci v42 S 1992 p629; SBF v28 Mr 1992 p37)

Wilson, Edward O. *Sociobiology: The New Synthesis*. Belknap/Harvard University Press, 1975. Wilson's study of the biological basis of animal and human behavior raised a firestorm of controversy when it was first published and is still being debated today. (BRD 1975 p1366; Choice v12 N 1975 p1196; Sci v190 O 17, 1975 p261)

(*Note*: See also entries in Chapter Eleven under "Evolution—General.")

Zoology—General

Ackerman, Diane. *The Moon by Whale Light*. Random House, 1991. 249 p. 0–394–58574.

When Ackerman writes about an animal, she first gets close to it—close enough to swim among a pod of killer whales and to allow a bat to roost in her hair. Her writing style is inimitable. (BL v88 O 1, 1991 p227; KR v59 S 1, 1991 p1125)

The Atlas of Endangered Species. John Burton, ed. Macmillan, 1991. 256 p., photos, index. 0–02–897081–0.

The leading cause of animal extinction is the degradation of their habitats by human activities. Thus, the problem is both local and global. This encyclopedic work looks at both aspects, and also provides species-by-species reports. (BL v88 F 1, 1992 p1051; LJ v116 D 1991 p130; SLJ v38 My 1992 p30)

Chadwick, Douglas. *The Kingdom: Wildlife in North America*. Sierra Club, 1990. 197 p., photos. 0–87156–617–6.

A brilliant visual tour of animal life in the remaining North American wildernesses, this is an oversize book where the text is as intelligent as the photographs are beautiful. (LJ v116 Mr 1, 1991 p59)

Kevles, Bettyann. *The Female of the Species*. Harvard University Press, 1986. 270 p., illus., biblio., index. 0–674–29865–9.

The role of the female animals in sexual behaviors and reproduction, and other sex-related characteristics. Incorporates sociobiological theory and suggests feminist interpretations. (BL v82 Mr 15, 1986 p1048; Choice v24 S 1986 p156; LJ v111 Ap 1, 1986 p155)

Quammen, David. *Natural Acts: A Sidelong View of Science and Nature*. Schocken Books, 1985. 221 p. 0–8052–3967–7; *The Flight of the Iguana*. Doubleday, 1989. 296 p. 0–385–26327–9.

Quammen confesses himself to be a dilettante, "a haunter of libraries and a snoop." The exuberant essays and expositions in these books are full of strange bits of biological trivia, strung together by a strong environmental ethic and the author's unique literary style. (BL v81 Mr 15, 1985 p1016; LJ v114 Mr 1, 1989 p43)

Tudge, Colin. *Last Animals at the Zoo: How Mass Extinctions Can Be Stopped*. Island Press, 1992. 266 p., notes, index. 1–55963–158–9.

Modern zoos are much more than just entertainment facilities. In an era when many species are faced with extinction in their natural habitats, zoos have become sanctuaries where they can be preserved and rehabilitated. Tudge of the Zoological Gardens of London explains the need to maintain biodiversity and argues that zoos are essential in doing so. (BL v88 Mr 1, 1992 p1187; LJ v118 Mr 1, 1993 p42; Nature v354 N 21, 1991 p185)

Also Recommended

California Center for Wildlife. *Living with Wildlife*. Sierra Club, 1994. How to enjoy and peacefully coexist with wildlife in a variety of human environments. (BL v90 Ap 1, 1994 p1414; LJ v119 My 1, 1994 p133)

Grzimek's Animal Life Encyclopedia. Bernard Grzimek, ed. Van Nostrand, 1975. In thirteen volumes, this remains the standard world zoology reference.

Macmillan Illustrated Animal Encyclopedia. Phillip Whitfield, ed. Macmillan, 1984. Characteristics and diversity of living vertebrates. (BL v81 S 15, 1984 p97; LJ v109 S 1, 1984 p1663)

Sinclair, Sandra. *How Animals See*. Facts on File, 1985. Eye structures, brain activities, and animal vision. Innovative photography simulates what animals actually see. (BL v81 Jl 1985 p1487; LJ v110 Je 15, 1985 p67)

Zoology—Animal Behavior

Agosta, William. *Chemical Communication: The Language of Pheromones*. Scientific American, 1992. 180 p., gloss., index. 0–7167–5036–8.
   Pheromones are biochemicals that send instinctive signals within and among species to elicit various behavioral reactions. Beginning with how pheromones operate in simple organisms, the author proceeds to the more elaborate systems found in insects and mammals. (Choice v30 D 1992 p644; SBF v29 D 1992 p14)

Benyus, Janine. *Beastly Behaviors*. Addison-Wesley, 1992. 366 p., index. 0–201–57008–4.
   The author, a zookeeper, offers this as a "field guide" to the behavior and body language of some of the most popular zoo animals in the world. The range of behaviors includes the most subtle, such as preening or scratching, to the most conspicuous, like mating or feeding. Hundreds of species are covered. (BL v89 O 1, 1992 p222; LJ v117 S 1, 1992 p208; SBF v29 Mr 1993 p41)

Morton, Eugene, and Jake Page. *Animal Talk: Science and the Voices of Nature*. Random House, 1992. 275 p., biblio., index. 0–394–58337–x.
   The "Dr. Dolittle Myth" is that the various sounds that animals make constitute a language akin to that spoken by humans. In dispelling this myth, the authors—an ornithologist and a science journalist—describe how animal vocalizations communicate environmental information within a species and how certain kinds of sounds seem to have universal meanings. (BL v88 Ap 15, 1992 p1490; LJ v117 My 15, 1992 p122; NYTBR v97 Jl 15, 1992 p10)

Also Recommended

Attenborough, David. *The Trials of Life: A Natural History of Animal Behavior*. Little, Brown, 1991. Derived from a television series, the chapters in this book explore basic behaviors such as mating, feeding, and home building in hundreds of different species. (New Sci v128 D 1, 1990 p62; TES O 12, 1990 pR3)

Griffin, Donald R. *Animal Minds*. University of Chicago Press, 1992. The author makes the controversial argument that animals do think. Cites a good deal of original research, but not so scholarly that it cannot be read by a generalist. (KR v60 Jl 15, 1992 p896; PW v239 Jl 27, 1992 p53)

Zoology—Birds

Harrison, Colin, and Alan Greensmith. *Birds of the World*. Dorling Kindersley, 1993. 415 p., photos, gloss., index. 1–56458–295–7.
A versatile book that could be used by birdwatchers, students, and general readers, or for reference. (SBF v30 Ap 1994 p77)

Katz, Barbara. *So Cranes May Dance*. Chicago Review Press, 1993. 279 p., index. 0–55652–171–5.
Cranes, with their long, graceful bodies, are favorites among birdwatchers. Seven of the fifteen modern species are extinct. Katz of the International Crane Foundation writes of the causes of the species' plight and the foundation's efforts to preserve them. (BL v89 Jl 1993 p1932; KR v61 My 15, 1993 p644; LJ v118 Je 1, 1993 p182)

Newton, Ian. *Birds of Prey*. Facts on File, 1990. 240 p., photos. 0–8160–2182–1.
Raptors—falcons, eagles, vultures, and buzzards—are among the most admired and feared of avian species. This well-illustrated, oversize book shows these birds in nature and discusses their biology, habitats, behaviors, and relations with human beings. (LJ v115 Jl 1990 p123)

Short, Lester L. *The Lives of Birds: Birds of the World and Their Behaviors*. Holt, 1993. 256 p., illus., index. 0–8050–1952–9.
Many of us may never see a Mexican chalaca or a jungle babbler, but this book can help us understand their behaviors through analogies with more commonly seen birds. The author gives a far-ranging view of basic bird behaviors like migration, egg-laying, and nest-making. (BL v89 Je 1, 1993 p1754; LJ v118 My 15, 1993 p93)

Stap, Don. *A Parrot Without a Name: The Search for the Last Unknown Bird on Earth*. Knopf, 1990. 224 p. 0–394–55596–1.
Deep in the Amazonian rainforests, there still exist plant and animal species undiscovered by science. As the title suggests, this book tells the story of one of them. (BL v86 My 15, 1990 p1765; LJ v115 My 1, 1990 p110)

Also Recommended

Cambridge Encyclopedia of Ornithology. Michael Brooke and Tim Birkhead, eds. Harvard University Press, Cambridge University Press, 1991. A complete reference on the anatomy, physiology, ecology, and behavior of birds. (BL v88 Mr 1, 1992 p1304; Choice v29 Mr 1992 p1108; LJ v116 D 1991 p132)

Gill, Frank. *Ornithology.* 2nd ed. Freeman, 1994. Entirely up-to-date edition of a text and reference.

Hume, Robe. *Owls of the World.* Running Press, 1991. An illustrated guide to the species within two families of owls. (BL v88 Ja 1, 1992 p800; LJ v117 Ja 1992 p169)

Johnsgard, Paul A. *Cormorants, Darters, and Pelicans of the World.* Smithsonian, 1993. Comparative biology, anatomy, behavior, zoogeography and population dynamics of these marine and wetland birds. (SBF v29 D 1993 p266)

Proctor, Noble S. *Manual of Ornithology: Avian Structure and Function.* Yale University Press, 1993. A combination lab manual and textbook. (Choice v31 S 1993 p158; LJ v118 Jl 1993 p72; SBF v29 Ag 1993 p169)

Robbins, Chandler S. *Birds of North America: A Guide to Field Identification.* Golden Press, 1983. A guide for birdwatchers.

Zoology—Fish, Aquatic Life, and Marine Mammals

Bulloch, David. *The Underwater Naturalist.* Lyons and Burford, 1991. 250 p., illus. 1–55821–108–x.
   The author, a marine naturalist and an avid scuba diver, introduces readers to the diversity of life found in the coastal regions of the United States. The book is arranged by marine environment types—for example, deep water, coral reefs, tropical waters, and so on. (BL v87 F 15, 1991 p1167; LJ v116 Mr 1, 1991 p114)

Carroll, David M. *Trout Reflections: A Natural History of the Trout and Its World.* St. Martin's, 1993. 143 p. 0–312–09464–7.
   The trout, a highly prized game fish, is also recognized by scientists as an indicator of the health of inland waters. In journal format, the author shares observations and gives scientifically accurate observations of trout species. Fishing also figures prominently. (LJ v118 Mr 15, 1993 p103; PW v240 F 15, 1993 p226)

Ellis, Richard, and John E. McCosker. *Great White Shark.* HarperCollins, 1991. Photos, notes, index. 0–06–016451–4.
   Human lore holds the great white to be perhaps the most terrifying creature in the ocean. It is, indeed, a hungry predator, but humans have taken a much greater toll on it than vice versa. This book gathers research, historical records,

and personal accounts in presenting a balanced view of the species. Superbly illustrated. (BL v88 D 15, 1991 p738; LJ v116 D 1991 p186)

Ellis, Richard. *Men and Whales*. Knopf, 1991. Illus., index. 0–394–55839–1.

As fearsome as whales are, they have almost always fared poorly in confrontations with human beings. Ellis describes whale families and behaviors, then introduces human beings to the equation in order to show how we have interfered with their basic species needs and, in some cases, systematically slaughtered them. Gives all the shameful and frightening details. (BL v88 O 15, 1991 p389; LJ v116 O 15, 1991 p114; Nature v356 Ap 16, 1992 p632)

Gormley, Gerard. *A Dolphin Summer*. Taplinger, 1985. 196 p. 0800822641.

Gormley's style of writing could be called scientific novelization. In this book, he narrates a tale of the first eight months of a female dolphin's life, its behaviors, its social networks, and its physical environment. Fictional episodes such as an encounter with killer whales and brushes with humanity are compelling. (BL v82 S 1, 1985 p12; KR v53 O 15, 1985 p1115; LJ v110 Ag 1985 p106)

Gormley, Gerard. *Orcas of the Gulf*. Sierra Club, 1990. 216 p. 0–87156–601–x.

Gormley creates a fictional pod (a family) of orcas and, through them, explores the full range of species behaviors, from their predations to their close family bonds. The content is based on information culled from more than 200 whale observations. (BL v86 Ap 15, 1990 p1594; LJ v115 My 15, 1990 p92)

Moyle, Peter. *Fish: An Enthusiast's Guide*. University of California Press, 1994, 272 p., index. 0–520–07977–9.

Fish anatomy and behaviors, as well as ecological concerns on various species. Gives lots of ideas about how to express an active interest in fish, such as developing and stocking a pond. The concluding chapter gives resources. (SBF v30 My 1994 p104)

Also Recommended

Moyle, Peter B. *Fishes: An Introduction to Ichthyology*. Prentice-Hall, 1988. 0133192113. A textbook.

Vermeij, Geraat J. *A Natural History of Shells*. Princeton University Press, 1993. An illustrated guide to malacology for students and collectors. (SBF v30 My 1994 p105)

Zoology—Insects and Land Invertebrates

Brackenbury, John. *Insects in Flight*. Blandford, 1992. Photos, index. 0–7137–2301–7.

Flying insects are capable of some extraordinary feats of aviation. In words and spectacular photos, this book examines the anatomies of various flying insects, the incredible mechanical means by which they fly, and even ways by which some species simulate flight (by hopping, for example). The last chapter is called "New Challenges." (SBF v29 Ag 1993 p175)

Holldobler, Bert, and E. O. Wilson. *The Ants.* Harvard University Press, 1990. 736 p. 0–674–04075–0.
This huge book contains a glut of detail on a very specialized subject, but it is also a Pulitzer Prize winner that is guaranteed to give anybody an appreciation of these insects that are so small they are often overlooked. In fact, ants are incredible creatures who live in close, complex societies. This will remain the standard reference on all ant species for generations to come. A more popularized account is the authors' *Journey to the Ants.* Harvard University Press, 1994. (Choice v28 O 1990 p334; LJ v116 Mr 1, 1991 p60; Nature v344 Ap 26, 1990 p894)

Imes, Rick. *The Practical Entomologist.* Fireside: Simon and Schuster, 1992. Photos, gloss., index. 0–671–74695–2.
This is an introduction to entomology for those with an interest in collecting or studying insects. Organized by taxonomic order, the book has sections on the major classes of insects and the families contained within them. Throughout, there are suggestions about what you can do to channel intellectual curiosity about insects into an active hobby. (LJ v117 D 1992 p180; SBF v29 Ja 1993 p14)

McGavin, George. *Bugs of the World.* Facts on File, 1993. 192 p., photos. 0–8160–2737–4.
The title, a bit flippant, might actually attract the attention of general readers. The tone of the writing is light, but no-nonsense. Chapters are arranged by topic, such as "The Structure of Bugs," "Diseases and Enemies," and "Mating and Egg Laying." (Am Ent v40 Spr 1994 p58)

Preston-Mafham, Rod, and Ken Preston-Mafham. *Spiders of the World.* Facts on File, 1984.
Human revulsion notwithstanding, spiders are ecologically important and seldom dangerous to people. This book begins with spider anatomy and structure, then goes into more depth when discussing behaviors. (ARBA v16 1985 p534; BL v81 O 15, 1984 p271; Choice v22 N 1984 p449)

Winston, Mark L. *Killer Bees: The Africanized Honey Bee in the Americas.* Harvard University Press, 1991. 162 p., index. 0–674–50352–x.
The "killer bees" invasion of the United States has been the subject of jokes and sensationalization. In fact, the Africanized honey bee is a resilient species which, while physically dangerous to humans, is of more concern because of the damage it can do to crops. (LJ v117 F 1, 1992 p119; SBF v28 Ap 1992 p73)

Wootton, Anthony. *Insects of the World*. Facts on File, 1984. 224 p., illus., biblio., index. 0–87190–99102.

This book emphasizes the diversity, adaptability, and sheer numbers of insect species on Earth. There are fairly standard discussions of anatomy, biology, and behavior. The author also attempts to put human/insect interactions in a positive light. (BL v81 D 1, 1984 p467; Choice v22 Ja 1985 p706; SBF v20 Mr 1985 p204)

Zoology—Land Mammals

Adamson, Joy. *Born Free: A Lioness of Two Worlds*. Pantheon, 1987. 220 p. 0394561414.

Originally published in 1960, this was one of the first books to depict the threats that human beings pose to large wild animals and the trials faced by conservationists. After nearly thirty-five years, it remains popular. (BRD v1961 p7; BL v56 Je 1, 1960 p595; LJ v85 Je 15, 1960 p2445)

Chadwick, Douglas. *The Fate of the Elephant*. Sierra Club, 1992. 492 p., biblio., index. 0–87156–635–4.

The magnificent African elephant, hunted avariciously and illegally for its ivory tusks, is near extinction. Wildlife biologist Chadwick tours several elephant habitats in studying the factors that have led to its endangerment. The threats to it are complex, and equally sophisticated measures will be required to ensure its survival. (KR v60 Ag 15, 1992 p1029; PW v234 S 7, 1992 p88)

Cheney, Dorothy L., and Robert Seyfarth. *How Monkeys See the World*. University of Chicago Press, 1991. 377 p., index. 0–226–10245–9.

The authors begin by asking, "What is it like to be a monkey?" Based on their research with East African vervet monkeys, they offer their ideas on animal cognition. Among their more provocative speculations is that monkeys are capable of certain kinds of abstract thought. (Choice v28 Jl 1991 p1805; Nature v35 Ap 18, 1991 p565; TLS N 30, 1990 p1299)

Fenton, M. Brock. *Bats*. Facts on File, 1993. 207 p., photos, index. 0–8160–2679–3.

This book discusses this greatly misunderstood mammal in words and pictures. The chapter on echolocation—the bat's amazingly sensitive form of sonar "vision"—is especially fascinating. (BRpt v12 My 1993 p56; Choice v30 My 1993 p1492; LJ v118 F 1, 1993 p105)

Goodall, Jane. *In the Shadow of Man*. Houghton Mifflin, 1983; *The Chimpanzees of Gombe*. Houghton Mifflin, 1986; *Through a Window: My Thirty Years with the Chimpanzees of Gombe*. Houghton Mifflin, 1990. 320 p. 0–395–50081–8.

Goodall is that rare combination, a leading researcher who has also displayed a commitment to popularizing her work. Her observations of the behaviors and personalities of chimpanzees have forced human beings to take a close look at themselves. These three books make for enlightening, progressive reading. In the earlier works, the focus is on chimp behavior and how it parallels that of humans. The most recent books are retrospective, with thoughts on animal activism and ecology issues. (BL v87 Mr 15, 1991 p1469; LJ v116 Mr 1, 1991 p60; Nature v348 N 22, 1990 p371)

Lopez, Barry. *Of Wolves and Men*. Scribner's, 1978. 309 p. 0684156245.

To northern native peoples in the Americas, the wolf is a highly symbolic creature that features prominently in myth and ritual. Whites, however, have treated the wolf as a nuisance and hunted it to elimination in all but a few areas in the lower forty-eight states. In this very popular book, Lopez gives a true picture of wolves in the wild and how they interact with humans. (BRD 1978 p815; LJ v103 O 15, 1978 p2123)

McNamee, Thomas. *The Grizzly Bear*. Knopf, 1984. 276 p. 0394529987.

That the huge and powerful grizzly could become endangered in the wild is a sad statement on how severely human beings have ravaged even remote wilderness areas. McNamee, a poet and writer, displays a humanistic muse and a scientific eye for detail in his writing. (KR v52 S 1, 1984 p851; LJ v109 O 15, 1984 p1952)

Schaller, George B. *The Last Panda*. University of Chicago Press, 1993. Photos, index. 0–226–73628–8.

The exotic giant panda of China's Sichuan province is one of the most captivating of the world's mammals. Schaller, a leading panda researcher, finds human activity largely to blame for the species' endangerment. With human resolve, however, he believes that the panda can be saved. (BL v89 Mr 15, 1993 p1283; LJ v119 Mr 1, 1994 p54; NYTBR v98 Mr 28, 1993 p1)

Thapar, Valmik. *Tiger: Portrait of a Predator*. Facts on File, 1986. Illus. 0816012385.

Thapar, an Indian zoologist, describes the biology and natural history of the tigers of the Indian subcontinent, particularly those he observed at the Ranthambhore Tiger Preserve. (LJ v111 S 1, 1986 p207; SBF v22 Ja 1987 p171; SLJ v33 S 1986 p157)

Walker's Mammals of the World. 5th ed. 2 vols. Johns Hopkins University Press, 1991. 670 p., photos, index. 0–8018–3970–x.

Describes the over 4,000 mammal species in the world. Entries give genera and species, physical descriptions, life histories, and biological/ecological characteristics. Clear writing and sharp photographs make this a good book for high school and up. (SBF v28 Ap 1992 p78)

Also Recommended

Kinkead, Eugene, and John Hamberger. *The Squirrel Book*. Dutton, 1980. The natural history of the gray squirrel. (BL v77 O 15, 1980 p292; KR v48 O 1, 1980 p1332; LJ v105 O 1, 1980 p2094)

Mammal Species of the World. Don E. Wilson and Deeann Reeder, eds. Smithsonian, 1993. A taxonomic and geographic reference, somewhat technical in nature. (Choice v31 S 1993 p84; SBF v29 O 1993 p198; SciTech v17 My 1993 p18)

Preston-Mafham, Rod. *Primates of the World.* Facts on File, 1992. A general text and photos. (BRpt v11 Mr 1993 p54; Choice v30 F 1993 p988; SLJ v38 D 1992 p155)

Thomas, Elizabeth Marshall. *The Hidden Life of Dogs.* Houghton Mifflin, 1993; *The Tribe of the Tiger: Cats and Their Culture.* Simon and Schuster, 1994. Although admittedly anthropocentric, the author has some popular and insightful observations on the behaviors of these common domestic animals. (BRD My 1994 p320; BL v89 Ap 15, 1993 p1480; LJ v118 Ap 15, 1993 p123)

Whitehead, Kenneth. *The Whitehead Encyclopedia of Deer.* Voyageur Press, 1993. The definitive encyclopedia covering all world deer species. (BL v90 S 1, 1993 p91; LJ v118 Je 1, 1993 p110)

Zoology—Reptiles and Amphibians

Mattison, Christopher. *Lizards of the World.* Facts on File, 1989. 192 p. 0816019002.
   A systematically classified volume, with emphasis at the level of family, then subfamily. For general readers and amateur naturalists. (ARBA v21 1990 p661; Choice v27 Ap 1990 p1345; SBF v25 Mr 1990 p204)

Mattison, Christopher. *Snakes of the World.* Facts on File, 1986. 190 p. 081601082x.
   Chapters are divided according to topics; for example, physical appearance, food and feeding, reproduction, and so on. The concluding chapter is a synopsis of commonalities and differences among species. (BL v82 My 15, 1986 p1346; Choice v23 Jl 1986 p1697; TLS Jl 4, 1986 p745)

Sprackland, Robert G. *Giant Lizards.* T.F.H. Publications, 1992. 288 p. 0866226346.
   Large, in this case, is defined as one meter. Information given includes common and scientific name, habitat, geographic distribution, and various other facts. Excellent photography brings readers closer to these reptiles than many would probably like in real life. (ARBA v24 1993 p662)

Also Recommended

Carroll, David. *The Year of the Turtle.* Camden House, 1991. Swamp and marsh life in New Hampshire. Superb drawings by the author. (LJ v116 Ja 1, 1991 p140)

Ernst, Carl H. *Venomous Reptiles of North America.* Smithsonian, 1992. Identification of snakes, lizards, and so on. (LJ v117 Jl 1992 p115; SBF v28 D 1992 p265)

Preston-Mafham, Rod, and Ken Preston-Mafham. *The Encyclopedia of Land Invertebrate Behavior.* MIT Press, 1993. Summarizes research on species in proportion to the amount that is known about them. Avoids jargon.

Smith, Hobart Muir. *Amphibians of North America: A Guide to Field Identification.* Golden Press, 1978; *Reptiles of North America: A Guide to Field Identification.* Golden Press, 1982. Reliable guides in the Golden series.

OTHER RESOURCES

Periodicals for General Readers

American Biology Teacher. National Association of Biology Teachers, 8/yr., revs. 0002–7685.
   The articles are written for high school and college teachers of the biological sciences and cover topics of current interest. The "How-to-Do-It" section gives ideas on pedagogical and field activities that work with students, but might also be adopted by interested amateurs.

American Birds. National Audubon Society, 5/yr., revs.
   A field journal that gives information of concern to bird watchers and enthusiasts.

American Entomologist. Entomological Society of America, q., revs. 1046–2821.
   A professional publication with relatively nontechnical articles of interest to the entire discipline. The reviews and editorials are useful.

BioEssays. International Council of Scientific Unions. Council of Biologists. m., revs.
   Current topics in genetics and molecular biology covered in usually lively prose. Gives reviews and summaries in a readable way for students, academics, and informed generalists.

BioScience. American Institute of Biological Sciences, m., revs. 0006–3568.
   The leading general biology magazine for amateur naturalists and interested readers, this glossy and intelligent publication covers all fields of the biosciences. Opinion pieces, essays, and miscellaneous news announcements also appear.

National Wildlife. National Wildlife Federation, bi-m. 0028–0402.
   The editorial creed says it all: "To create and encourage an awareness among the people of the world of the need for conservation and proper management of those resources of the Earth upon which our lives and welfare depend."

(*Note*: See also periodicals entries in Chapter Eleven.)

Audiovisual Resources

BIOLOGY—GENERAL TOPICS

Edward O. Wilson: Reflections on a Life in Science. Harvard University Press, 1993 (videocassette, 60 minutes, color).

Wilson's scientific career has been both heralded and controversial. This video introduces the man in a very personal way as he recounts his life, career, and personal fascination with living things. Enlightening for generalists and a possible career motivator for young adults. (SBF v30 Mr 1994 p56)

The Gene—Future Quest. Hawkhill Associates, 1991 (videocassette, 34 minutes, color).

Scientists active in the fields of biotechnology and genetic research describe the techniques by which genes are manipulated and the potential benefits of this new, often misunderstood technology. For balance, there is also an interview with Jeremy Rifkin, a noted opponent of gene research. (SBF v27 N 1991 p249)

Perspective: Bacteria—The New Alchemy. Chip Taylor Communications, 1993 (videocassette, 24 minutes, color).

The analogy between contemporary biotechnology and alchemy is intriguing in that biotechnology's objectives are to genetically transmute biological entities. With bacteria, for example, the aim of biotechnological manipulation is to enhance useful properties that they already possess. (SBF v30 Ap 1994 p87)

The Secret of Life: How Molecular Biology Is Transforming Medicine, Agriculture, and Our Understanding of Life. WGBH TV, 1993 (videocassette).

Originally broadcast on television as a special production of a Boston station, this fine video looks at molecular biology with an emphasis on social aspects and the practical uses and related issues of technology.

ANATOMY AND PHYSIOLOGY

Inside Information: How the Brain Works. Films for the Humanities, 1990 (videocassette, 50 minutes, color).

The human brain has been likened to a biological computer. Taking that analogy as its point of departure, this video studies the ways by which the brain processes information. A running subtopic is cognitive evolution and how human thought and perception came to be. (SBF v27 N 1991 p249)

Living Body (series). Films for the Humanities, 1985 (videocassettes, 26 minutes ea., color).

From the twenty-two-segment television series of the same name, this series covers all aspects of human anatomy, growth, development, reproduction, and physiology.

Also Recommended

Bodyworks: Anatomy of the Human Body in Action. Media Productions International, 1990 (videocassette, 60 minutes). This presentation is derived in part from Gerard Tortora's standard text *Principles of Anatomy and Physiology.*

Body in Question (series). British Broadcasting Corp.–KCET Television (Los Angeles), 1980 (videocassettes, 60 minutes each). This thirteen-part series, originally done for the BBC, looks at the human body and physiology from social and medical perspectives.

ANIMAL LIFE

If Dolphins Could Talk. PBS Video, 1990 (videocassette, 60 minutes, color).
    The central theme of this video recognizes dolphins as highly intelligent animals and thus wonders what they would tell us about the declining health of marine environments if they could talk. (SBF v228 N 1992 p248)

Orca: Killer Whale or Gentle Giant. Video Project, 1990 (videocassette, 26 minutes, color).
    The sleek black and white killer whale is one of the most beautiful and feared of marine mammals. Its reputation for viciousness notwithstanding, it is also a highly intelligent and family-centered creature. This video uses some impressive underwater photography to show how the orca lives and, in doing so, gives a balanced view of its nature. (SBF v27 My 1991 p122)

The Panda. Landmark Films, 1991 (videocassette, 47 minutes, color).
    The giant panda of the People's Republic of China has been a major attraction at several American zoos. In its natural habitat, however, it is greatly endangered. Currently, naturalists are hard at work studying the species in the hopes of developing a plan for its survival. (SBF v28 My 1992 p120)

The Secret World of Bats. Bat Conservation International, 1992 (videocassette, 46 minutes, color).
    Originally shown on CBS in 1992, this video argues for the conservation of bat species by showing them to be not the fearsome predators of vampire movies, but complex and fascinating species. The slow-motion photography and other up-close footage are spectacular. (SBF v28 N 1992 p248)

Tracks of the Grizzly. Churchill Media, 1991 (videocassette, 27 minutes, color).
    Information about grizzlies, with excellent footage, and their interactions with humans. The efforts to preserve grizzlies are featured significantly. (SBF v28 My 1992 p120)

The Whales that Came Back. Landmark Films, 1991 (videocassette, 26 minutes, color).
    The California gray whale and its migratory behaviors. The near-catastrophic effects of whaling are also discussed. (SBF v28 N 1992 p248)

Wolves. PBS Video, 1990 (videocassette, 60 minutes, color).

From the National Audubon Society, this video dispels some of the myths about wolf biology and behavior, all within the context of wolf/human interactions. The issue of reintroducing wolves to certain habitats in the lower forty-eight states is given informed discussion. (SBF v27 Ag 1991 p185)

Also Recommended

Land of the Tiger. WQED TV–National Geographic Video, 1986 (videocassette, 60 minutes). The biology of the tiger and other topics in the zoology and biogeography of India.

Nature's Kingdom (series). Landmark Films, 1993 (videocassettes, 8 minutes each, color). The brief installments in this series give live footage of numerous species and basic facts about their lives and habitats. (SBF v29 D 1993 p278)

A Passion for Birds. Churchill Media, 1991 (videocassette, 26 minutes). From the *Northwest Wild* series produced in Seattle, this segment looks at birds in that region and the ongoing efforts of biologists to understand their lives and environments. (SBF v28 My 1992 p120)

CD-ROMs

Anatomist 2.1. Folkstone Design (one CD).

A versatile classroom tool using hypermedia to teach human anatomy; for middle school through general adult levels.

Dictionary of the Living World. Media Design Interactive, 1991 (one CD).

An award-winning resource that has many classroom applications. Features text entries on over 3,000 species, 450 color images, and even animated movies and clips.

Encyclopedia of Animals. Applied Optical Media Corporation, 1992.

General zoology for high school students.

Life Story. Wings for Learning, 1992–Sunburst (one CD).

Enables students at the secondary, college, or general adult level to explore the discovery and significance of DNA.

Mammals: A Multimedia Encyclopedia. National Geographic Association, 1990.

Up-to-date research on over 200 species; includes full screen visuals and programmed animal vocalizations. Also contains forty-five motion picture clips from National Geographic specials.

Multimedia Encyclopedia of Mammalian Biology. McGraw-Hill (one CD).

Contains access to all five volumes of *Grzimek's Encyclopedia of Mammals*, as well as 4,000 color photos and information not published elsewhere.

The Nature of Genes. MedMedia, 1992 (one CD).
  Genetics and medical applications of gene therapy in multimedia.

Ornithology. CEDROM.
  A multimedia encyclopedia of Western bird species, with artistic illustrations.

Chapter Eight

Chemistry

A chemist is a poet who took a wrong turn.
John Read, Humor and Humanism in Chemistry

By example, [Robert Boyd] taught us the arts of judgment, of scientific appraisal, of assessing possible significance of scientific insight; and he showed us how continually one has to exercise these mature qualities of thought. . . . I did not get such good examination results in chemistry. But what did that matter? I did receive the spirit of science, and so, for that much better reason, I elected to specialize in chemistry.
Christopher Kelk Ingold

Everything is a chemical.
American Chemical Society

Chemists argue that their discipline is "the central science." That claim is legitimate. The American Chemical Society's observation that "everything is a chemical" is indubitably true, and it makes an important point about how chemists view their science. Although chemistry has deep roots of its own, it also has branches reaching into almost all other disciplines. In fact, a liberal grounding in chemistry is often a prerequisite for entrance into fields such as physics, biology, and the health sciences.

Despite its centrality, chemistry is vastly underrepresented in popular literature. While books on hot topics in physics and biology, for example, seem to proliferate on commercial bookstore shelves, often few or no chemistry titles can be found. This is due partly to what some in the field have identified as chemistry's image problem. The word "chemical" can conjure up repellant visions of toxic substances stored in rusty steel drums at desolate dump sites. Chemistry may also have been comparatively ignored in popular literature precisely because of its centrality—it is sometimes viewed as rather mundane, the realm of the commonplace. This perception is unfortunate, for chemistry affects daily life probably more than any other discipline, and, to an open mind, even common chemical reactions can be deeply fascinating. Possessing a bedrock knowledge of general chemistry is both a practical and a rewarding aspect of science literacy.

Although new developments in chemistry do not grab headlines, there have been some truly revolutionary discoveries in recent years. For example, in 1985 "buckyballs," so named because of their physical resemblance to Buckminster Fuller's dome structures, were discovered, and their superconducting and other properties are being actively researched. In 1990 three MIT chemists announced that they had created a self-replicating molecule that seems to reproduce in a manner similar to living systems. These and other subjects need to be popularized, and it might be expected they will be in future works.

Because few chemistry popularizations have been published in recent years, many of the titles listed in this chapter are textbooks or semi-scholarly, academic press publications. While perhaps not ideal sources for general information needs, they do provide solid information on important topics. In recent years, the American Chemical Society has launched a vigorous publications campaign aimed at filling this gap, and some commercial publishers have also shown a new willingness to invest in chemistry popularizations. It can be hoped that many new books and videos of high quality will appear in the years to come.

CHEMISTRY—GENERAL

Reference

Emsley, John. *The Elements*. 2nd ed. Oxford University Press, 1991. index. 0–19–855568–7.

Probably the most convenient reference source on the chemical elements available. For each element, it gives data on chemical, physical, environmental, and other properties. A useful addition to reference collections, or for use as a supplement with other readings. (SBF v28 Ap 1992 p71)

The Facts on File Dictionary of Chemistry. Facts on File, 1988. 249 p. 0–8160–1866–9.

Many librarians have high trust in the Facts on File line of reference books. As a science literacy resource, this dictionary is effective because of its broad coverage and straightforward presentation. Suits high school and most public libraries perfectly; academic and larger public libraries might want to consider a more advanced, technical dictionary. (RSSEMA p69; SBF v24 Ja 1989 p156)

McGraw-Hill Encyclopedia of Chemistry. 2nd ed. McGraw-Hill, 1993. 1236 p. 0070454558.

The scope, currency, and authority of encyclopedias in the McGraw-Hill series make them good choices for science literacy collections. This is the best one-volume general encyclopedia for chemistry available. As always, though, libraries owning the multivolume parent need not purchase this spinoff. (Choice v31 O 1993 p268; WLB v68 S 1993 p119)

Also Recommended

A Concise Dictionary of Chemistry. Oxford University Press, 1990. Culled from Oxford's *Concise Science Dictionary.* Contains around 3,000 entries. (ARBA v18 1987 p661; TES My 23, 1986 p28)

Concise Encyclopedia of Chemistry. Mary Eagleson, ed. De Gruyter, 1994. Contains around 12,000 entries from general, inorganic, physical, and technical chemistry.

Grant and Hackh's Chemical Dictionary. 5th ed. Roger Grant and Claire Grant, eds. McGraw-Hill, 1987. Especially strong for covering international chemical terminology. (SBF v23 My 1988 p282)

Kirk-Othmer Concise Encyclopedia of Chemical Technology. Martin Grayson et al., eds. Wiley, 1985. Even at 1,318 pages, this is just a fraction of its multivolume parent. The concise version is more than adequate for most science literacy collections. Public and college libraries. (RSSEMA p128)

Lange's Handbook of Chemistry. McGraw-Hill, 1992. 0748–4584. Of principal value to chemists and chemistry students, but also useful to anybody whose business involves solving problems that deal with chemical products. (RSSEMA p73)

Van Nostrand Reinhold Encyclopedia of Chemistry. 4th ed. Douglas Considine and Glenn Considine, eds. Van Nostrand, 1984. If past publishing history holds true, a new edition of this standard work is due soon. (BL v81 S 1984 p36; SciTech v8 My 1984 p5)

Autobiography and Biography

Djerassi, Carl. *The Pill, Pygmy Chimps, and Degas' House: The Autobiography of Carl Djerassi.* HarperCollins, 1992, 318 p. 0–465–05759–4.

The autobiography of a chemist who was in the forefront of the development of oral contraceptives (the Pill). As much a history of social attitudes toward reproductive and women's rights as of the scientific discovery. (BL v88 Mr 1, 1992 p1187; KR v60 F 1, 1992 p156; LJ v117 Ap 1, 1992 p126)

Nobel Laureates in Chemistry, 1901–1992. American Chemical Society, 1994. 0–8412–2459–5.

This latest installment in ASC's fine History of Modern Chemical Sciences series is from an excellent lineage and covers a topic that is unique and important. The series is very good and the need for a book covering this subject is great, but since the final book is unavailable for review at this time, only a qualified recommendation can be made.

Patterson, Elizabeth Chambers. *John Dalton and the Atomic Theory.* Doubleday, 1970. 348 p.

Not only a readable and authoritative biography, but also an excellent period piece. The author depicts Dalton's great achievements within the full context of his social and scientific times. (BL v67 D 1, 1970 p282; Choice N 1970 p1248; LJ v95 Jl 1970 p2460)

Pioneers in Polymer Science. Raymond B. Seymour, ed. Kluwer, 1989. 272 p. 0–7923–0300–8.

Great advances have been made in polymer science in this century. This title serves as both a general history of the field and a biography of the most influential figures in its development. Most of the contributors are well known in the discipline. (Choice v27 Je 1990 p1705)

Serafini, Anthony. *Linus Pauling: A Man and His Science.* Paragon, 1991. 310 p., photos. 0913729884.

Pauling is the winner of two Nobel prizes: one for chemistry, and a second for peace. Until his recent death, he was also a leading proponent of the medical benefits of vitamin C. The style in this biography is casual and openly deferential to Pauling, whom the author describes as "hard-driving and boundlessly energetic," "the American cowboy of science." (BL v85 Mr 15, 1989 p1229; KR v57 F 15, 1989 p280; LJ v114 Mr 15, 1989 p83)

Also Recommended

Morselli, Mario. *Amedeo Avogadro.* Reidel, 1984. The intellectual and political life of an independent thinker. (Nature v311 O 25, 1984 p766; Sci v226 O 26, 1984 p432)

Pearce, Williams. *Michael Faraday: A Biography.* Da Capo, 1987. Faraday was a chemist and physicist, a pioneer of research into electricity, and a scientist with a strong sense of public mission. (NYTBR v92 N 22, 1987 p50)

Profiles, Pathways, and Dreams: Autobiographies of Eminent Chemists. Jeffrey I. Seeman, series ed. American Chemical Society, 1990– . This ACS series is projected to comprise twenty-two autobiographies.

Todd, Alexander R. *A Time to Remember: The Autobiography of a Chemist.* Cambridge University Press, 1983. 257 p. Very personal in places, but sometimes technical. (Choice v21 Jl 1984 p1629; Sci v224 Je 15, 1984 p1230; TLS Ag 3, 1984 p867)

History of Chemistry

Brock, William H. *The Norton History of Chemistry.* Norton, 1993 (from the Norton History of Science series). c. 768 p., illus., biblio., index. 0–393–03536–0.

Chemistry emerged as a scientific discipline from the mysterious practices of alchemy, with contributions coming also from medieval medicine and technology. This general history could serve as a course reference, but it also makes for interesting reading. No better one-volume history of the discipline is likely to appear for many years. (LJ v119 Mr 1, 1994 p54; SBF v30 Ja 1994 p8)

Knight, David. *Ideas in Chemistry: A History of the Science.* Rutgers University Press, 1992. 213 p. 0–8135–1835–0.

Characterized by the author as a ''biography'' of chemistry, this is actually more of a history of currents of thought and doctrines within the discipline. At the same time, the experimental side of the field is not neglected. This is an academic history, with all of the obligatory scholarly appurtenances, but the author's liberal and sometimes incisive commentaries liven it up. For readers with a basic knowledge of the field. (SBF v29 Ja 1993 p8; TLS Ag 28, 1992 p23)

Also Recommended

Salzberg, Hugh W. *From Caveman to Chemist.* American Chemical Society, 1991. A world history of chemistry with the ACS's imprimatur. (Choice v29 F 1992 p920; LJ v117 Mr 1, 1992 p45)

General Topics

Heiserman, David L. *Exploring Chemical Elements and Their Compounds.* TAB Books, 1992. 376 p., index. 0–8306–3025–5.

Succinct chapters describe each of the chemical elements, including their atomic structures, historical backgrounds, chemical properties, compounds and isotopes, and uses. Handy as a reference, but the author's inclusion of tidbits of chemical trivia also makes this book a pleasure to read. (Choice v29 Ap 1992 p1256; SBF v28 Mr 1992 p44; SLJ v38 My 1992 p153)

Hoffman, Roald, and Vivian Torrence. *Chemistry Imagined: Reflections on Science.* Smithsonian, 1993. 168 p., index. 1–56098–214–4.

Nobel laureate Hoffman wrote the descriptive essays, and visual artist Torrence painted the imaginative collages that make this an intelligent and stimulating collaboration. The text discusses chemistry at a general level, with chapters on ''The Periodic Table,'' ''Energy and Form,'' and ''The Philosopher's Stone.'' Throughout, the science is interspersed with humanistic expressions from art, literature, and philosophy. (BL v89 Ap 15, 1993 p123; LJ v118 Ap 15, 1993 p123; SBF v30 Ja 1994 p8)

Jacques, Jean. *The Molecule and Its Double*. McGraw-Hill, 1993. 128 p. 0–07–032399–2.

Pasteur, famous for his contributions to microbiology, was also the discoverer of the "right-" and "left-handedness" of molecules. This work laid the groundwork for the development of the field of stereochemistry. There are no mathematical equations and few formulae in this book, but there is intelligent conceptual depth in the discourse. The author is a French chemist, historian of science, and popularizer. (J Ch Ed v71 p A53; New Sci v139 S 18, 1993 p43; SBF v29 D 1993 p264)

Levi, Primo. *The Periodic Table*. 1st American ed. Originally published in Italy as *Sistema Periodico*. Translated by Raymond Rosenthal. Schocken Books, 1984. 233 p. 0–8052–3939–4.

Organized around twenty-one chapters with the titles of chemical elements ranging from argon to zinc, Levi's moving and much honored book tells of how he survived imprisonment at Auschwitz by working as a chemist for the Germans. This book has been read, studied, and appreciated by scientists, social scientists, and humanists alike. (BL v81 D 1, 1984 p476; LJ v110 Ja 1985 p81; NYTBR v89 D 28, 1989 p84)

Lewis, Grace Ross. *1,001 Chemicals in Everyday Products*. Van Nostrand, 1991. 344 p., index. 0–442–01458–9.

An alphabetical listing of chemicals used in preservatives, pesticides, flavorings, colorants, antiseptics, solvents, coatings, and so on, that can be found in almost any household. Gives data on uses and possible health effects.

Richards, W. Graham. *The Problems of Chemistry*. Oxford University Press, 1986. 104 p., index. 0–19–219191–8.

Richards, an Oxford lecturer, steps down from the lectern and distills the major principles of the entire discipline into a work of just 104 pages. (SBF v22 Mr 1987 p218)

Scott, Andrew. *Molecular Machinery: The Principles and Powers of Chemistry*. Blackwell, 1989. 192 p. 0631164413.

A painless introduction to basic chemistry. In the first seven chapters, the author outlines foundation concepts: atomic structure, bonding, chemical equilibrium and reaction rates, and so on. The remaining chapters illustrate applications of these concepts. (New Sci v125 F 1990 p60)

Snyder, Carl H. *The Extraordinary Chemistry of Ordinary Things*. Wiley, 1992. 634 p., illus., notes, gloss., index. 0–471–62971–5.

Developed as a text for the general education requirements of nonscience majors, this book is extremely successful at making the world of chemistry seem vital and relevant. The point made by the title is well taken: chemically speaking, ordinary things are anything but. Anybody with a desire to learn about chemistry and apply that knowledge would find this book very rewarding. (New TB v77 N 1992 p1360)

Timmreck, Roy S. *Power of the Periodic Table: The Secret of Change in the Universe—The Chemical Reaction.* Royal Palm, 1991. 203 p., index. 1–878862–00–6.

While chemistry is perhaps most known as an experimental science, its theoretical side is rich and intriguing. Some of the theoretical vistas depicted in this book relate to atomic structures, quantum mechanics, ionic and covalent bonding, and periodicity. Best for readers with some prior knowledge of chemistry. (SBF v27 N 1991 p231)

Also Recommended

American Chemical Society. *What's Happening in Chemistry?* ACS, 1987. The hot topics in chemistry (circa 1987) are covered in nineteen succinct, to-the-point articles. Although this was written for journalists wishing to bone up on their chemistry, it will appeal to all general readers. (SBF v23 My 1988 p300)

Brown, Theodore, et al. *Chemistry: The Central Science.* Prentice-Hall, 1991. Each chapter in this text gives a helpful summary and lists key terms. (SciTech v15 Mr 1991 p8)

Goldberg, David. *Theory and Problems of Chemistry Foundations.* McGraw-Hill, 1991. An outline including basic calculations and terminology.

Hess, Fred. *Chemistry Made Simple.* Doubleday, 1984. 208 p. 0–385–18850–1. The revised edition of a classic.

Oxford Chemistry Primers (series). Oxford University Press, 1991– . This very useful series includes about a dozen titles to date. Each covers a specific topic in under 100 pages. Not all titles are of general interest, but the series merits watching.

Pimentel, George. *Opportunities in Chemistry: Today and Tomorrow.* National Academy Press, 1987. Based on a prior report of the National Research Council, this book divides career opportunities into seven categories. (SBF v23 My 1988 p300)

Wilbraham, Antony. *Addison-Wesley Chemistry.* 2nd ed. Addison-Wesley, 1990. An excellent high school text.

CHEMISTRY—SPECIAL TOPICS

Analytical Chemistry

Baiulescu, G. E., et al. *Education and Teaching in Analytical Chemistry.* Horwood, 1982. 190 p., index. 0–0–85312–384–5.

Although this is written for teachers and their interests, the authors' central contention, that analytical chemistry can best be understood through a multidisciplinary approach, makes this book highly informative for general readers as well.

Also Recommended

(*Note*: Textbooks provide the only general information on the field as a whole; the following are some recent titles.)

Hargis, Larry. *Analytical Chemistry: Principles and Techniques.* Prentice-Hall, 1988.

Rubinson, Kenneth A. *Chemical Analysis.* Little, Brown, 1987.

Biochemistry

Fruton, Joseph S. *A Skeptical Biochemist.* Harvard University Press, 1992. Biblio., index. 0–674–81077–5.

In the spirit of Robert Boyle's famous *A Skeptical Chymist* (1661), the author outlines methods and presuppositions of the discipline. The main topic is the fusion of biology with chemistry and how that has affected both disciplines. The commentary on contemporary practices is thoughtful. (LJ v118 Mr 1, 1993 p43; Nature v359 S 17, 1992 p203)

Lehninger, Albert L., David L. Nelson, and Michael Cox. *Principles of Biochemistry.* Worth, 1993. 1090 p., illus.

There are four parts to this exhaustive but not encyclopedic text: (1) Foundations of Biochemistry, (2) Structure and Catalysis, (3) Bioenergetics and Metabolism, and (4) Information Pathways (*genetic* information, that is). For undergraduate courses. (J Ch Ed v70 Ag 1993 pA223)

Also Recommended

Campbell, Mary K. *Biochemistry.* Harcourt, Brace, 1991. A text for nonchemistry majors from other fields of science and engineering.

Kuchel, Philip, and Gregory Ralston. *Theory and Problems of Biochemistry.* McGraw-Hill, 1988. An outline. Covers mammalian biochemistry principally.

(*Note*: See also entries in Chapter Seven under "Molecular Biology" and in Chapter Thirteen under "Biotechnology and Genetic Engineering.")

Chemical Technology

Aftalion, Fred. *A History of the International Chemical Industry.* University of Pennsylvania Press, 1991. 411 p., photos, index. 0–8122–1297–5.

The scope of this history is monumental—from prehistory to modern times. The aim of the book (and, more generally, of the series in which it was published) is to establish chemistry as a vital science for today's society and as the intellectual equal of other sciences. To quote the author: "It has been written to the glory of scholars and industrial leaders who by their daring, their shrewdness and their tenacity have provided us with the means for a better life." (Choice v29 Ap 1992 p1256; Sci v254 O 25, 1991 p588)

Crone, Hugh. *Chemicals and Society: A Guide to the New Chemical Age.* Cambridge University Press, 1987.
    The list of problems that many people associate with chemicals includes pollution, disease, herbicides, unhealthy food additives, and various others. Crone objectively and reasonably assesses the nature of chemical risks. (Choice v24 Je 1987 p1570; New TB v72 F 1987 p55)

Spitz, Peter H. *Petrochemicals: The Rise of an Industry.* Wiley, 1988. 588 p. 0–471–85985–0.
    Union Carbide, Shell Development Corporation, Dow Chemical—the stories of these and other giant players in the international, billion dollar petrochemical industry are fascinating and complex. Although the industry has been bashed a great deal in recent years, this sympathetic but objective account puts some of the criticisms into perspective. (Am Sci v77 Jl 1989 p409; Nature v334 Ag 25, 1988 p659)

Also Recommended

Hounshell, David A., and John Kelly Smith, Jr. *Science and Corporate Strategy: Du Pont R&D, 1902–1980.* Cambridge University Press, 1988. A history of the famous American chemical corporation. (Choice v26 Ap 1989 p1366; Nature v339 Je 22, 1989 p589; Sci v244 My 19, 1989 p840)

Morris, Peter J. T. *The American Synthetic Rubber Research Program.* University of Pennsylvania Press, 1989. This book explores the vicissitudes of the synthetic rubber industry, with special emphasis on its largely successful research initiatives. (Choice v27 Je 1990 p1702; Nature v346 Jl 26, 1990 p328; Sci v248 My 18, 1990 p892)

(*Note*: See also entries in various categories in Chapter Thirteen.)

Crystallography

Hammond, C. *Introduction to Crystallography.* Rev. ed. Oxford University Press, 1992. 129 p., illus., index. 0–19–865433–3.
    Step-by-step through the basic concepts of crystallography, beginning with simple lattice structures. Includes problems and useful illustrations.

Mercer, Ian. *Crystals.* Harvard University Press, 1990. 60 p., photos, index. 0–674–17914–5.
    The strength of this book is the over 150 illustrations and photographs. The

text deals with topics related to crystal growth, symmetry, structure, occurrences, and properties. (Kliatt v24 S 1990 p46)

Experiments

Shakhashiri, Bassam Z. *Chemical Demonstrations*. 4 vols. University of Wisconsin Press, 1989–1992.
   The four volumes in this well-regarded and widely used set provide explicit information for conducting classroom demonstrations of important concepts in chemistry. Each section has specific directions (along with precautions), lists materials needed, and gives tips for making presentations. Collectively, the demonstrations span the discipline. The level is young adult through advanced undergraduate. (SBF v25 Ja 1990 p124; SBF v28 Ap 1992 p78)

Also Recommended

Chemical Safety Matters. International Union of Pure and Applied Chemistry. Cambridge University Press, 1992. Every institution where chemical experiments are performed should have a safety manual.

Coyne, Gary. *The Laboratory Handbook*. Prentice-Hall, 1992. Procedures, standards, and safeguards for all labs. (Choice v30 D 1992 p647)

Newton, David E. *Consumer Chemistry Projects for Young Scientists*. Watts, 1991. Practical experiments for high school students. (BL v87 Je 15, 1991 p1950; SBF v27 My 1991 p107)

Inorganic Chemistry

Asimov, Isaac. *The Noble Gases*. Basic, 1966. 171 p.
   Still probably the only popular book on the noble gases available. Covers their history from discovery to the preparation of the first noble gas compounds.

Also Recommended

(*Note*: Textbooks provide the only information available on the field as a whole; the following are some current titles.)

Ebsworth, E.A.V., et al. *Structural Methods in Inorganic Chemistry*. CRC, 1992. The strength of this book is the case history approach to techniques and problems that, in the abstract, can be difficult to grasp. (Nature v 332 Mr 10, 1988 p185)

Owen, S. M., and A. T. Brooker. *A Guide to Modern Inorganic Chemistry*. Longman Scientific, 1991. (SciTech v15 Jl 1991 p18)

Sharpe, A. G. *Inorganic Chemistry*. 3rd ed. Longman Scientific, 1992. (J Ch Ed v70 N 1993 pA304)

Organic Chemistry

Elias, Hans-Georg. *Mega Molecule: Tales of Adhesives, Bread, Diamonds, Eggs, Fibers. . . .* Springer-Verlag, 1987. 202 p., illus.

A sequential approach to molecular science, from the simplest molecules to so-called mega-molecules (a.k.a. polymers), which make up many common domestic and industrial products. Discusses the bonding mechanisms and structures of these molecules. Lucid writing makes this a highly recommended choice. (SBF v23 N 1987 p77)

Morawetz, Herbert. *Polymers: The Origins and Growth of a Science.* Wiley, 1985. 306 p., illus., index. 0–471–89638–1.

Polymers are long, tenacious molecules that can be linked into durable yet flexible chains. This trait makes them unique and useful. General readers might appreciate them because they are the chemical stuff of many household products—rubber, nylon, and plastics, to name just a few. The history of polymer research straddles the pure and applied branches of the field. (Choice v23 N 1985 p472; Nature v317 O 31, 1985 p773)

Nickson, Alex, and Ernest Silversmith. *Organic Chemistry: The Name Game.* Pergamon, 1987. 347 p., illus., index. 0–08–03441–x.

"Undecacyc . . . eicosane!" The full, systematic name for this compound is too long to reproduce here, but the point is that, in chemistry, terminology is a very big deal indeed. Making sense of the multisyllabic alphabet soup requires a bit of work. The authors consulted voluminous literature and hundreds of scientists in writing this book, which, despite the scholarly effort involved in its creation, is not a chore to read—in fact, a gentle sense of humor pervades it. (Choice v25 My 1988 p1429; Nature v333 J3 9, 1988 p510; Sci v239 Mr 4, 1988 p1184)

Streitwieser, Andrew, et al. *Introduction to Organic Chemistry.* 4th ed. Macmillan, 1992. 1256 p., illus., index. 0–02–418170–6.

A well-crafted textbook that is markedly student friendly. The latest edition includes new information in the fields of biochemistry and related biological sciences.

Also Recommended

Mandelkern, Leo. *An Introduction to Macromolecules.* 2nd ed. Springer-Verlag, 1983. Begins with the structure of molecular chains and proceeds through discussions of natural and synthetic polymers, crystalline polymers, and macromolecules with biological importance. (SciTech v8 Ap 1984 p11)

Most, Clark F., Jr. *Experimental Organic Chemistry.* Wiley, 1988. The ins, outs, and experiences of an organic chemistry lab. (SBF v24 N 1988 p75)

Sharp, J. T., I. Gosney, and A. G. Rowley. *Practical Organic Chemistry: A*

Student Handbook of Techniques. Chapman and Hall, 1989. Laboratory methods. Emphasizes safety. (SBF v25 My 1990 p245)

Physical Chemistry

Laidler, Keith. *The World of Physical Chemistry.* Oxford University Press, 1993. 476 p. 0–19–855597–0.
A very thorough approach to physical chemistry, with chapters on thermodynamics, electrochemistry, quantum mechanics, colloid and surface chemistry, and other topics. The references and reading lists enhance the book's utility. College level. (SBF v30 Ap 1994 p71; Nature v365 O 14, 1994 p615)

Pauling, Linus, and Roger Hayward. *Architecture of Molecules.* Freeman, 1970; Linus Pauling. *The Chemical Bond.* Cornell University Press, 1967. 267 p.
In between his two Nobel Prizes, Pauling wrote about chemistry in a number of popular venues. Of his work during that period, these two books are still read and can be found in a great many libraries. Although probably not relevant for classroom use any longer, they remain good supplemental readings or references.

Prigogine, Ilya, and Isabelle Stengers. *Order out of Chaos.* Bantam, 1984. 349 p., illus., index.
One of the first accounts of chaos theory, before it was called chaos theory. Covers technical and philosophical aspects of thermodynamics and chemical equilibrium. Prigogine, a chemist and physicist, is something of a Renaissance man of science, and his opinions are influential. (Choice v22 Ja 1985 p703; LJ v110 Ja 1985 p52; SBF v20 Mr 1985 p196)

Salem, Lionel. *Marvels of the Molecule.* Translated by James D. Wuest. VCH, 1987. 104 p., illus., gloss. 0–89573–345–5.
Dedicated to ''chemists and non-chemists alike,'' this is probably one of the bestsellers of the field. There are five parts, each consisting of eight two-page chapters. On the first page of the chapter is a description of a molecule or some chemical principle, and on the facing page is a drawing that illustrates that principle. The formula is simple and the content easy to digest, but the lessons learned stick with the reader. (Choice v25 D 1987 p647; SBF v23 My 1988 p300; SciTech v11 O 1987 p14)

Also Recommended

Matthews, Peter. *Experimental Physical Chemistry.* Clarendon, 1986. Experiments and exercises divided into three types: ''Equilibrium,'' ''Structure,'' and ''Change.'' (Nature v326 Mr 12, 1987 p214)

Vemulapalli, G. K. *Physical Chemistry.* Prentice-Hall, 1993. A massive, up-to-date text. (J Ch Ed v70 N 1993 pA304)

OTHER RESOURCES

Periodicals for General Readers

Accounts of Chemical Research. American Chemical Society, m. 0001–4842.

Publishes concise, critical reviews of current research and investigation, as well as occasional summaries or review articles. College level.

Chemical Week. Chemical Week Associates, w. 0009–272x.

News and features on the global chemical industry, for those in business or with broad interests in the field.

ChemTech. American Chemical Society, m. 0009–2701.

The magazine of chemical sciences, technology, and innovation. Regular departments include "Resources," "Research," "Development," and "The Marketplace." Includes lively editorials and commentary.

Journal of Chemical Education. American Chemical Society, m., revs. 0021–9584.

Includes articles of general interest, features, and sections on secondary school chemistry issues and experiments.

Audiovisual Materials

The Atom that Makes the Difference (Carbon); Who Found the Missing Link (Uranium); A Restoration Drama (Sulfur); Science in the Saddle (Nitrogen). Films for the Humanities, 1993 (all in color, approx. 30 minutes).

These offerings from the *Periodic Table and the Human Element Series* emphasize the scientists who were important in the discovery and study of the titular elements. Excellent for high school or college classrooms, or general viewing. (SBF v29 O 1993 p214)

Catalysis: Technology for a Clean Environment. Council for Chemical Research, 1993 (32 minutes, color).

Chemical technologists make use of various enzymes in reducing water and air pollution. The processes of enzymatic reactions are not simple, though, and their use can create new problems if unchecked. This would be a good video for high school or college students, or anybody who wishes information about the real-life difficulties of environmental rehabilitation. (SBF v29 D 1993 p278)

Experiment: Chemistry (series). Films for the Humanities, 1986 (20 minutes each, color).

Each entry in this series demonstrates a different analytical chemical experiment; for example "Experiment: Mass Spectrometry," "Experiment: Spectrophotometry," and "Experiment: Magnetochemistry."

Linus Pauling: A Century of Science and Life. Carolina Biological, 1987 (28 minutes).

The preeminent American figure in chemistry on his scientific vocation and commitments to social activism. From the Century of Science and Life series.

Organic Chemistry. TV Ontario (60 minutes, color).

There are six lively segments in this video, three that describe carbon's bonding properties and three on industrial hydrocarbon processes using enzymes. Excellent visual effects show what happens when carbon bonds. (SBF v24 My 1989 p319)

Organic Molecules in Action. Media Guild, 1989 (24 minutes, color).

At one level, this video is a scientific detective story where researchers work to identify a specific molecule. At a second level, it demonstrates sophisticated organic processes used in the manufacture of an organic chemical. The two stories are effectively linked in a way that yields a "big picture." (SBF v25 My 1990 p282)

Also Recommended

Dmitri Mendeleyev: Father of the Periodic Table. Films for the Humanities, 1991. Produced in Russia, this video gives insight into Mendeleyev and is also a good introduction to the periodic table. (J Ch Ed v69 Ap 1992 pA140)

An Industrial Chemist. Hawkhill Associates, 1986 (12 minutes, color). The development of post-it notes, the peel-off notepads that have become omnipresent in the modern office. (SBF p118)

People Who Took Chemistry, That's Who. American Chemical Society, 1992. (15 minutes, color; includes 46 p. notebook). Practicing chemists are shown in fields from academe to ecology. (J Ch Ed v 70 Ja 1993 pA25)

The Periodic Table. Media Guild, 1989 (24 minutes, color). Originally produced in 1978, this reissue has gained popularity with a new generation. (SBF v25 My 1990 p282)

Physical Chemistry: Principles of Chemical Change. Media Guild, 1987. A BBC/Open University series of videos at the high school and general college level. (SBF v24 Ja 1989 p182)

CD-ROMs

CHEM-Bank. National Institute for Occupational Safety and Health–Silver Platter (one CD).

Databases containing information on potentially hazardous chemicals; includes "Registry of Toxic Effects of Chemical Substances," "Oil and Hazardous Materials—Technical Assistance Data System," and "Chemical Hazard Response Information System."

Dialog Ondisc: Polymer Encyclopedia. Dialog Information Services–Wiley, 1990 (one CD).

Contains the complete *Encyclopedia of Polymer Science and Technology.* Materials, methods, and advances in polymer sciences. Specialized.

Fine Chemicals Database. Chemron, Inc.

A chemical supplies database for over fifty international manufacturers.

Chapter Nine

Mathematics

There is a story about two . . . aristocrats who decided to play a game in which the one who calls the largest number wins.

"Well," says one of them, "you name your number first."

After a few minutes of hard mental work the second aristocrat finally named the largest number he could think of. "Three," he said.

Now it was the turn of the first one to do the thinking, but after a quarter of an hour he finally gave up. "You've won," he agreed.

George Gamow, One, Two Three . . . Infinity

Nature is written in the language of mathematics.

Galileo

Mathematics is written for mathematicians.

Copernicus

Throughout history, advances in all scientific fields have been parallelled by, and in some cases triggered by, the development of new mathematical tools. Today, all scientific and technical disciplines—and, increasingly, many in the social sciences—use higher mathematics to gather and document data, describe and analyze findings, explore relationships, solve problems, and make predictions. Dependence on mathematics is especially strong in the physical sciences, where the objects of research are often practically unseeable and unmeasurable entities, knowable only through mathematical expression. For example, the existence of whole families of subatomic particles has been hypothesized because, mathematically, they need to be there. Thus, there is extensive interplay between mathematics and all sciences, and a student entering any scientific profession generally begins by learning its mathematical foundations. In some respects, all scientists and engineers are mathematicians.

Mathematical Reviews, the chief index covering the discipline, currently lists ninety-four main subject categories, each of which is further divided into any number of subcategories. The professional literature of mathematics is thus extremely broad. While numerous specialties exist, at the most general level mathematics is divided into pure and applied branches. These might be thought of as the theoretical and practical sides of mathematics, respectively, although here too there is overlap. In pure mathematics, theorems are posited, then computational proofs are worked out. At the simplest level, "proof" means that the theorem is mathematically correct, but researchers also aim for such elusive and subjective traits as elegance and beauty in their calculations. Pure mathematics can legitimately be called an art. Applied mathematics addresses real-life problems, such as in the fields of mechanics, thermodynamics, numerical analysis, computer science, and all engineering disciplines.

The observation by Copernicus at the beginning of this chapter is valid. Mathematics is the most difficult of all scientific fields to popularize, because so few laypersons speak its language. Mathematical ideas are communicated by symbolic notation, and not a single word of any spoken language is necessarily required for the transmission of information. Thus, that which *can* be translated into lay language at all must cover topics that have some correlates, such as numbers or shapes, within the realm of common experience. Computer imaging technologies are capable of taking mathematical models, like fractal images, and representing them visually in ways that are vivid and attract the attention of popular audiences. Much mathematical research, however, simply cannot be translated, except in extremely general terms.

Many prolific areas of mathematical study are not represented in popular literature. Many probably will never be popularized. Among current areas of active research, concepts that can be mathematically modeled and illustrated using computer generated imagery will continue to find an audience in the popular media. Among nonspecialists with fairly high degrees of mathematical know-how, problem solving books and recreational math workbooks are always popular. Finally, in recent years adult math phobics have been coming out of the closet, and several primers and remedial books have been written for them. It is likely that more and more specialized titles in this category will be published.

The word "numeracy" has been suggested as the mathematical equivalent of literacy. Simply put, numerate individuals possess a basic facility with numbers and measurements; they can make reasonable assessments of such things as risk factors and their chances of winning the lottery. Numerate people also know how essential mathematics is to the practice of science and technology, and thus how it affects their lives. Of course, numerate people are likely to possess, or at least strive for, higher than average mathematical aptitudes. Some basic texts are included in this chapter as a means by which such individuals can learn or enhance those skills, either in class or through self-directed study. In all, the books listed here are dedicated to the promotion of numeracy.

MATHEMATICS—GENERAL

Reference

Webster's New World Dictionary of Mathematics. William Karush, ed. Webster's New World, 1989. 0–131–92667–5.

Several fields are served by mathematics, so a good dictionary can serve broad reference needs. In all, this work contains over 14,000 entries, some fairly lengthy. Diagrams and charts enhance the text. Works well in high school, public, undergraduate, or community college libraries. (BL v86 D 15, 1989 p860; RSSEMA p96)

Also Recommended

Encyclopedic Dictionary of Mathematics. 2nd ed. Kiyosi Ito, ed. Mathematical Society of Japan/MIT Press, 1987. In four volumes, this source represents the totality of mathematics at a level appropriate for undergraduate math majors and up.

Facts on File Dictionary of Mathematics. Carol Gibson, ed. Facts on File, 1988. Another good general dictionary for the discipline. (RSSEMA p86)

Wells, David. *Penguin Dictionary of Curious and Interesting Numbers.* Penguin, 1986. A good source for hard-to-find information on number theory and the properties of numbers.

Autobiography and Biography

Albers, Donald, and G. L. Alexanderson. *Mathematical People.* Birkhauser, 1985. 372 p., photos. 0–8776–3191–7; Donald Albers, G. L. Alexanderson, and Constance Reid. *More Mathematical People.* Harcourt Brace Jovanovich, 1990. 375 p., photos. 0–15–158175–4.

Anybody who thinks of mathematicians as dull and squint-eyed will find a revelation in these books. Here you will meet forty-three of the finest contemporary mathematicians in the world, and, in addition to learning about their lives and work, you will also discover who among them flunked algebra, reads *Mad Magazine*, and always wanted to be a magician. (BL v81 Je 15, 1985 p1423; Choice v22 Je 1985 p1519; NYTBR v90 Je 30, 1985 p26)

Halmos, Paul R. *I Want to Be a Mathematician.* Springer, 1985. 421 p., photos, index. 0–387–96078–3.

A bright student considering a career in mathematics might be well advised to read this book by a creative mathematician who conveys a genuine passion for his work. Few research mathematicians of Halmos's caliber have also matched his devotion to study and teaching. (NETB v71 F 1986 p357)

Hollingdale, Stuart. *Makers of Mathematics*. Penguin, 1989. 433 p. 0–14–014922–8.

Reworked for a general audience from material that first appeared in technical publications of the Institute of Mathematics and Its Applications, this book takes as its central premise that the history of mathematics is the history of great mathematicians. (TES F 16, 1990 p27)

Kanigel, Robert. *The Man Who Knew Infinity: A Life of the Genius of Ramanujan*. Scribner's, 1991. 438 p., photos, index. 0–684–19259–4.

Srinivasa Ramanujan is one of the tragic geniuses in the history of science. In 1913 Ramanujan, a self-educated Brahmin clerk in Madras, posted a letter to G. H. Hardy, the preeminent British mathematician of the day, in which he begged Hardy's opinion on some of his ideas related to number theory. Hardy quickly recognized brilliance in Ramanujan's work. Through Hardy's efforts, Ramanujan was brought to England and shown great honors. There he also despaired for the life he had left behind, attempted suicide, and, seven years later at the age of thirty-five, died of a mysterious illness. (BL v87 Je 1, 1991 p 1848; KR v59 Ap 1, 1991 p456; Sci v253 Jl 19, 1991 p334)

Ulam, S. M. *Adventures of a Mathematician*. Scribner's, 1976. 317 p., photos, index. 0–684–14391–7.

Perhaps the preeminent mathematician to work on the Manhattan Project, Ulam developed so-called Monte Carlo methods, which are used to estimate probabilities based on artificial sampling. This is a personal, reflective, entertaining, and readable biography. (BL v72 Ap 1, 1976 p1081; Choice v13 S 1976 p848; Sci v193 Ag 13, 1976 p568)

Also Recommended

Dick, Auguste. *Emmy Noether, 1882–1935*. Translated by H. I. Blocher. Birkhauser, 1970. Noether, an algebraist, has been called the greatest female mathematician in history.

Kac, Mark. *Enigmas of Chance: An Autobiography*. Harper and Row, 1985. Personal and professional recollections. (BL v82 S 1, 1985 p12; LJ v110 Ag 1985 p88; PW v228 Jl 12, 1985 p42)

Reid, Constance. *Hilbert*. Springer, 1986. Reid accomplished what some scholars claimed was impossible by writing an accurate yet readable biography of David Hilbert. (Choice v7 S 1970 p864; LJ v95 My 15, 1970 p1851)

Simmons, George. *Calculus Gems: Brief Lives and Memorable Mathematics*. McGraw-Hill, 1992. 0–0705–7566–5. Biographies of various mathematicians, not confined to those working with calculus.

History of Mathematics

Boyer, Carl. *A History of Mathematics.* 2nd ed. Wiley, 1989. 762 p., index. 0–471-09763-2.

The best available general history of mathematics, significantly revised from the first edition (1985). Serves multiple purposes—readable enough to help the generalist, but technical enough to aid the student or scholar. (Choice v27 N 1989 p518; Choice v24 Ap 1987 p1186; New Tech B v 75 Mr 1990 p363)

Dunham, William. *Journey Through Genius: The Great Theorems of Mathematics.* Wiley, 1990. 300 p., illus., index. 0–471-50030-5.

A surprising number of the greatest mathematical constructs in history can be appreciated by anybody with a high school level grasp of algebra and geometry. With this as his basic premise, Dunham explores the works of such mathematical giants as Archimedes, Newton, Euler, and Cantor. (KR v58 My 1, 1990 p644; LJ v101 Mr 1, 1991 p62; SciTech v14 My 1990 p4)

Also Recommended

Beckman, Peter. *A History of Pi.* Dorset, 1989. Explores the history of pi as ''a little mirror of the history of man.''

Burton, David M. *The History of Mathematics: An Introduction.* Allyn and Bacon, 1985. A text that could serve general readers.

The History of Mathematics: A Reader. John Fauvel and Jeremy Gray, eds. Macmillan, 1987. The most accessible book available containing important, original mathematical writings.

General Topics

Darling, David. *Equations of Eternity: Speculations on Consciousness, Meaning, and the Mathematical Rules that Orchestrate the Cosmos.* Hyperion, 1993. 190 p. 1–5682-875-4.

Why physical reality can be so accurately described by mathematics is one of the mysteries of science. Darling explores this relationship in a mind-boggling, speculative tour of the mind and the universe. (Choice v31 Ja 1994 p825; NYTBR v98 D 5, 1993 p74; SBF Ja 1994 p8)

Dudley, Underwood. *Mathematical Cranks.* Mathematical Association of America, 1992. 372 p., illus., index. 0–88385-507-0.

Mathematical ''cranks,'' as Dudley defines them, are smatterers and tinkerers who come to believe that they have done something impossible, or something they have not. The author wittily reveals the fallacies of various favorite crank endeavors, like trisecting the angle, finding a simple proof of the four color theorem, or proving Fermat's last conjecture. (Choice v30 My 1933 p1505; SBF v29 Ag 1993 p174)

Ekeland, Ivar. *Mathematics and the Unexpected.* University of Chicago Press, 1988. 146 p. 0–226–19989–4.

"Catastrophe theory" is not quite the disastrous area of scholarship that its name implies. In this award winning popularization, Ekeland shows how it applies to the mathematics of time, which is central to many science disciplines. He sprinkles relevant quotes throughout from Homer, Proust, Shakespeare, and others. (KR v56 My 1, 1988 p664; LJ v113 Ag 1988 p165; Nature v335 S 1, 1988 p2)

Guillen, Michael. *Bridges to Infinity: The Human Side of Mathematics.* Tarcher, 1983. 205 p., gloss., index. 0–87477–233–8.

Math anxiety, the author claims, is actually several afflictions, each the result of misperceptions about what math can and cannot do. Thus, he sets out to clarify those misperceptions. The brief chapters in this book describe, in lucid prose, the nature and purposes of the major fields of mathematics. (Choice v21 Je 1984 p1500; LJ v108 D 1, 1983 p2256; PW v224 O 28, 1983 p63)

Halmos, Paul R. *Problems for Mathematicians, Young and Old.* Mathematical Association of America, 1991. 338 p. 0–88385–5.

Textbooks usually present problems within the context of a particular lesson, which sometimes makes it difficult to conceptualize real-life approaches to identifying and solving problems. Halmos invites readers to attack problems systematically by grouping them into fourteen categories, such as "calculus," "combinatorics," "set theory," "matrices," and so on. He also mercifully provides a section of hints. (Choice v30 S 1992 p163; SBF v28 Je 1992 p141)

Hoffman, Paul. *Archimedes' Revenge: The Joys and Perils of Mathematics.* Norton, 1988. 285 p., illus., index. 0–393–02522–5.

According to legend, Archimedes, in a fit of rage, concocted an impossibly complex problem about grazing cattle just to confound his contemporaries. This book aims to alleviate some of the confusion inherent in mathematical niceties so that generalists can get a grasp of the subject. The author sketches the scope of mathematics and what mathematicians do. (NYTBR v94 S 10, 1989 p42; SBF v24 Ja 1989, p138)

Peterson, Ivars. *Islands of Truth: A Mathematical Mystery Cruise.* Freeman, 1990. 325 p., illus., biblio., index. 0–7167–2113–9.

Peterson, a reporter for *Science News*, claims that it is characteristic of the discipline of mathematics that more is unknown than known. There are, however, "islands of truth," and by taking the reader on visits to several of them, he maps the frontiers of contemporary mathematics. The illustrations are very entertaining. (BWatch v11 D 1990 p4; LJ v116 Mr 1, 1991 p62; SciTech v14 S 1990 p8)

Peterson, Ivars. *The Mathematical Tourist: Snapshots of Modern Mathematics.* Freeman, 1988. 204 p., illus. 0–7167–1953–3.

Contemporary mathematical topics discussed include fractal geometry, topology and differential geometry, number theory (with applications to computer science), and various others. (Nature v336 N 17, 1988 p292; NYTBR v94 S 10, 1989 p42; SBF v24 Ja 1989 p138)

Salem, Lionel, Frederic Testard, and Coralie Salem. *The Most Beautiful Mathematical Formulas*. Wiley, 1992. 141 p., index. 0–471–55276–3.
In forty-nine short chapters, the authors present playful and entertaining looks at various mathematical formulae, from "Fermat's Last Theorem" to "The Chances of Winning a Lottery." Cartoon illustrations enhance the book's whimsical style.

Also Recommended

Blocksma, Mary. *Reading the Numbers: A Survival Guide to the Measurements, Numbers and Sizes Encountered in Everyday Life*. Penguin, 1989. Could be read for pleasure or used for reference. (LJ v115 Mr 1, 1994 p44)

Davis, Philip, and Reuben Hersh. *The Mathematical Experience*. Birkhauser, 1981. This book was written to explain just what it is that mathematicians do. Strong on historical topics. (Choice v19 S 1981 p120; LJ v106 My 1, 1981 p982)

Flato, Moshe. *The Power of Mathematics*. McGraw-Hill, 1990. A summary of the work of mathematicians, with observations on mathematics education.

Hardy, G. H. *A Mathematician's Apology*. Introduction by C. P. Snow. Cambridge University Press, 1967. 151 p. A classic exposition on the significance of pure mathematics.

Kasner, Edward, and James R. Newman. *Mathematics and the Imagination*. Microsoft, 1989. Illus. 1–55615–104–7. A reprint of a classic mathematics popularization. (LJ v115 Mr 1, 1990 p44)

Pappas, Theoni. *The Joy of Mathematics*. Tetra, 1989; *More Joy of Mathematics*. Tetra, 1991. Essays for the uninitiated. (NYTBR v94 S 10, 1989 p42)

Savant, Marilyn Vos. *The World's Most Famous Mathematics Problem*. St. Martin's, 1993. Fermat's last theorem and others, succinctly described in this slim book.

Staszkow, Ronald, and Robert Bradshaw. *The Mathematical Palette*. Saunders, 1991. To be used with a general math course for liberal arts students.

Wolfram, Stephen. *Mathematica: A System for Doing Mathematics by Computer*. 2nd ed. Addison-Wesley, 1992. A remarkable program for "machine assisted human thought." (NETB v76 N 1991 p1483)

(*Note*: Most of the books listed here deal with topics in both pure and applied mathematics.)

MATHEMATICS—SPECIAL TOPICS

Algebra

Asimov, Isaac. *The Realm of Algebra.* Fawcett, 1981; Ballantine, 1984. 143 p. 0-449-24398-2.

Part of a series in which Asimov explains various areas of mathematical thought, this book is a popular choice for readers seeking a painless introduction to algebra. Practical applications are emphasized throughout. (BRD 1962 p48; BL v58 D 1, 1961 p227; LJ v 86 D 15, 1961 p4371)

Cuoco, Albert. *Investigations into Algebra.* MIT Press, 1990. 623 p. 0-262-53071-6.

Written by a high school math teacher, this book is an exploration of algebraic concepts, including discrete mathematics, combinatorics, and higher algebras. Students generate data, make conjectures, and establish proofs. Could be used as a text or for self-directed study. Requires access to an IBM compatible running the Logo programming language, which is not unusual in high school or college math labs. (Choice v28 My 1991 p1520; SBF v27 Ap 1991 p73; SciTech v15 F 1991 p9)

Also Recommended

(*Note*: The following are texts and outlines that could be used either with classes or by anybody wishing to brush up on algebraic fundamentals. Any of several other standard texts could also work well.)

Aufman, Richard N. *Algebra for College Students.* Houghton Mifflin, 1992.

Baley, John D. *Algebra: A First Course.* Wadsworth, 1990.

Dykes, Joan. *Elementary Algebra.* HarperPerennial, 1992.

Orr, Bill. *College Algebra.* HarperPerennial, 1992.

Applied Mathematics

For All Practical Purposes: Introduction to Contemporary Mathematics. Solomon Garfunkel, project director. COMAP, 1988. 250 p., illus., index. 0-7167-1830-8.

Companion to the PBS video series of the same name, this book covers all fields of applied mathematics in sections titled "Management Science," "Statistics," "Social Choice," and "On Size and Shapes." An excellent text for high school students or undergraduate nonscience majors. Full of illuminating sidebars and real-life problems (BL v84 Ap 1, 1988 p1296; R&R BK N v3 Ap 1988 p23; SciTech v12 Ap 1988 p4)

(*Note*: See also various sections in Chapter Thirteen.)

Arithmetic/Number Theory

Humez, Alexander, Nicholas Humez, and Joseph Maguire. *Zero to Lazy Eight: The Romance of Numbers.* Simon and Schuster, 1993. 0–671–74282–5.

This is a playful approach to numbers that combines exposition with arithmetical games, with a good dose of speculation thrown in. The authors are adept at exposing the cultural connections between mathematics and language. (BL v89 Je 1, 1993 p1754; KR v61 Je 15, 1993 p768; LJ v118 Jl 1993 p70)

Ifrah, Georges. *From One to Zero: A Universal History of Numbers.* Viking, 1985. 503 p., biblio., chrono. 0–6070–37395–8.

When a student asked the author the innocent question, "How did numbers start?," he embarked upon the research that culminated in this book. Beginning with humankind's first primitive notions of counting, the author details the development of the complete number systems that we use today. (BL v81 Jl 1985 p1487; KR v53 Jl 1985 p696; LJ v110 Jl 1985 p80)

Reid, Constance. *From Zero to Infinity.* Mathematical Association of America, 1992. 200 p. 0–88385–505–4.

A classic that has remained in print over thirty-five years. Reid covers several topics in number theory by dealing with the numbers zero through nine in separate chapters (i.e., Chapter Zero, Chapter One, etc.). She discusses each digit's special properties and, from that foundation, digresses into other, more complex realms of numbers.

Van Dyke, James, et al. *Prealgebra.* Saunders, 1990. 868 p., illus., index. 0–03–014832–4.

A text designed to take students from arithmetic to algebra. Covers basic operations, whole numbers, decimals, rational numbers, graphing, and other topics.

Also Recommended

Julius, Edward. *Rapid Math: Tricks and Tips. Thirty Days to Number Power.* Wiley, 1992. A quick fix for innumerates.

McLeish, John. *Number: The History of Numbers and How They Shape Our Lives.* Fawcett Columbine, 1992. Number systems throughout history. (New Sci v133 Ja 11, 1992 p47; TES Ja 3, 1992 p17)

Calculus

(*Note*: Standard calculus texts could be used in any science literacy collection. Several good ones exist. The following are examples.)

Anton, Howard. *Calculus.* 3rd ed. Wiley, 1988.

Text created to ease the average student into calculus. (SBF S 1989 p11)

Lang, Serge. *A First Course in Calculus*. 5th ed. Springer, 1986.
From the respected Undergraduate Texts in Mathematics series.

Pauling, Edward. *Calculus for Business, Economics and Social Sciences*. Mc-Graw-Hill, 1990; Ayers, Frank, and Elliott Mendelson. *Differential and Integral Calculus*. 3rd ed. McGraw-Hill, 1990.
Two entries from an outline series.

Chaos Theory—Mathematical Aspects

Stewart, Ian. *Does God Play Dice? The Mathematics of Chaos*. Blackwell, 1989. 349 p., illus., refs., index. 1557861064.
In debunking quantum physics, Einstein declared that "God does not play dice." In modern science, though, there is ample evidence that chaotic systems do indeed exist. Writing in his usual entertaining style, Stewart expounds the mathematical constructs by which irregularities in nature can be studied. His conclusion: "If God played dice, he'd win."

(*Note*: See also entries in Chapter Twelve under "Chaos and Complexity Theories.")

Computational Guides

Howard, W. J. *Doing Simple Math in Your Head*. Coast, 1992. 130 p., index. 0–9627341–5–2.
No mathematics background (or even a facility for numbers) is required for this handy guidebook. The author teaches readers how to simplify mathematics problems by reordering factors, making good estimates and approximations, and coming to workable solutions. (BWatch v13 Mr 1992 p9; LJ v117 F 1, 1992 p119; SBF v28 Ap 1992 p76)

Kogelman, Stanley, and Barbara Heller. *The Only Math Book You'll Ever Need*. Rev. ed. Facts on File, 1994. 268 p., index. 0–8160–2767–6.
Practicality is paramount in this book of everyday mathematics. Step by step, the authors show you how to balance your checkbook, determine whether it is cheaper to buy in bulk quantities, figure out how much you need to save for retirement, and perform hundreds of other important calculations that can help you better manage your affairs and understand the significance of numbers. (BL v90 O 15, 1993 p401)

(*Note*: Various other books of this type exist.)

Geometry and Topology—General

Sved, Martha. *Journeys into Geometries*. Mathematical Association of America, 1991. 182 p. 0–88385–500–3.

The problems in this book are presented in dialogue fashion, the intent being to invite students to explore geometric landscapes. The author claims that imaginative mathematics is as much process as product, and thus emphasizes methodology and encourages students to develop and use their own intuitions. (Choice v29 S 1992 p164; SBF v28 Ap 1992 p77)

Also Recommended

Francis, George K. *A Topological Picture Book.* Springer, 1987. Shows the mathematical aspects of surfaces. (Nature v330 N 19, 1987)

Grunbaum, Branko, and G. C. Shepard. *Tilings and Patterns.* Freeman, 1987. Tilings are solid, two-dimensional figures. Excellent illustrations. Relevant to all fields of science, as well as some in art and design. (Choice v24 Mr 8, 1987 p1104; Nature v326 Ap 9, 1987 p553; Sci v236 My 22, 1987 p996)

Lipschutz, Seymour. *General Topology.* HarperPerennial, 1992. An outline of the field.

Wells, David. *Penguin Dictionary of Curious and Interesting Geometries.* Penguin, 1991. Not terribly useful as a dictionary, but a treasure trove of geometrical oddities. (New Sci v132 1991 p56)

Geometry—Fractals

Lauwerier, Hans. *Fractals: Endlessly Repeated Geometrical Figures.* Princeton University Press, 1991. 209 p., biblio., index. 0–691–08551–x.
   This, the author says, is "a book to work with." Novice and more advanced students are given instructions on how to create fractal images using a personal computer. Sample programs are included. General readers wishing only a narrative overview of the subject will find that in Chapters 1 to 7. Assumes only high school level mathematics proficiency. (Choice v29 Mr 1992 p1118; SBF v23 Ap 1992 p77)

Mandelbrot, Benoit. *The Fractal Geometry of Nature.* Freeman, 1983. 468 p., illus., biblio., index. 0716711869.
   An early account of fractal geometry by the scientist who developed the concept, this work sensitized a generation of mathematicians to the importance of fractals and captured the attention of the lay public. Widely read and cited. (Nature v302 Mr 3, 1983 p91)

McGuire, Michael. *An Eye for Fractals: A Graphic and Photographic Essay.* Addison-Wesley, 1991. 176 p., illus. 0–201–55440–2.
   Fractals are repeated, self-similar visual patterns that can be seen throughout nature, as in the leaves of a tree or the movements of clouds across the sky. With a Ph.D. in physics and an impressive resumé as a photographer, the author has perfect qualifications for this well-illustrated view of the geometries of the

natural world. (Choice v28 Jl 1991 p1800; Nature v350 Ap 11, 1991 p524; SBF v27 My 1991 p103)

Also Recommended

Feder, Jens. *Fractals*. Plenum, 1988. A well-received popularization of fractals. (Am Sci v77 S 1989 p490; Sci v245 S 29, 1989 p1515)

Mathematics Education

Dewdney, A. K. *200% of Nothing: An Eye-Opening Tour Through the Twists and Turns of Math Abuse and Innumeracy*. Wiley, 1993. 182 p. 0–471–57776–6.
In response to a *Scientific American* article about how mathematics is misunderstood and misused, Dewdney was inundated with examples sent by readers, whom the author calls his "math detectives." This book could be read for pleasure or used in a classroom. (LJ v118 Ap 15, 1993 p122; Nature v364 Ag 5, 1993 p498; PW v240 Ap 5, 1993 p60)

Engel, Arthur. *Exploring Mathematics with Your Computer*. Mathematical Association of America, 1993. 301 p., index. 0–88385–639–5.
This book is designed to use with microcomputer and Turbo Pascal compiler to explore the breadth of mathematics, including number theory, statistics, probability, combinatorics, and numerical analysis. There are over 180 sample programs and hundreds of exercises.

On the Shoulders of Giants: New Approaches to Numeracy: Lynn Arthur Steen, ed. National Research Council, 1990. 232 p., index. 0–309–04234–8.
The need for reform in mathematics education has been cited as a priority among science literacy activists. Much of the problem is caused by a common perception that mathematics has little to do with daily life and is thus more bother to learn than it is worth. The essays in this book put forth a new, integrated vision of mathematics as a vital part of life that, if cultivated, can facilitate a deeper understanding of the world. (Choice v28 My 1991 p1522; SBF v27 Ja 1991 p6)

Paulos, John Allen. *Innumeracy: Mathematical Illiteracy and Its Consequences*. Hill and Wang, 1989. 135 p. 0–8090–7477–8.
The author was inspired to write this book, in part, because over the years he had heard so many people admit defeat where mathematics is concerned. Symptomatic of innumeracy is an inability to deal rationally with large numbers and the probabilities associated with them. Paulos uses jokes, puzzles, and real-life scenarios to show how to achieve and apply a better understanding of numbers. (BL v85 N 1, 1988 p439; LJ v114 My 1, 1989 p48; Sci v245 Ag 11, 1989 p654)

Tobias, Sheila. *Overcoming Math Anxiety*. Norton, 1994. 260 p. 0–39–303577–8.

Math anxiety is probably the single greatest cause of adult innumeracy. Many people happily write off math after graduating from high school or college, only to have it come back to haunt them later in life. Tobias argues that this affliction can be overcome, and this book is designed to help.

Also Recommended

A Challenge of Numbers: People in the Mathematical Sciences. Mathematical Sciences Education Board, 1990. Describes deficiencies and recommended reforms in math education. Gives many pertinent statistics.

Philosophy of Mathematics—General

Barrow, John. *Pi in the Sky: Counting, Thinking, and Being*. Oxford University Press, 1992. 317 p., illus., biblio., index. 0–19–853956–8.

In his preface, Barrow asks, "How can our inky squiggles on pieces of paper possibly tell us how the world goes round?" This book explores human mathematical intuition in a broad cultural and historical context. The subject is conveyed as both art and science. (BL v89 O 15, 1992 p387; Choice v30 Ap 1993 p1348; SBF v29 Mr. 1993 p39)

Davis, Philip, and Reuben Hersh. *Descartes' Dream: The World According to Mathematics*. Houghton Mifflin, 1987. 321 p., illus., index. 0–15–12526–02.

In a mystic vision, the French philosopher and mathematician René Descartes conceived of a world where science and nature became unified by mathematical reason. The lively essays in this volume explore the ways by which that vision has become today's reality. Above all, this book explores how applied mathematics affect culture and society. The technical level ranges from popular to professional, so browsing is encouraged. (Choice v24 Mr 1987 p1104; LJ v111 O 15, 1986 p102; SBF v23 Mr 1987 p231)

Hofstadter, Douglas. *Godel, Escher, Bach: An Eternal Golden Braid*. Basic, 1978. 777 p., refs., index. 0–465–02685–0.

A cult classic among the mathematically and philosophically inclined. Hofstadter claims that "reality is a system of interconnecting and interrelating braids that are endlessly folding upon each other." He develops this thesis by examining the work of these three creative geniuses from different fields. To finish this book is considered a badge of honor among fans. (BRD 1979 p587; LJ v104 O 1, 1979 p108; NYTBR Ap 29, 1979 p13)

Maor, Eli. *To Infinity and Beyond: A Cultural History of the Infinite*. Birkhauser, 1986. 275 p., photos, index. 0–8176–3325–1.

Conceptually, the idea of infinity appears in art, religion, philosophy, and many other institutions of human thought. Although it can be expressed in these ways, only mathematics can truly describe infinity. The mathematization of infinity began with the Greeks and today is at the fore of many sciences. (Choice v24 Jl 1987 p1714; New Tech B v72 F 1987 p59; Sci v237 Ag 7, 1989 p666)

Paulos, John Allen. *Mathematics and Humor.* University of Chicago Press, 1980. 116 p. 0–226–65624–3.
    What makes a joke funny? Often, it is some unexpected twist, or a "discontinuity." Mathematicians have developed techniques for dealing with various types of discontinuities, and these can actually be used to create a mathematical structure for jokes. Requires moderate mathematical facility. (BRD 1980 p1102; Choice v18 F 1981 p811; LJ v105 S 1980 p1744)

Also Recommended

Lehman, Hugh. *Introduction to the Philosophy of Mathematics.* Rowman and Littlefield, 1979. One of the more accessible introductions to the subject available. (Choice v16 F 1980 p1596)

Russell, Bertrand. *Principles in Mathematics.* Norton, 1937. An influential work by the famous scholar.

Philosophy of Mathematics—Logic

Copi, Irving, and Carl Cohen. *Introduction to Logic.* 9th ed. McGraw-Hill, 1990. 728 p., gloss. index. 0–02–325041–0.
    Designed as a textbook for a beginning course in logic. The purpose of this book is to teach the student reasoning skills that can be applied to legal, social, political, and personal situations.

McNeill, Dan, and Paul Freiberger. *Fuzzy Logic.* Simon and Schuster, 1993. 319 p. 0–671–73843–7.
    Two computer journalists popularize this emerging branch of mathematical logic, which has promising computer applications and also raises intriguing philosophical questions. Fuzzy logic refutes the existence of absolutes; when this core principle is adopted, new and innovative tangents of thought arise. (BL v89 F 15, 1993 p1021; Byte v18 Ap 1993 p195; LJ v118 F 1, 1993 p108)

Recreational Mathematics

Gardner, Martin. *Fractal Music, Hypercards, and More . . .* Freeman, 1992. 328 p., index. 0–7167–2188–0; *The Mathematical Circus.* Rev. ed. Mathematical Association of America, 1992. Illus., biblio. 0–88385–506–2; *The Unexpected Hanging and Other Mathematical Diversions.* Rev. ed. University of Chicago Press, 1991. 0–226–28256–2.
    Scientific American magazine's prolific mathematical wizard has enchanted and perplexed readers for decades with his popular "Mathematical Games"

feature. These collections mix the best of the old with new insights and diversions. With his quirky sense of humor and his keen intuition for mathematical irony, Gardner is the almost unchallenged master of the recreational mathematics genre. His work should be included in any science literacy collection.

Shasha, Dennis. *Codes, Puzzles and Conspiracy.* Freeman, 1992. 241 p. 0–7167–2275–5.
   A book of puzzles, designed so that insight and creativity are just as important in their solution as mathematical expertise. Most of the puzzles are entwined with interesting stories. (SBF v28 O 1992 p203)

Stewart, Ian. *Another Fine Math You've Gotten Me Into.* Freeman, 1992. 269 p. 0–7167–2342–5.
   Stewart is an author and a frequent contributor to *New Scientist* magazine, Britain's leading periodical vehicle for popular science. The sixteen chapters in this book introduce the reader to characters such as the Worm family and Rock-chopper Rocknuttersson, whose off-the-wall problems range from designing the largest sofa that can be fit around a corner to using group theory to figure out the best way to transport a lion, a llama, and a head of lettuce across a river. The author demonstrates how each of these problems fits into various schools of mathematical thought.

(*Note*: Aficionados of recreational mathematics materials have a voracious appetite. Numerous other books of collected puzzles and brainteasers are available. Also, some works are available on specific problems, such as the four color problem and the traveling salesman problem.)

Statistics and Probabilities

Gonick, Larry, and Woollcott Smith. *The Cartoon Guide to Statistics.* HarperPerennial, 1993. 230 p.
   This is a sort of hybrid comic book/textbook in which a narrative presentation of statistical concepts is accompanied by lighthearted line drawings, generally of people whose expressions and thought balloons comment on the text. In twelve chapters the authors cover probabilities, random variables, sampling, distributions, regression, and experimental designs. (R&R BK N v8 D 1993 p50)

Huff, Darrell, and Irving Geis. *How to Lie with Statistics.* Pictures by Irving Geis. Norton, 1954. 142 p.
   This slim, pleasant book has been required reading in many introductory statistics classes for two generations. Gives humorous examples of how statistics can be misleading and suggests ways to recognize sound data analysis. (Atl v193 F 1954 p83; BRD 1954 p443; LJ v79 Ja 1, 1954 p57)

Moore, David S., and George McCabe. *Introduction to the Practice of Statistics.* Freeman, 1989. 839 p. 0–71671989–4.

A voluminous text, but very well organized. This extremely thorough work could serve anybody with ongoing interests in any field of statistics for a lifetime. There are also hundreds of problems and examples.

Phillips, John L., Jr. *How to Think about Statistics*. Freeman, 1992. 256 p., illus. 0–7167–2288–7.
    To the layperson, it sometimes seems as though statistics can prove anything. This book aims to help make sense of statistical claims by discussing where statistics come from, how they can be interpreted, and the proper ways to apply them. Each chapter describes a different statistic and gives specific examples. College level. (SBF v28 Je 1992 p136)

Also Recommended

Moore, David S. *Statistics: Concepts and Controversies*. 3rd ed. Freeman, 1991. A text.

Wagner, Susan. *Introduction to Statistics*. HarperPerennial, 1991. An outline with problems.

Symmetry—Mathematical Aspects

Bunch, Bryan R. *Reality's Mirror: Exploring the Mathematics of Symmetry*. Wiley, 1989. 286 p., index. 0–471–50127–1.
    Symmetries have been discovered in many areas of study—art, anthropology, biology, and physics, for example; but it is only through mathematics that they can be described and understood. Bunch puts forth multifarious examples and writes in a personal, anecdotal style in order to illustrate the complexities behind symmetrical and asymmetrical patterns. (BRD 1990 p255; Choice v27 Ja 1990 p836; SBF v25 Mr 1990 p202)

OTHER RESOURCES

Periodicals for General Readers

American Mathematical Monthly. Mathematical Association of America, 10/yr., revs. 0002–9890.
    Publishes articles, notes, and features about mathematics; the readership is intended to include anybody who is mathematically inclined. Accessibility is a key editorial criterion. The "Telegraphic Reviews" section is an excellent collection development resource.

Chance: New Directions for Statistics and Computing. Springer-Verlag, q. 0093–2480.
    For those who use statistics in any field of endeavor. Has a casual tone.

Journal of Recreational Mathematics. Baywood Publishing, q. 0022–412x.
Mathematical games, puzzles, and brainteasers.

Mathematical Gazette. Mathematical Association, q., revs. 0025–5872.
A British general interest mathematical journal. Has an excellent review section.

Mathematical Intelligencer. Springer-Verlag, q., revs. 0343–6993.
Explores the culture of mathematics in intriguing, entertaining ways.

Mathematics Magazine. Mathematical Association of America, bi-m., revs. 0025–570x.
An undergraduate level journal of mathematical miscellanea, with a special interest in historical topics.

Mathematics Teacher. National Council of Teachers of Mathematics, m., revs. 0025–5769.
Junior high through two-year college math. Pedagogy aside, there is a lot here for the mathematically curious.

Audiovisual Materials

Chaos, Fractals and Dynamics: Computer Experiments in Mathematics. Robert Devaney–Science Television Co., 1991 (videocassette, 60 minutes in two parts).
An easy introduction to an exciting new field of mathematical thought, with some exquisite pictures.

Discrete Mathematics: Cracking the Code. COMAP, 1993 (videocassette, 33 minutes).
In mathematics, elaborate codes are used to protect, manipulate, and compress data. The science of codes—cryptography—has many social, economic, and technological dimensions. This well-illustrated video provides a good introduction (SBF v30 My 1994 p120)

For All Practical Purposes. COMAP, 1986–1987 (26 video installments, c. 30 minutes each).
The videotapes in this series, developed by the Consortium for Mathematics and Its Applications and originally broadcast on PBS, collectively cover the entire universe of contemporary mathematics for nonspecialists. The series could serve as the basis for a semester class or for structured independent study, or individual installments could be used to introduce various topics.

Futures 2. PBS Videos, 1992. (videocassettes, 15 minutes each).
Jaimie Escalante, the real-life mathematics teacher who gained fame from the movie *Stand and Deliver*, hosts this series of videos designed to show young adults how mathematics is used in various professions. Among the titles are installments showing how math relates to fields from architecture and engineer-

ing to fashion and cartography. The pace is rapid and the content is infectious for students, but probably inappropriate for adults.

The Power of Algebra. Louisiana Public Broadcasting/Great Plains National Instructional Television Library, 1989 (15 minutes each).

The ten brief installments in this series are designed to provide a pedagogical overview to the methods and significance of modern algebra for high school students. Sections include "Inverse Operations," "Order of Operations," "Basic Properties," "Positive and Negative Numbers," "Using Positive Exponents," "Polynomials and Equations," "Factoring I, II," "Fractions," and "Words into Symbols." Perfectly appropriate for adults in need of a brushup, too.

Statistics: Decisions Through Data. COMAP, 1992 (60 minutes each).

A statistics professor developed this series under the auspices of the Consortium for Mathematics and Its Applications and with funding from the Annenberg Foundation. The five videos move progressively through twenty-one units, each about ten to twenty minutes long. Throughout, examples are taken from real life. Among the many topics are the demise of baseball's .400 hitters and the relationship between aspirin consumption and heart attack risk. Titles include "Basic Data Analysis," "Data Analysis for One Variable," "Data Analysis for Two Variables," "Introduction to Inference," and "Planning Data Collection."

Also Recommended

Algebra 1. Video Tutor, 1987 (c. 50 minutes each; series includes instruction on algebraic operations in six volumes).

History of Mathematics. Media Guild, 1984 (c. 30 minutes each). A unique series on the history of mathematics for young adult viewers. (SBF v27 Ag 1991 p183)

CD-ROMs

Advanced Maths Workshop. Format PC Ltd., 1991 (one disc).

Coursework designed in modular format covering advanced tutorials in pure and applied mathematics, and statistics. From advanced secondary to beginning professional difficulty levels. In all, over 1,300 worksheets.

Complete Math Workshop. Global Learning System, Ltd., 1990 (one disc).

A fully interactive math syllabus with over 700 student worksheets covering foundational, intermediate, and higher level modules. For high school, college, general adult, and beginning professional instruction.

Exploring Mathematics with Mathematica. Addison-Wesley (one disc).

Derived from Stephen Wolfram's book *Mathematica: A System for Doing Mathematics*, this fully interactive program works through inventive and challenging mathematical data processing problems. Creative animations, and has sound too.

Mathematics, Volumes 1 and 2. Xploratorium Anglia Polytechnic, 1991 (one disc each).

Hypercard, multimedia products that explore mathematics fundamentals (e.g., basic algebra, coordinate geometry) and proceed to calculus and mathematical modeling.

Chapter Ten

Medicine and Health Sciences

Medicine is the most distinguished of all the arts, but through ignorance of those who practice it and of those who casually judge such practitioners, it is now of all the arts far the least esteemed.

Hippocrates

When it comes to your health, I recommend frequent doses of that rare commodity among Americans—common sense.

Vincent Askey

Take this quickly, while it is still a cure.

Armand Trousseau

Decisions about individual health care are among the most personal and practical that consumers must make; they may quite literally have life or death consequences. Some examples of weighty decisions that might face any teenager in America include:

- Will this new chocolate and celery diet really work?
- How can I protect myself from the AIDS virus?
- What are the known health risks of marijuana?
- Should I be concerned about my friend's depression?
- Will we still be able to afford my college tuition if my father has his surgery?

To answer questions like these, consumers need very specialized kinds of information. There are many appropriate and useful information resources, as well as many biased, superficial, and even dangerous ones. Locating and identifying information that will enable them to make effective decisions calls for well-developed evaluative skills.

Likewise, whole societies require information in order to confront complex public health problems. At the local, national, and even international levels, these issues are extremely complicated and often abstract, but they nonetheless have undeniably real impacts on individual lives. Politically, they can give rise to fiery debate, for example, about the cost of providing universal health insurance. Economics is an almost ubiquitous factor in public health decision making. Finally, in a market economy, medical ethics become even more problematic. When taking the Hippocratic oath, the physician swears to treat patients humanely and compassionately. Does that, however, extend to assisting a terminally ill patient plan for suicide? Does that include aborting a fetus that is known to have grave genetic defects? Questions such as these resonate not just in the conscience of the personal physician, but throughout society as well.

Thus, science literacy information resources for medicine and the health sciences must give much more than just quantitative data and the results of laboratory tests. Science is just one part of medical practice and progress. Making informed health care decisions can involve acting in numerous other arenas, where scientific expertise might be less important in bringing about positive outcomes than political savvy, economic fitness, and/or ethical persuasiveness. Practical health sciences information must give the "big picture" even when the subject is very specific. Many times, health care consumers will look outward for information, but find that they must look inward for answers. Good health science information resources give consumers the knowledge that empowers them to make sound choices.

Representing the popular literature of the medical and health sciences in anything approaching comprehensive fashion would require an entire book. There are innumerable books and other resources of the self-care and self-help varieties. For example, scores of books are available on dietary regimens for coping with everything from allergies to AIDS. This bibliography is quite selective, focusing on general works that discuss the profession of medicine, social and technological issues related to health sciences, basic authoritative reference materials, and also a smattering of titles on specific maladies and themes of particular current interest. Many subjects are not directly represented by citations listed here. Various other health sciences collection development tools could be used to develop a more comprehensive health sciences collection; for example, the Medical Library Association periodically publishes its "Brandon List," a core bibliography of materials for a basic medical library.

When offering consumer health sciences information, health sciences librarians often tell patrons that they should address any questions to their doctors. The same proviso applies here.

MEDICINE AND HEALTH SCIENCES—GENERAL TOPICS

Reference

Columbia College of Physicians and Surgeons. *Complete Home Medical Guide.* Columbia University Press, 1989. 930 p. 0517572168.

This full-service resource to health care and maintenance comes in eight sections: (1) The Nature of Health and Medicine, (2) What to Do until the Doctor Comes, (3) Your Body and How It Works, (4) Personal Health, (5) Disease Treatment and Prevention, (6) Drugs and Their Use, (7) Directory of Resources, and (8) Medical Terms and Their Meanings. (BL v85 D 1, 1988 p622; LJ v113 N 1, 1988 p49)

Current Medical Diagnosis and Treatment. Marcus A. Krupp and Milton Chatton, eds. Lange Medical (annual). 0092–8682.

A thorough reference on diseases, symptoms, and treatments in thirty-four general areas. Heavily used by consumers and physicians alike. (RSSEMA p196)

The HarperCollins Illustrated Medical Dictionary. Ida C. Dox, B. John Melloni, and Gilbert Eisner, eds. HarperCollins, 1933. 533 p., illus. 0–06–273142–4.

Formerly *Melloni's Illustrated Medical Dictionary,* this is especially recommended for science literacy collections on the strength of the magnificent illustrations that adorn every page. (WLB v68 O 1993 p86)

Merck Manual of Diagnosis and Therapy. 16th ed. Robert Berkow et al., eds. Merck and Co., 1992. 0–911910–16–6.

For general practitioners, but widely used for health reference by librarians and consumers as well. Expert authors contributed articles which were editorially reviewed "to ensure accuracy, adequate and relevant coverage of each subject, and clean and simple exposition." Gives information on 290 diseases and health care topics in all. (RSSEMA p194)

The Oxford Companion to Medicine. John Walton et al., eds. 2 vols. Oxford University Press, 1986. 0–19–261191–7.

A good general purpose medical reference, this could serve as a dictionary, a guidebook to information on diseases and illnesses, and even a source for philosophical or biographical material. Contains brief definitions and articles, often with bibliographies, on numerous subjects. (ARBA v18, 1987 p626; Choice v24 D 1986 p608; TLS S 19, 1986 p1024)

Also Recommended

Conn's Current Therapy. Saunders (annual). Treatments for a variety of maladies.

Harbert, John C. *Doctor in the House: An Accessible, Practical Guide to over 400 Medical Problems Facing You and Your Family Today.* Humana, 1994. A well-done home health handbook.

Health United States. U.S. Department of Health, Education and Welfare, National Center for Health Services Statistics (annual). Statistical reports on the health status of and trends in the United States population.

International Dictionary of Medicine and Biology. 3 vols. Wiley, 1986. These three volumes, compiled by a first-class editorial board, were ten years in the making. As comprehensive as any resource of this kind can be. (BL v83 O 1, 1986 p209; Choice v23 Jl 1986 p1658; LJ v111 Je 15, 1986 p61)

Physicians' Desk Reference. Medical Economics (annual). The definitive source for drug information in the United States. Indicates drug effects, dosages, routes, methods, hazards, side effects, and precautions. (RSSEMA p205)

World Health Statistics Annual. World Health Organization (annual). Presents global statistical information designed to provide country and world overviews on trends in health and causes of mortality.

(*Note*: Several excellent medical dictionaries exist. Every library should have at least one. Other reference-style resources are listed in various categories throughout this chapter.)

Autobiography and Biography

Austrian, Robert. *Life with Pneumococcus.* University of Pennsylvania Press, 1985. 168 p. 0812279778.
    In his memoirs, the author pays his respects to the bacterial pathogen that he has dedicated his career to investigating. Profiles the fascinating intellectual relationship that develops between a researcher and the subject of his study. The author very unselfconsciously accounts for his own contributions to modern developments in pneumococcal diseases. (SciTech v9 N 1985 p15)

Bendinger, Jessica, and Elmer Bendinger. *Biographical Dictionary of Medicine.* Facts on File, 1990. 284 p., index. 0–8160–1864–2.
    The "cast of characters involved in the history of medicine," described in essay entries of 1,000–5,000 words each, including selected bibliographical entries. Also lists a smattering of non-Western physicians and medical figures. (BL v87 Mr 1, 1991 p1420; Choice v28 Ap 1991 p1284)

Cournand, Andre F. *From Roots . . . to Late Budding: The Intellectual Adventures of a Medical Scientist.* Gardner Press, 1987. 232 p. 0–89876–108–5.
    The author, a Nobel Prize winner for his work in cardiology, recounts his career in medicine and medical research. There is also substantial information on his influence as a social activist after retirement. (LJ v112 Mr 1, 1987 p33)

Gay, Peter. *Freud: A Life for Our Time.* Norton, 1988. 810 p., index. 0–393–02517–9.

Gay, a German writer who has won a Pulitzer Prize, displays a critical insight and empathy into Freud's thinking that other biographers have lacked. The author focuses on Freud's environmental and personal influences and, in doing so, establishes a convincing historical and intellectual framework for making sense of his work. Impeccable scholarship and masterful storytelling. (BL v84 Ap 1, 1988 p1292; Choice v26 N 1988 p572; Time v131 Ap 18, 1988 p85)

Koop, C. Everett. *Koop: The Memoirs of the Former Surgeon General*. Random House, 1991. 320 p., illus., index. 0–394–57626–8.

During his tenure as surgeon general, Koop consistently put science and public health concerns above politics. He emerged from those years as one of the heroes of the Reagan administration. These personal memoirs look at his influences, his convictions, and his opinions on current health issues. (BL v88 S 1, 1991 p2; LJ v116 S 15, 1991 p88; NYTBR v96 S 8, 1991 p9)

Medawar, Peter. *Memoirs of a Thinking Radish*. Oxford University Press, 1986. 209 p. 0–19–217737–0.

The title reflects this book's casual and somewhat facetious tone. Medawar, a Nobelist for his work in immunology and a dedicated science popularizer throughout his career, tells of his early influences, his educational background, and his landmark research. (BL v82 My 1, 1986 p1271; LJ v111 Jl 1986 p77; NYTBR v91 Je 15, 1986 p15)

Also Recommended

Hoffman, Edward. *The Right to Be Human: A Biography of Abraham Maslow*. Tarcher, 1989. Biography of the champion of "humanism," whose ideas have become core in psychiatry and other caring professions. (SBF v25 S 1989 p 6)

Luria, S. E. *A Slot Machine, a Broken Test Tube*. Harper and Row, 1984. The autobiography of a groundbreaking medical researcher who contributed to knowledge of bacteriophages and DNA. (BL v80 Ja 1, 1984 p655; LJ v109 Ja 1984 p82; Sci v225 Jl 6, 1984 p47)

Morantz, Regina, et al. *In Her Own Words: Oral Histories of Women Physicians*. Greenwood Press, 1982. Stories of nine female physicians from three distinct historical eras, beginning with the late nineteenth century. Well edited to underscore social and cultural influences. (LJ v107 S 1, 1982 p1669)

Paul, Olgesby. *Take Heart: The Life and Prescription for Living of Dr. Paul Dudley White*. Harvard University Press, 1986. A biography of the physician who cared for Eisenhower after his heart attack and became a champion of the benefits of exercise for cardiac health. (BL v83 O 15, 1986 p315)

History of Medicine

Magner, Lois. *A History of Medicine*. Dekker, 1992. 393 p., illus., index. 0–8247–8673–4.

The best concise general history of medicine available. Included are broad sketches of the history of Western medicine from the Paleolithic era to modernity, as well as introductions to the varied traditions of the Near and Far East. Each chapter contains a list of further readings. (Choice v30 D 1992 p652; SciTech v16 My 1992 p18)

Starr, Paul. *The Social Transformation of American Medicine*. Basic, 1982. 514 p. 0–465–07934–2.

A Pulitzer Prize winner in 1984, this important book traces the evolution of medical practice in America from its decentralized beginnings to today's mega-industry. Starr gives full consideration to the complex socioeconomic and political factors that have shaped health care institutions and how attitudes toward them have changed over time. Identifies mistakes that were made along the way and cautions that they must not be revisited. (KR v50 S 1, 1982 p1054; LJ v107 N 1, 1982 p2084)

Also Recommended

Nuland, Sherwin. *Doctors: The Biography of Medicine*. Knopf, 1988. A history of medicine that focuses on physicians and scientists as agents of technological advance. (BL v84 My 15, 1988 p1563; Choice v26 O 1988 p350; LJ v113 Ag 1988 p166)

General Topics

Beasley, Joseph. *The Betrayal of Health: The Impact of Nutrition, Environment, and Lifestyle on Illness in America*. Random House, 1991. 274 p., index. 0–8129–1897–5.

Beasley, a physician and medical clinic administrator, sees a direct relationship between personal and social illnesses. We live in a toxic society where the environment is unhealthy, personal health habits are poor, and health care is often expensive or inaccessible. It is imperative, he argues, that the medical profession address these problems. (BL v84 Ag 1991 p2082; KR v59 Jl 15, 1991 p899; LJ v116 S 15, 1991 p105)

Cooper, Robert K. *Health and Fitness Excellence: The Scientific Action Plan*. Houghton Mifflin, 1989. 523 p., illus., index. 0–395–47589–9.

A health strategy that takes into account the whole body. Another reviewer called this a "Wellness Bible" for the nineties. (BL v85 F 1, 1989 p904; KR v56 D 15, 1988 p1803; LJ v114 F 1, 1989 p78)

Dutton, Diana B. *Worse than the Disease: Pitfalls of Medical Progress*. Cambridge University Press, 1992. Index. 0–521–34023–3.

Through four case studies, Dutton relates the consequences of medical mistakes and the political pressures that sometimes cause them. Her analyses of these cases make it clear that decision makers must be better informed and held to higher standards. This book is serious reading with a social conscience. (SBF v28 N 1992 p229)

Encyclopedia of Health (series). Chelsea House, 1990–

A proposed seventy-nine-volume series in which each monograph gives a lay introduction to a specific topic, for example, "The Nervous System," "Emergency Medicine," "Substance Abuse," "Sexually Transmitted Diseases," and so on. Collectively, these works provide a thorough overview of health care topics; individual titles can be selected for coverage of specific topics. (SBF v27 My 1991 p109; SBF v27 O 1991 p206)

Harris, Jeffrey. *Deadly Choices: Coping with Health Risks in Everyday Life.* Basic, 1993. 269 p., notes, index. 0–465–02889–6.

To a degree, we are as healthy as we choose to be. Unfortunately, many of us make poor choices. Our choices regarding habits, exercise, weight control, nutrition, and sexuality all affect our health. This book is about making rational individual choices on those matters. (BL v90 S 15, 1993 p113; KR v61 S 1, 1993 p1118)

Weisse, Allen B. *Medical Odysseys: The Different and Sometimes Unexpected Pathways to Twentieth-Century Medical Discoveries.* Rutgers University Press, 1991. 250 p., notes, index. 0–8135–1616–1.

How do medical discoveries happen? Often, through equal measures of painstaking research and ordinary luck. Weisse, a physician, looks at how major breakthroughs were reached in several areas, such as the discovery of penicillin, the development of the first artificial kidney, and the study and treatment of Legionnaire's disease. (BL v87 F 15, 1991 p1169; Choice v28 Jl 1991 p1807; LJ v116 F 1, 1991 p95)

Also Recommended

Smolen, Rick, et al. *The Power to Heal.* Prentice-Hall, 1990. 224 p. A timeless and multicultural exposition of healers, along with their stories and thoughts. (NW v116 S 24, 1990 p45)

Thomas, Lewis. *The Youngest Science: Notes of a Medicine-Watcher.* Bantam, 1984; *The Fragile Species.* Scribner's, 1992. Thoughts on biomedical topics from the physician and essayist. (BL v88 F 1, 1992 p986; SBF v19 S 1983 p24; SLJ v30 O 1983 p24)

Weissmann, Gerald. *They All Laughed at Christopher Columbus: Tales of Medicine and the Art of Discovery.* Times Books, 1987. Essays on medicine, physicians, and the social forces that shape health care practices. (BL v83 Mr 15, 1987 p1085; KR v55 Ja 15, 1987 p123; LJ v112 Ap 1, 1987 p157)

MEDICINE AND HEALTH SCIENCES—SPECIAL TOPICS

Diseases—AIDS

AIDS in the World, 1992. Jonathan Mann et al., eds. Harvard University Press, 1992. 1037 p. 0674012658.

A global summary of the AIDS pandemic; includes information on the extent of the disease's penetration and responses. (LJ v118 Mr 1, 1993 p98; SBF v29 O 1993 p199; SciTech v17 F 1993 p21)

Langone, John. *AIDS: The Facts*. Rev. ed. Little, Brown, 1991. 266 p., notes. 0–316–51414–4.

This book has a direct, no-nonsense question and answer format. Most chapters are given interrogatory titles, such as "What Is AIDS?," "What Is the AIDS Virus?," "Where Did the Virus Originate?," and so on. The author's answers are honest and comprehensible, and he is very clear about stating when there is ongoing debate on an issue. A sound, reasoned approach to AIDS education for a variety of audiences. (BL v84 F 1, 1988 p895; KR v55 D 15, 1987 1725)

Shilts, Randy. *And the Band Played On*. St. Martin's, 1987. 630 p. 0312009941.

The publication of Shilts's book was a landmark in AIDS awareness in America. The now-deceased author studied the early years of AIDS in America, before it became a pandemic, and suggested that swift, dedicated efforts to control the disease could have curtailed its spread. The opportunity, however, was lost due to a lack of political will, bureaucratic shortsightedness, and plain bigotry. (BL v84 S 15, 1987 p92; KR v55 S 1, 1987 p1303; PW v232 S 11, 1987 p72)

Verghese, Abraham. *My Own Country: A Doctor's Story of a Town and Its People in an Age of AIDS*. Simon and Schuster, 1994. c.352 p. 0–671–78514–1.

A compassionate and insightful account of a small town's coming of age in the face of this new and misunderstood disease. This book represents the dilemma of many whose values have been challenged when confronted by AIDS victims. (BL v90 Ap 1, 1994 p1412; KR v62 Ap 1, 1994 p471; LJ v119 Ap 1, 1994 p126)

Also Recommended

Arno, Peter S., and Karyn L. Feiden. *Against the Odds: The Story of AIDS Drug Development, Politics, and Profits*. HarperCollins, 1992. At present there are just two drugs available to combat AIDS. This book tells of how bureaucracy, unspoken homophobia, and bottom-line capitalism have hindered AIDS drug research. (KR v60 Mr 15, 1992 p363; LJ v117 Ap 15, 1992; Sci v257 S 25, 1992 p1975)

Corea, Gena. *The Invisible Epidemic: The Story of Women with AIDS*. HarperCollins, 1992. The author, associate director of the Institute of Women and Technology, bases her work on in-depth interviews with women affected by AIDS and writes a chronology of their despair and hope. (BL v89 O 5, 1992 p382; LJ v118 Ja 1993 p68)

Joseph, Stephen C. *Dragon Within the Gates: The Once and Future AIDS Epidemic*. Carroll and Graf, 1992. A former health commissioner of New York City tells of politics and science in public health AIDS initiatives. (KR v60 O 5, 1992 p382; PW v239 S 14, 1992 p96; NYTBR v97 N 15, 1992 p26)

Diseases—Alcoholism and Addictions

Beasely, Joseph. *How to Defeat Alcohol.* Times Books, 1990. 224 p. 0–8129–1807–x.

By a physician experienced with alcoholics, this book gives helpful information about the disease, including genetics, social factors, and treatments. (PW v236 D 15, 1989 p65; SBF v25 My 1990 p260)

Blum, Kenneth, and James Payne. *Alcohol and the Addictive Brain.* Free Press, 1991. 320 p., index. 0–02903701–8.

Having conducted personal visits to treatment centers nationwide and drawn upon the latest scientific research, the authors define alcoholism as a physiological disease process with psychological and social dimensions. Part One of this book is a historical summary of the problem and methods for coping with it, while Part Two—the more significant—examines the mechanisms and biochemical bases of alcoholism. The authors then establish a scientific model for alcoholic craving and suggest the possible existence of an "alcogene." (SBF v28 Ag 1992 p172; SLJ v38 Je 1992 p95)

Goldstein, Avram. *Addictions: From Biology to Drug Policy.* Freeman, 1994. 321 p., index. 0–7167–2384–0.

Goldstein develops the view that addictions to specific substances are unique diseases, with distinct symptoms and requiring special treatments. He draws upon biological research and applies it toward enlightened public policy formulation. (Choice v31 Je 1994 p1614)

Goode, Erich. *Drugs in American Society.* 4th ed. McGraw-Hill, 1993. 434 p. 0070239231.

The fully revised edition of a reliable source of information on drugs and drug abuse in the United States. Expanded by over 100 pages from the previous edition.

Weil, Andrew, and Winifred Rosen. *From Chocolate to Morphine: Everything You Need to Know about Mind Altering Drugs.* Houghton Mifflin, 1993. 0–395–66079–3.

Addictive substances are more numerous and their effects more pervasive than many know. Taking a humanistic approach to coping with drugs in society, the authors argue that better public education is necessary in order to combat all aspects of addictive behaviors and their consequences.

Also Recommended

Blum, Kenneth. *Handbook of Abusable Drugs.* Gardner Press, 1984. Information on a spectrum of drugs and addictive substances. (Choice v22 Ap 1985 p1191)

Diseases—Anorexia

Brumberg, Joan Jacobs. *Fasting Girls: The Emergence of Anorexia Nervosa as a Modern Disease.* Harvard University Press, 1988. 366 p. 0–674–29501–3.

Eating disorders have long been misdiagnosed and misunderstood. This is a history of cultural and medical attitudes toward appetite disorders, especially as they pertain to women's health. (Atl v262 Jl 1988 p82; BL v84 Mr 15, 1988 p1210; Choice v26 O 1988 p348)

Diseases—Cancer

McAllister, Robert, et al. *Cancer.* HarperCollins, 1993. 329 p., index. 0–465–00845–3.

Divided into three parts, this book tells of the history of scientists' efforts to understand and cure cancer, the latest laboratory findings on diagnosis and treatment, and the National Cancer Institute's protocols for dealing with the ten most common forms of the disease. (LJ v118 Ap 15, 1993 p118; PW v240 Mr 29, 1993 p53)

Rosenberg, Steven, and John Barry. *The Transformed Cell: Unlocking the Mysteries of Cancer.* Putnam, 1992. 398 p., index. 0–399–13749–1.

What triggers the uncontrolled metastasis of healthy cells is a great mystery of oncology. While research has yet to discover cancer's causes, there is great hope that cures can be developed that will take advantage of the body's natural immunities. The authors look at various of these biological treatments and other possibilities. (BL v88 Ag 1992 p1985; KR v60 Jl 1, 1992 p833; PW v239 Jl 13, 1992 p40)

Also Recommended

The American Cancer Society Cancer Book: Prevention, Detection, Diagnosis, Treatment, Rehabilitation, and Cure. Arthur Holleb, ed. Doubleday, 1986. Written by an oncologist to give the public information on what cancer is, how it can be detected, and current therapeutics. (BL v82 Je 1, 1986 p1425)

Diseases—Chronic Fatigue Syndrome

Kenny, Timothy. *Living with Chronic Fatigue Syndrome.* Thunder's Mouth, 1994. 320 p., index. 1–56025–075–5.

Chronic fatigue syndrome has just recently been identified as a serious disease that attacks adults in their prime. In graphic, sometimes disturbing terms the author, a broadcast journalist, recounts his experiences with the disease. Especially enlightening are his frustrations with the skepticism that he frequently encountered regarding the seriousness of his illness. (LJ v119 My 1, 1994 p133)

Diseases—Diabetes

Rayfield, Elliot. *Diabetes: Beating the Odds.* Addison-Wesley, 1992. 158 p. 0201577844.
  The diagnosis, treatment, and prevention of diabetes in a style written for a broad, general audience. (SBF v29 Mr 1993 p47)

Raymond, Mike. *The Human Side of Diabetes.* Noble Press, 1992. 350 p., index. 1–879360–09–8.
  A first person account of living and coping with a common form of diabetes. (LJ v117 Mr 1, 1992 p112)

Diseases—Heart Diseases

Selzer, Arthur. *Understanding Heart Disease.* University of California Press, 1992. 211 p. 0520065603.
  This book covers the basic physiologic and pathologic processes of heart disease. Gives descriptions of common problems such as hypertension, coronary artery disease, congenital heart defects, and so on, and offers solutions for prevention and care. (Choice v31 S 1993 p166; SBF v29 My 1993 p104)

Yale University School of Medicine Heart Book. Barry L. Zaret et al., eds. Hearst Books, 1992. 432 p. 068097197.
  A comprehensive sourcebook, this could serve as a reference for patients or anybody with an interest in how the heart functions in health and disease. Strong coverage on preventative measures. Special sections elaborate on heart health issues for women and ethnic groups. (BL v88 Ap 1, 1992 p1450; LJ v117 Mr 1, 1992 p112; PW v239 Ap 27, 1992 p264)

Also Recommended

Cohan, Carol, et al. *A Patient's Guide to Heart Surgery: Understanding the Practical and Emotional Aspects of Heart Surgery.* HarperPerennial, 1991. Especially interesting for its perspective on the oft-neglected psychological aspects of heart disease and recovery.

Diseases—Lung Diseases

Krogh, David. *Smoking: The Artificial Passion.* Freeman, 1992. 170 p.
  Without moralizing, the author explains the nature and etiology of the smoking habit. This is a book for anybody who wishes to quit smoking or who is curious about the extent and impact of smoking on the general public. (Kliatt v27 Mr 1993 p34; SBF v28 D 1992 p263)

Diseases—Neurological Disorders

Alzheimer's, Stroke, and Twenty-Nine Other Neurological Disorders. Frank E. Bair, ed. Omnigraphics, 1993. 600 p., illus., index. 1–55888–748–2.

A source with information on over thirty common, and some not so common, neurological maladies, with information on diagnosis, progression, and therapies, as well as statistical and demographic material. Also gives a good summary of the brain and how it relates to general body health. Much of the information is taken from publications of the National Institute for Neurological Disorders and Stroke. (A Lib v24 S 1993 p719; BL v90 S 1, 1993 p80)

The Healing Brain: A Scientific Reader. Robert Ornstein and Charles Swencionis, eds. Guilford Press, 1990. 262 p., index. 0–89862–394–4.

Contributors to this volume come from various disciplines and write on broad topics related to the physiological/psychological roles of the brain in healing. The theme is aptly conveyed by the editors' description of the brain as "the body's health maintenance organization." (Choice v28 F 1991 p1006; SBF v26 N 1990 p113; SciTech v15 My 1991 p19)

Sacks, Oliver. *The Man Who Mistook His Wife for a Hat.* HarperPerennial, 1990. 243 p., biblio., 0060970790.

Clinical yarns of a practicing neurologist, entertainingly told. Although the disorders that Sacks describes seem odd, he conveys compassion for his patients and demonstrates that we all can learn from their experiences. (LJ v111 F 15, 1986 p189; KR v53 N 15, 1985 p1253; Nature v318 D 19, 1985 p609)

Also Recommended

Dippel, Raye L. *Caring for the Alzheimer Patient: A Practical Guide.* Prometheus, 1991. A necessary book for anybody who cares for somebody who has Alzheimer's, or who simply wishes to learn more about this disease as well as other forms of dementia. (SBF v27 My 1991 p104)

Levinson, Harold. *Dyslexia: A Scientific Watergate.* Stonebridge, 1994. New insights into a disease that science is just beginning to understand. (*Note*: No reviews are available for this book.)

Martin, Russell. *Matters Gray and White: A Neurologist, His Patient, and the Mysteries of the Brain.* Holt, 1987. The author recounts his experiences during a year spent with a hospital neurologist. (A Lib v18 Ja 1987 p24; KR v54 D 1, 1986 p1782; LJ v111 N 1, 1986 p105)

Restak, Richard. *The Brain Has a Mind of Its Own.* Harmony, 1991. Restak is a witty essayist from the relatively new field of neuropsychiatry. (BL v88 O 15, 1991 p392; NYTBR v97 Ja 12, 1992 p14; SBF v28 My 1992 p104)

Restak, Richard. *Receptors.* Bantam, 1994. The functions of neurotransmitters in the brain.

Diseases—Pain

Wall, Patrick. *Defeating Pain.* Plenum, 1991. 285 p. 0306439646.

Pain is often considered merely a symptom of a disease, and thus is not treated directly. To sufferers, though, pain is a real disease with real consequences. Fortunately, research into the nature and mechanisms of pain is beginning to yield results that could change the lives of afflicted patients. (BL v88 S 15, 1991 p106; PW v238 S 13, 1991 p69)

Diseases—Polio

Smith, Jane C. *Patenting the Sun: Polio, the Salk Vaccine, and the Children of the Baby Boom.* Morrow, 1990. 320 p., photos, index. 0–688–09494–5.

Salk's development of the polio vaccine was one of the first triumphs of laboratory medicine. The story of his research, which took place within a fantastically supportive social and political environment, makes for fascinating reading and contains lessons for medical researchers today. (BL v86 My 1, 1990 p1677; LJ v115 Ap 15, 1990 p120; NYTBR v95 My 6, 1990 p15)

Diseases—Tuberculosis

Ryan, Frank. *The Forgotten Plague: How the Battle Against Tuberculosis Was Won—and Lost.* Little, Brown, 1993. 460 p., photos, index. 0–316–76380–2.

In the late nineteenth and early twentieth centuries, efforts by several famous medical researchers—Paul Ehrlich, Robert Koch, René Dubos, and others—led to what seemed like a victory over tuberculosis. Today, however, TB is back with a vengeance, and combined with AIDS it threatens to become "an alliance of terror." (BL v89 My 15, p1665; LJ v118 My 15, 1993 p90; NYTBR v98 Ag 1, 1993 p1)

Also Recommended

Rothman, Sheila. *Living in the Shadow of Death: Tuberculosis and the Social Experience of Illness in American History.* HarperCollins, 1994. A history that focuses on the experiences of patients and how society viewed those with this deadly disease. (LJ v119 Ja 1, 1994 p113)

Emergency Medicine

Heimlich, Henry, and Lawrence Galton. *Dr. Heimlich's Home Guide to Emergency Medicine.* Simon and Schuster, 1981. 350 p., index. 0–671–24947–9.

The physician whose name is famous for the Heimlich maneuver, which can save choking victims, collects hundreds of practical techniques for dealing with common (and some rarer) medical emergency situations. Part One of this book is given the rather urgent title "Read This First." (BL v76 Je 1, 1980 p1396; KR v48 F 1, 1980 p184; LJ v105 Ap 1, 1980 p869)

Environmental Health

Moeller, Dade. *Environmental Health*. Harvard University Press, 1992. 332 p., illus., index. 0–674–25858–4.

Where we live and work can be hazardous to our health. This source examines various aspects of the environment, such as air, water, and food, and how they can enhance or diminish one's health. From there, broader topics of institutional, local, regional, and even global health are considered. (LJ v117 F 15, 1992 p191)

Also Recommended

Critical Condition: Human Health and the Environment. Eric Chivin, ed. MIT Press, 1993. Examines the nature of human health risks related to such environmental factors as air pollution, radiation, occupational environments, and so on. (SBF v30 Ja 1994 p10)

Geriatrics

Caroline, Rob, and Janet Reynolds. *The Caregiver's Guide: Helping Elderly Relatives Cope*. Houghton Mifflin, 1991. 458 p., index. 0–395–50086–9.

Decisions about how to care for elderly loved ones can be among the most difficult that a person faces in life. This sensitive and ultimately positive book gives practical advice for dealing with such situations in a loving way. In addition to offering general information about the aging process, the authors also provide suggestions for handling specific problems.

Hayflick, Leonard. *How and Why We Age*. Ballantine, 1994. 0–345–33918–5.

A gerontologist explains the biological mechanisms of aging and the effects of time on the body and mind. He passionately debunks many of the myths of aging, with references to the scientific literature. (KR v62 Je 1, 1994 p753; PW v241 Je 27, 1994 p73)

Keeton, Kathy. *Longevity: The Science of Staying Young*. Viking, 1992. 332 p., index. 0–670–83961–2.

The strength of this book is the author's treatment of various theories of aging and research into the causes of longevity. Keeton, a founder of *Longevity* magazine, summarizes diverse opinions and also gives her own "recipe" for living longer and healthier. An excellent list of resources appears in the appendices. (BL v88 Ap 15, 1992 p1492; KR v60 Mr 1, 1992 p301; SBF v28 O 1992 p200)

Also Recommended

Aging and Public Health. Harry T. Phillips and Susan Gaylord, eds. Springer, 1985. As American society grays, various new public health and policy issues must be addressed. (Choice v23 O 1985 p323)

Ebersole, Priscilla, and Patricia Hall. *Toward Healthy Aging.* Mosby, 1990. Humanistic psychology applied toward health and wellness of senior citizens. (SciTech v14 Ap 1990 p18)

Fries, James F. *Aging Well.* Addison-Wesley, 1989. Answers seniors' questions on aging and disease. (LJ v115 Mr 1, 1990 p44; PW v236 Jl 21, 1989 p55)

Holistic Health

Kastner, Mark, and Hugh Burroughs. *Alternative Healing.* Halcyon, 1993. 346 p., biblio., index. 0–9635597–1–2.
   Many people dissatisfied with conventional medical treatments have turned to alternative therapies for health care. This is an A to Z guide to over 160 of the most popular and common. Each section gives information on applications and case studies. Could also be used for reference. (BL v90 N 15, 1993 p644; LJ v118 O 15, 1993 p60)

Moyers, Bill. *Healing and the Mind.* Doubleday, 1993. 369 p. 0385468709.
   Written to accompany Moyers's television program, *Healing and the Mind* is about how mental qualities and attributes affect a patient's physical well-being. This principle, accepted in other cultures but somewhat alien to Western medicine, is now the subject of much research and is the basis for reforms in the health care profession. (BL v89 D 15, 1992 p698; NYTBR v98 My 23, 1993 p29)

Weil, Andrew. *Natural Health, Natural Medicine.* Houghton Mifflin, 1990. 342 p., biblio., index. 0–395–49340–4.
   Weil is a fervent believer in the rewards of a salubrious lifestyle and preventative medicine, which, he claims, are factors often overlooked by conventional practitioners. He gives a good introduction to ''naturopathic'' alternatives to health promotion and well-being. (BL v86 My 15, 1990 p1767; LJ My 1, 1990 p108)

Also Recommended

Murray, Michael, and Joseph Pizzorno. *Encyclopedia of Natural Medicine.* Prima Publications, 1991. A valuable reference in three parts: an introduction to natural medicine, a section on basic principles, and an extensive alphabetical section on appropriate treatments of various diseases. (BL v87 Je 1, 1991 p1899; LJ v116 My 1, 1991 p72)

Hospitals and Medical Care Facilities

Colen, B. D. *OR: The True Story of Twenty-Four Hours in a Hospital Operating Room.* Dutton, 1993. 214 p. 0–525–93518–5.
   The real-life workings of an operating room are in some ways rather mundane,

but, as shown on television, it can also be the scene of some high drama. The author, a journalist, goes into some detail on the roles of orderlies, support staff, surgeons, and others who work in the operating room. Also considers patient issues and concerns. (BL v89 Mr 15, 1993 p1286; LJ v118 Mr 1, 1993 p100; SBF v29 Ag 1993 p170)

Rosenberg, Charles E. *The Care of Strangers: The Rise of America's Hospital System.* Basic, 1987. 0–465–00877–1.
A sensitive and well-researched history of hospitals in America, from their beginnings in the nineteenth century, through the changes that occurred within the medical profession in the early twentieth century, up to today's age of technology. Gives attention to the significance of the nursing profession. (BL v84 O 15, 1987 p353)

Stevens, Rosemary. *In Sickness and in Wealth: American Hospitals in the Twentieth Century.* Basic, 1989. 432 p., index. 0–465–03223–0.
As institutions, hospitals are uniquely chartered in that they are private institutions that exist for the public good. Thus, many times hospital administrators face quandaries that arise when what is good for business is not in the best interests of the patients. This book traces the rise of the hospital system and its management today. (BL v85 Mr 15, 1989 p1233; Choice v27 S 1989 p168; NYTBR v94 Ag 20, 1989 p14)

Also Recommended

Shohen, Sandra. *Emergency! Stories from the Emergency Department of a New York City Hospital.* St. Martin's, 1989. An account of the management of emergency services at New York's Roosevelt Hospital, written by a former administrator. (BL v86 O 1, 1989 p245; LJ v114 O 1, 1989 p112; SBF v25 S 1990 p53)

Immunology

Davis, Joel. *Defending the Body: Unraveling the Mysteries of Immunology.* Atheneum, 1989. 250 p., gloss., index. 0–689–11946–1.
Relying to a large degree on information gleaned from interviews with scientists and physicians, Davis explores the complex and marvelous workings of the human immune system. Immunology touches upon virtually all aspects of health and disease; thus, the author ranges far afield in his expositions. (BL v85 Mr 15, 1989 p1232; LJ v114 My 1, 1989 p95; SBF v25 Mr 1990 p205)

Also Recommended

Desowitz, Robert S. *The Thorn in the Starfish: How the Human Immune System Works.* Norton, 1987. The history of immunology, how antigens and antibodies work in the body, and how habits and lifestyle affect the immune system. (BL v83 Je 15, 1987 p1548; KR v55 My 1, 1987 p689; LJ v112 My 1, 1987 p75)

Medical Ethics—General

Annas, George J. *Judging Medicine.* Humana, 1988. 438 p., index. 0–89603–132–2.

Most of the articles in this thoughtful compilation originally appeared in the author's column in the *Hastings Center Report.* Covers issues on eclectic topics such as patients' rights, pregnancy and birth, mental illness, government regulations, death and euthanasia, and transplantation. (BL v84 Je 15, 1988 p1698; SBF v25 S 1989 p7)

Callahan, Daniel. *What Kind of Life? The Limits of Medical Progress.* Simon and Schuster, 1990. 319 p. 0671670964.

In the health care field, the cost and complexity of procedures are two factors that can lead to troubling ethical dilemmas, such as when an expensive, risky operation is the sole hope for a patient who, even with the surgery, has poor odds of survival. The author attempts to come to grips with this and other similarly difficult cases by developing cost/benefit and benefit maximization models of health care. (BL v86 D 1, 1989 p706; LJ v115 Ja 1990 p139; NYTBR v94 D 24, 1989 p1)

Jonsen, Albert R. *The New Medicine and the Old Ethics.* Oxford University Press, 1990. 171 p., index. 0–674–61725–8.

The author believes that discussions of ethics are not doomed to subjectivity and ambiguity; rather, enduring ethical principles still apply in today's high-tech, high-cost world of health care. Examining the history of medicine, with reference to related fields of thought and human culture, he identifies and underscores these principles. This book helps define a debate that is current and urgent. (Choice v28 Ap 1991 p1342; New Sci v131 Ag 17, 1991 p44; SBF v27 Ap 1991 p71)

Also Recommended

Ethical Issues in the New Reproductive Technologies. Richard Hull, ed. Wadsworth, 1990. Principally about infertility and related legal matters. (SBF v25 S 1990 p32)

Medical Ethics—Abortion

Morowitz, Harold, and James Trefil. *The Facts of Life: Science and the Abortion Crisis.* Oxford University Press, 1992. 175 p., index. 0–19–507927–2.

Seeking objectivity, the authors, both excellent science popularizers, concentrate on the biological side of fetal development and, from there, offer views about the more philosophical concepts of when human life begins. Raises heady questions and invites readers to ponder them. (BioSci v43 Jl 1993 p495; Choice v30 My 1993 p1501; SBF v29 D 1992 p15)

Also Recommended

Mohr, James C. *Abortion in America: The Origins and Evolution of National Policy, 1800–1900.* Oxford University Press, 1978. A historical study of abortion with an implicit pro-choice slant. (BRD 1978 p922; Choice v15 Je 1978 p582; NYTBR Ap 2, 1978 p7)

Olasky, Marvin. *Abortion Rights: A Social History of Abortion in America.* Crossway, 1992. Argues against abortion from a social morality point of view. (BL v89 N 15, 1992 p567)

Medical Ethics—Death and Euthanasia

Hill, T. Patrick, and David Shirley. *A Good Death: Taking More Control at the End of Your Life.* Addison-Wesley, 1992. 176 p., index. 0–201–06223–2.
A respectful, sensitive treatment of a very difficult subject, this book looks at the human issues involved in death and the choices that must be faced. Honestly examines a wide range of practical topics, such as making a living will and how to communicate your desires if you are incapacitated. Should be of comfort to individuals and families. (SBF v29 D 1992 p9)

Kevorkian, Jack. *Prescription—Medicide: The Goodness of Planned Death.* Prometheus, 1991. 268 p.
The controversial gadfly of medicine offers his thoughts on physician-assisted suicide. Kevorkian's writing is as iconoclastic as his public personality, but he insists that euthanasia is a moral choice and that physicians have the responsibility to abide by their patients' wishes. This book is an important statement in a complicated debate. (BL v88 S 1, 1991 p7; LJ v116 S 1, 1991 p220)

Kübler-Ross, Elisabeth. *On Death and Dying.* Macmillan, 1991. 0025671111.
Originally published in 1969, this classic discusses the fear of death and the psychological stages through which the terminally ill pass. Interviews with dying patients and their caregivers make this book as intimate and compassionate as it is informative.

Medical Genetics

Milunsky, Aubrey. *Choice, Not Changes: An Essential Guide to Your Heredity and Health.* Little, Brown, 1989. 488 p., index. 0–316–57423–6.
Knowing your genetic heritage is important for self-care and personal health decision-making. This book provides a framework for understanding how an individual's health is affected by genetics, and what can and cannot be done. Considers medical, legal, and ethical questions. (BL v85 F 1, 1989 p904; LJ v114 F 15, 1989 p171; SBF v25 S 1989 p18)

Pollen, Daniel. *Hannah's Heirs: The Quest for the Genetic Origins of Alzheimer's Disease.* Oxford University Press, 1993. 296 p., refs., index. 0–19–506809–2.

Hannah, a nineteenth century Ukrainian woman who developed Alzheimer's in her forties, was the matriarch for an intense study of a family's genetic history. Today, Hannah's descendants are the heirs to a "personal biological holocaust." In studying the transmission of Alzheimer's across those generations, however, scientists have made some remarkable discoveries, and Hannah's heirs can now make more informed decisions about their own lives. (BL v89 My 15, 1993 p1665; KR v61 My 1, 1993 p578; LJ v118 Je 1, 1993 p184)

Thompson, Larry. *Correcting the Code.* Simon and Schuster, 1994. 384 p., index. 0–671–77082–9.

Throughout its brief history, the field of medical genetics has experienced several ups and downs. In 1990, however, a breakthrough was reached when a team of National Institute of Health doctors successfully applied genetic therapy for a four-year-old girl suffering from a rare immune system disease. The author is a great believer in the future of medical genetics; his optimism comes through resoundingly. (BL v90 Mr 15, 1994 p1312; KR v62 Ja 1, 1994 p55)

Also Recommended

Neel, James. *Physician to the Gene Pool.* Wiley, 1994. An "in the trenches" reflection on fifty years in genetic research by a leader in the field. Covers social and technological issues. (BL v90 F 1, 1994 p987; KR v62 F 1, 1994 p124)

(*Note*: See also entries in Chapter Seven under "Molecular Biology" and in Chapter Thirteen under "Biotechnology and Genetic Engineering.")

Medical Technology

Bronzino, Joseph, et al. *Medical Technology and Society.* MIT Press, 1990. 570 p., illus, index. 0–262–02300–8.

This sweeping survey of the current state of the art in medical technology is an impressively complete single-volume treatment of the subject. Explores both the promises and the problems associated with medical technology. (Choice v28 N 1990 p514; LJ v116 Mr 1, 1991 p62)

Nursing

Bullough, Vern L., and Bonnie Bullough. *The Care of the Sick: The Emergence of Modern Nursing.* Watson, 1978. 0–88202–183–4.

This history of nursing begins by exploring the origins of home care in ancient times, proceeds through the Middle Ages, when nursing was often the sole form of health care available, then surveys the growing professionalization of nursing with the rise of modern medicine in this century. Gives consideration to social and gender issues related to the profession. (Choice v16 Ap 1979 p256)

Heron, Echo. *Intensive Care: The Story of a Nurse.* Atheneum, 1987. 0–689–11808–2.

Engrossing, at times both funny and sad, this book is a registered nurse's personal account of the trials and dilemmas she has faced in a hospital emergency room. The author succeeds at being honest, realistic, and hopeful. (BL v83 Je 15, 1987 p1548; KR v60 Ap 15, 1987 p616; PW v231 Ap 24, 1987 p58)

Images of Nursing: Perspectives from History, Art, and Literature. Anne Hudson Jones, ed. University of Pennsylvania Press, 1988. 253 p., index. 0–8122–1254–1.

How society has viewed nurses and the nursing profession historically, and how those impressions have been conveyed in numerous media. Contributors are experts from various fields. (Choice v26 S 1988 p168)

Also Recommended

Carpineto, Jane. *R.N.* St. Martin's, 1992. The thoughts and experiences of three nurses. (BL v88 Ja 15, 1992 p896; KR v59 D 15, 1991 p1566; LJ v117 Ja 1992 p164)

Donahue, M. Patricia. *Nursing: The Finest Art.* Mosby, 1985. A lavishly illustrated and very informative history of nursing. (LJ v111 Jl 1986 p98; NYTBR v91 S 14, 1986 p31)

Nutrition

The Columbia Encyclopedia of Nutrition, Institute of Human Nutrition, Columbia University Press/Putnam, 1988. 349 p. 0–399–13298–8.

The over 100 alphabetically arranged articles contained here give useful and scientifically based information on various aspects of human nutrition. (BL v85 S 15, 1988 p132; LJ v113 Je 1, 1988 p102)

National Research Council. *Eat for Life: The Food and Nutrition Board's Guide to Reducing Your Risk of Chronic Disease.* National Academy Press, 1992. 0–309–04049–3.

Written for anybody with an interest in improving personal diet and nutrition, this book gives guidelines for implementing a basic dietary plan for life. Emphasizes how proper diet reduces the risks of certain chronic diseases. (SBF v28 Ag 1992 p172)

Also Recommended

Clayman, Charles. *Diet and Nutrition.* Reader's Digest, 1991. Straightforward, easy-to-read advice on basic topics in food and health. Part of a series bearing the American Medical Association's seal of approval.

Goodman, Sandra. *Vitamin C: The Master Nutrient.* Keats, 1991. 176 p. The complete story of vitamin C and its usefulness to human beings. (Choice v29 Jl 1992 p1712; SBF v28 Mr 1992 p45)

Operations and Surgery

Gutkind, Lee. *Many Sleepless Nights: The World of Organ Transplantation.* Norton, 1988. 368 p., index. 0–393–02520–9.

Donated organs have saved countless lives, but supplies are short and competition for organs fierce. The observations in this book are based on the author's experiences at a major American hospital and cover the gamut of patient and surgical issues. (BL v84 Je 15, 1988 p1699; LJ v113 S 1, 1988 p178; SBF v25 S 1989 p18)

Inlander, Charles. *Good Operations, Bad Operations.* Viking, 1993. 430 p., notes, index. 0–670–83778–4.

This book quotes some frightening statistics. For example, unnecessary surgery may result in some 86,000 deaths per year; 10 percent of hospital patients acquire some infection during their stay; and perhaps as many as 90 percent of hysterectomies are done for questionable reasons. Inlander of the People's Medical Society has compiled this as a practical guide to over 100 of the most common diagnostic and surgical procedures. It is designed to help consumers make more informed and intelligent choices. (LJ v118 Jl 1993 p70; PW 240 Jl 26, 1993 p69)

Also Recommended

Rutkow, Ira M. *Surgery: An Illustrated History.* Mosby Year Book, 1994. A large, well illustrated volume of the surgical art and science from prehistory to present. (NYTBR v99 Je 5, 1994 p34)

Sylvester, Edward. *The Healing Blade: Neurosurgery on the Edge of Life and Death.* Simon and Schuster, 1993. A history of neurosurgery. (BL v89 Ja 1, 1993 p782; LJ v118 Ja 1993 p156)

Wangensteen, Owen, and Sarah Wangensteen. *The Rise of Surgery.* University of Minnesota Press, 1979. Still the most popular one-volume history of medical surgery.

Pediatrics

Anderson, Peggy. *Children's Hospital.* Harper and Row, 1985. 532 p., biblio., 0–06–015089–0.

The author follows the desperate plight of six severely ill children "to discover how—or whether—the child, family, and the hospital staff members attempting to save or cure the child tolerate the intolerable." (BL v81 Je 1, 1985 p1360; BW v15 Je 9, 1985 p5; KR v53 Ap 1, 1985 p308)

Bergman, Abraham. *The "Discovery" of Sudden Infant Death Syndrome*. Praeger, 1986. 237 p. 0–275–92059–3.

The death of an infant is devastating to any parent, but despite the obvious human tragedy that it engenders, SIDS was not recognized as a medical condition until relatively recently. Although the causes of SIDS are still unknown, research is in progress and there is hope. (Choice v23 Jl 1986 p1700)

The Columbia University College of Physicians and Surgeons Complete Guide to Early Child Care. Crown, 1990. 514 p. 0517572176.

A comprehensive reference designed to give parents and caregivers quick information on a host of subjects. Contains sections on behavioral and medical conditions. (LATBR S 2, 1990 p6)

Also Recommended

Ashford, Janet Isaacs. *The Whole Birth Catalog: A Sourcebook for Choice in Childbirth*. Crown, 1985. Resources for women making choices about pregnancy and childbirth. (BL v79 Je 15, 1983 p1310; LJ v108 My 15, 1983 p1010; NYTBR v88 My 8, 1983 p36)

Brace, Edward R., et al. *Childhood Symptoms: Every Parent's Guide to Childhood Illnesses*. HarperPerennial, 1992. A popular handbook on children's diseases and their symptoms.

Pharmacology and Pharmaceuticals

Burger, Alfred. *Drugs and People: Medications, Their History and Origins and the Way They Act*. University Press of Virginia, 1986. 176 p., index. 0–8139–1101–x.

The author states that "this book is for persons who are interested in medicine and want to know where drugs come from." Having established a historical background on how drugs were discovered and synthesized, Burger proceeds to describe how specific pharmacological agents are applied toward diseases and various human conditions. (KR v54 Je 15, 1986 p902; LJ v111 My 15, 1986 p73)

Burkholz, Herbert. *The FDA Follies*. Basic, 1994. 228 p., index. 0–465–02369–x.

Throughout the 1980s, the FDA's corruption and inefficiency was almost like a joke to industry insiders—except that the consequences were no laughing matter. Since then, some much-needed reforms have been enacted. The whole story serves as a model of how a health care system went wrong, and how it can be fixed. (BL v90 Mr 1, 1994 p1166; LJ v119 Mr 1, 1994 p108; Nature v369 My 5, 1994 p27)

Fisher, Jeffrey. *The Plague Makers: How We Are Creating Catastrophic New Epidemics—and What We Must Do to Avert Them.* Simon and Schuster, 1994. 0–671–79156–7.

The excessive use of antibiotics can inadvertently encourage the development of new, resistant strains of microbes. The physician author describes the nature of the problem and what can be done about it. (LJ v119 Ap 15, 1994 p102)

Also Recommended

Graedon, Joe. *Aspirin Handbook: A User's Guide to the Breakthrough Drug of the Nineties.* Bantam, 1993. Moderate use of aspirin as a preventative measure.

Griggs, Barbara. *Green Pharmacy: A History of Herbal Medicine.* Inner Traditions, 1991. An introduction to ethnopharmacology. (BL v78 Ja 1, 1982 p574; Choice v19 Jl 1982 p1595)

Julien, Robert. *A Primer of Drug Action.* Wiley, 1985. Principles of drug actions and interactions in the human body.

Kramer, Peter D. *Listening to Prozac: A Psychiatrist Explores Antidepressant Drugs and the Remaking of the Self.* Viking, 1993. In popular literature, Prozac has been called the pill that can change your personality—that is, it can make you feel happy, more confident, and/or more assertive. First used for patients with more severe mental disorders, it is becoming a drug of choice for those with much milder problems. (BL v89 Je 1, 1993 p1758; LJ v118 My 1, 1993 p104; SBF D 1993 p266)

Maeder, Thomas. *Adverse Reactions.* Morrow, 1994. A history of the use and abuse of the antibiotic chloramphenicol. (LJ v119 Mr 15, 1994 p96)

Werth, Barry. *The Billion Dollar Molecule: One Company's Quest for the Perfect Drug.* Simon and Schuster, 1994. The business and science of Vertex, a biotechnology company dedicated to the development of new drugs. (BL v90 Ja 15, 1994 p886; Bus W F 14, 1994 p15; KR v61 D 1, 1993 p1514)

Practice of Medicine

Carpineto, Jane. *On Call: Three Doctors at the Frontlines.* St. Martin's, 1994. 224 p. 0–312–11041–3.

The physicians profiled in this book are from a hospital and an inner-city health clinic. In addition to looking at their daily routines, the author also extrapolates from their experiences to offer commentaries about the changing face of health care today. The concluding section includes interviews with the doctors, in which they offer their candid opinions about abortion, euthanasia, health care reform, and other critical issues of today.

Kaufman, Sharon R. *The Healer's Tale: Transforming Medicine and Culture.* University of Wisconsin Press, 1993. 324 p., photos, notes. 0–299–13550–0.

The opinion that, unlike in earlier times, contemporary medicine is impersonal and uncaring is, unfortunately, all too common. With that in mind, Kaufman interviewed seven eminent elderly physicians (all in their eighties) about how the practice of medicine has changed over the four decades of their careers. (LJ v118 Ap 1, 1993 p122; SciTech v17 My 1993 p20; WLB v68 S 1993 p112)

Klitzman, Robert. *A Year-Long Night: Tales of a Medical Internship.* Viking, 1989. 242 p. 0–670–81777–5.
In this memoir, the author tells of some of his unusual experiences during internship. Most poignant are those dealing with the dead or dying, which, he claims, is something that many physicians never learn to handle with grace and confidence. Additionally, Klitzman discusses his pre-med experiences in New Guinea, where he encountered some rare viral diseases. (BL v85 F 1, 1989 p904; LJ v114 F 1, 1989 p79; SBF v25 S 1989 p18)

Also Recommended

Empathy and the Practice of Medicine. Howard Spiro et al., eds. Yale University Press, 1994. Essays and papers on how medicine and doctors have changed since World War II. (SBF v30 Ap 1994 p73)

LeBaron, Charles. *Gentle Vengeance: An Account of the First Year at Harvard Medical School.* Penguin, 1982. A critical look at medical education, this book has remained extremely popular among pre-meds for over a decade.

Wolinsky, Howard. *The Serpent and the Staff: The Unhealthy Politics of the American Medical Association.* Putnam, 1994. A whistle-blower looks at the institutional goals and political clout of the AMA. (LJ v119 My 1, 1994 p133)

Psychology, Psychotherapy, and Behavioral Health

Cronkite, Kathy. *On the Edge of Darkness: Conversations about Depression.* Doubleday, 1994. Biblio., index. 0–385–42194–x.
Interviews with those who suffer from clinical depression and physicians who treat them. This book sympathetically addresses the deep personal feelings that are associated with depression, such as guilt, isolation, and thoughts of suicide. Also looks at the oddly similar phenomena of young adult and geriatric depressions. (KR v62 Ja 1, 1994 p31; LJ v119 F 15, 1994 p175)

Heston, Leonard L. *Mending Minds: A Guide to the New Psychiatry of Depression, Anxiety, and Other Serious Mental Disorders.* Freeman, 1992. Index. 0–7167–2158–9.
The "new psychiatry" emphasizes a full working partnership between the patient and the therapist. With that as his starting point, the author provides brief essays on psychiatric treatments in several categories: mood, thought, anxiety, drug abuse, and so on. Gives signs, symptoms, diagnostic information, and prog-

noses. Concludes with a section on the physiology and biochemistry of the brain. (SBF v28 Je 1992 p137)

Mondimore, Francis Mark. *Depression: The Mood Disease.* Johns Hopkins University Press, 1990. 226 p., index. 0–8018–3856–8.

"What causes depression?" is an intriguing question. Serious mood disorders can often seem indistinguishable from normal highs and lows, sometimes even to the sufferers. The author gives a far-ranging analysis of various types of mood complications, including those that affect teens, women, and elders. (SBF v25 N 1990 0113)

Weiss, Joseph. *How Psychotherapy Works: Process and Technique.* Guilford Press, 1993. 219 p., index. 0–89862–549–3.

In a remarkably easy-to-read fashion, Weiss expounds upon the basic principles and techniques of psychotherapy. Central to successful therapy is defining the individual's goals and how they meld with that person's perceptions of reality. The therapist's objective is to help the person come to an understanding of these goals and to assess his/her own psychological resources. Supported by over forty case studies. (SBF v30 Ap 1994 p73)

Yalom, Irvin. *Love's Executioner: And Other Tales of Psychotherapy.* Basic, 1989. 270 p.

In anecdotal essays the author tells of some odd and uplifting cases he has encountered during his years as a psychotherapist. (BL v85 Je 15, 1989 p1741; LJ v114 Ag 1989 p151; NYTBR v94 S 3, 1989 p5)

Also Recommended

Bruno, Frank J. *The Family Mental Health Encyclopedia.* Wiley, 1989. Approximately 700 definitions and brief articles on concepts related to the broad field of family mental health. (BL v86 S 1, 1989 p104; Choice v27 O 1989 p280; SBF v25 Ja 1990 p115)

Grob, Gerald. *The Mad Among Us: A History of the Care of America's Mentally Ill.* Free Press, 1994. A semi-scholarly history from the colonial period to the present. (LJ v119 F 15, 1994 p175)

Personality and Depression. Marjorie Klein et al., eds. Guilford Press, 1993. An outstanding source that looks at current models for understanding depression and research into the origins and uniqueness of personalities. (SBF v29 Ag 1993 p170)

(*Note:* See also entries in Chapter Seven under "Neurosciences.")

Reproductive Health

Cooper, Susan L., and Ellen S. Glazer. *Beyond Infertility: The New Paths to Parenthood.* Lexington Books, 1994. 376 p., index. 0–02–911813–1.

Medical technology has made it possible for couples previously considered infertile to have children, but at what cost? In this book, a psychologist and a social worker look at reproductive alternatives and the choices that they give couples who wish to have children.

Sexuality

Boston Women's Health Collective. *The New Our Bodies Ourselves.* Simon and Schuster, 1985; Paula Doress-Worters and Diana Laskin Siegal. *The New Ourselves, Growing Older: Women Aging with Knowledge and Power.* Simon and Schuster, 1994. 560 p., photos, index. 0–671–82797–4.

These resources by and for women are very popular and cover the gamut of women's health issues in an empathetic and scientific way. (A Lib v17 My 1986 p314; LJ v118 O 1, 1993 p58)

LeVay, Simon. *The Sexual Brain.* MIT Press, 1993. 162 p., gloss., index. 0–262–12178–6.

In 1991 LeVay published a controversial paper in *Science* in which he described physical differences between the brains of gay and straight men. This book explores the nature of those differences and broader issues related to the neurobiological basis for human sexuality. (Choice v31 O 1993 p317; Nature v363 Je 10, 1993 p505; SBF v29 Ag 1993 p164)

Michael, Robert et al. *Sex in America: A Definitive Survey.* Little, Brown, 1994. 300 p. illus., index. 0316911917.

An abridgment of "The Social Organization of Sexuality: Sexual Practices in the United States," which is the most far-reaching survey of American sexual behaviors since the Kinsey reports of two generations ago. (*Note*: No reviews are available for this book.)

Zilbergeld, Bernie. *The New Male Sexuality.* Bantam, 1992. 592 p., index. 0–553–08253–1.

Sequel to the author's popular *Male Sexuality*, this looks at male physical and psychological health care. (BL v88 Je 15, 1992 p1795; LJ Je 1, 1992 p156)

Also Recommended

Bullough, Vern. *Science in the Bedroom: A History of Sex Research.* Basic, 1994. Historical approaches to researching this basic, but extremely private human characteristic. (BL v90 Je 1, 1994 p1733; PW v241 My 9, 1994 p58)

Hamer, Dean, and Peter Copeland. *The Science of Desire: The Search for a Gay Gene and the Biology of Behavior.* Simon and Schuster, 1994. Cautious investigations into the possible genetic basis of homosexuality.

Social Aspects of Medical Practice

Beyond Crisis: Confronting Health Care in the United States. Nancy F. McKenzie, ed. New American Library, 1994. 0–452–01108–6.

These essays were collected as a project of the Health Policy Advisory Center for the purpose of surveying the state of health care in America today and focusing the debate on needed reforms. (LJ v119 Ja 1994 p150)

Califano, Joseph A., Jr. *America's Health Care Revolution: Who Lives? Who Dies? Who Pays?* Simon and Schuster, 1989. 0–671–68371–3.

Califano, a political insider, documents the public and private factors that contribute to the high costs of health care in America. The tone is reform-minded, and the author targets several areas of waste and inefficiency. (BL v82 Je 15, 1986 p1488; Choice v23 Jl 1986 p1300; LJ v111 Mr 15, 1986 p60)

Payer, Lynn. *Medicine and Culture: Varieties of Treatment in the United States, England, West Germany and France.* Holt, 1988. 204 p. 0805004432.

Payer compares and contrasts the cultural paradigms that underlie health care systems in four industrialized countries. In each, she finds that unique cultural and social values are at the root of differing practices and policies. (BL v84 Ap 15, 1988 p1379; Choice v26 S 1988 p168; LJ v113 My 15, 1988 p89)

Textbooks

(*Note*: Medical textbooks are written for the needs of students and practicing physicians. Even so, because they integrate vast bodies of knowledge on partic-ular subjects in an authoritative summary fashion, they are frequently consulted by lay readers and can represent whole subdisciplines in library collections. The following are the latest editions of several standard texts; they fall under the category of "Also Recommended.")

Adams, Raymond D., and Maurice Victor. *Principles of Neurology.* 5th ed. McGraw-Hill, 1993.

Allergy: Principles and Practice. 4th ed. 2 vols. Elliot Middleton, Jr., et al., eds. Mosby, 1993.

Balenger, John J. *Diseases of the Nose, Throat, Ear, Head and Neck.* 14th ed. Lea and Febiger, 1993.

Best and Taylor's Physiological Basis of Medical Practice. 12th ed. John B. West, ed. Williams and Wilkins, 1991.

Brocklehurst, John, et al. *Textbook of Geriatric Medicine and Gerontology.* 4th ed. Churchill Livingstone, 1992.

Cancer: Principles and Practice of Oncology. 2 vols. Vincent T. DeVita et al., eds. Lippincott, 1993.

Casarett and Doull's Toxicology. 4th ed. Mary O. Amdur et al., eds. McGraw-Hill, 1991.

Comprehensive Textbook of Psychiatry. 5th ed. 2 vols. Harold Kaplan and Benjamin Saddock, eds. Williams and Wilkins, 1989.

Cunningham, Gary, et al. *Williams Obstetrics.* 19th ed. Appleton and Lange, 1993.

Family Medicine: Principles and Practice. 4th ed. Robert Taylor, ed. Springer-Verlag, 1993.

Goodman and Gilman's Pharmacological Basis of Therapeutics. 8th ed. Alfred G. Gilman et al., eds. McGraw-Hill, 1990.

Griffith, H. Winter. *Instructions for Patients.* 4th ed. Saunders, 1989.

Harrison's Principles of Internal Medicine. 12th ed. 2 vols. Jean D. Wilson et al., eds. McGraw-Hill, 1991.

Heart Disease: A Textbook of Cardiovascular Medicine. 4th ed. Eugene Braunwald, ed. Saunders, 1992.

Modern Nutrition in Health and Disease. 8th ed. Maurice E. Shils and Vernon R. Young, eds. Lea and Febiger, 1993.

Nelson Textbook of Pediatrics. 14th ed. Richard E. Behrman et al., eds. Saunders, 1992.

Principles of Surgery. 6th ed. Seymour I. Schwartz et al., eds. McGraw-Hill, 1993.

Robbins Pathologic Basis of Disease. 4th ed. Ramzi Cotran et al., eds. Saunders, 1989.

Sexually Transmitted Diseases. 2nd ed. King K. Holmes et al., eds. McGraw-Hill, 1990.

Substance Abuse: A Comprehensive Textbook. 2nd ed. Joyce Lowinson et al., eds. Williams and Wilkins, 1992.

Zinsser Microbiology. 20th ed. Wolfgang Jolik et al., eds. Appleton and Lange, 1992.

Viruses

Garrett, Laurie. *The Coming Plague: Newly Emerging Diseases in a World Out of Balance.* Farrar Straus Giroux, 1994. 750 p., notes, index. 0–374–12646–1.
Fifty years ago, medical scientists felt that the total eradication of infectious diseases was imminent. Today, tuberculosis has returned, AIDS has already taken a deadly toll, and numerous exotic, extremely lethal microbial diseases are emerging rapidly, often as a result of human activity. A comprehensive and thoroughly documented study of biomedicine's new challenges. (*Note*: No reviews are available for this book.)

Radetsky, Peter. *The Invisible Invaders: The Story of the Emerging Age of Viruses*. Little, Brown, 1991. 406 p., index. 0–316–73216–8.

"Viruses cause more sickness than anything else on Earth," Radetsky explains. Some, such as the cold virus, are relatively benign, while others, like HIV, are not only deadly, but extremely difficult to understand and treat. This book offers a history of human encounters with viruses and a modern overview of their impact on health and society. The author is a science journalist with a background in creative writing. (BL v87 Ja 1, 1991 p897; Choice v28 Je 1991 p1659; LJ v115 N 15, 1990 p89)

(*Note*: See also entries in Chapter Seven under "Microbiology and Cell Biology.")

Also Recommended

Preston, Richard. *The Hot Zone*. Random, 1994. A riveting tale of how a team of scientists and doctors responded to the threat of a serious viral outbreak. (KR v62 Je 15, 1994 p829; PW v241 Je 27, 1994 p61)

Women's Health Issues

Nechaus, Eileen, and Denise Foley. *Unequal Treatment: What You Don't Know about How Women Are Mistreated by the Medical Community*. Simon and Schuster, 1994. 272 p., biblio., index. 0–671–79186–9.

The authors contend, with good evidence, that the medical profession has historically been a stronghold of male dominance. This has led to a common neglect of women's health issues and a lack of understanding about how women respond to treatments. (KR v62 My 15, 1994 p376)

Also Recommended

Dally, Ann G. *Women under the Knife: A History of Surgery*. Routledge, 1991. 0415905540. This history of gynecological surgery seems sometimes odd and barbaric. Dally examines the attitudes that permitted these conditions. (BL v88 Mr 1, 1992 p1187; LJ v117 Ap 1, 1992 p140)

Greer, Germaine. *The Change*. Knopf, 1992. Greer, the feisty and controversial feminist, examines an issue in women's health care that is on-target for many of today's aging baby boomers. (BL v89 S 1, 1992 p2; LJ v117 S 15, 1992 p86; PW v239 Ag 24, 1992 p66)

Nechas, Eileen, and Denise Foley. *The Women's Encyclopedia of Health and Emotional Healing*. Rodale Press, 1993. A popularized exposition of a host of women's health issues, with due consideration given to mental and emotional health. (BL v89 D 15, 1992 p698; LJ v118 F 1, 1993 p70)

Shephard, Bruce D., and Carroll Shephard. *The Complete Guide to Women's Health.* New American Library, 1985. Written by a husband and wife team of health care practitioners, this gives concise information on issues related to disease, sexuality, reproductive health, and others. (Kliatt v25 Ap 1991 p33; Choice v20 Mr 1983 p1020; SciTech v15 F 1991 p28)

Wadler, Joyce. *My Breast.* Addison-Wesley, 1992. A journalist recounts her experiences with breast cancer, which afflicts one in nine women. (BL v89 S 15, 1992 p107; KR v60 Ag 1, 1992 p980; LJ v117 S 15, 1992 p88)

OTHER RESOURCES

Periodicals for General Readers

American Health: Fitness of Body and Mind. RD Publications, 10/yr. 0730–7004.
  A glossy magazine with a number of regular columns on broadly defined health care topics and articles on timely, substantive issues. Targeted for a very general audience.

FDA Consumer. Food and Drug Administration, 10/yr. 0362–1332.
  Updates, research notes, and articles designed to help consumers make more informed decisions.

Harvard Health Letter. Harvard Medical School, m. 0884–3783.
  Each issue of this newsletter gives a fairly substantial "General Review" of a topic, offers "Insights" in the form of editorials, covers miscellaneous issues "In Brief," and presents, in question form, a "Forum."

Hastings Center Report. Hastings Center, bi-m., revs. 0093–0334.
  Published by a nonprofit association that "carries out educational and research programs on ethical issues in medicine, the life sciences, and the professions." Most articles and features take the form of opinion pieces.

Health Letter. Health Research Group, m. 0882–598x.
  Published to "fight for the public health and to give consumers more control over decisions that affect their health." Co-founded by Ralph Nader.

In Health: The Magazine of Health and Medicine. Hippocrates, Inc., bi-m.
  A relative newcomer to the field of popular health information resources, this magazine has attracted the attention of a wide consumer audience. Features articles and numerous regular columns. Continues *Hippocrates.*

JAMA: The Journal of the American Medical Association. American Medical Association, w., revs. 0098–7484.

This is a very prestigious vehicle for research, news, professional activities, and opinions on medicine in America. Other professional and research journals that, despite their specialization, would be useful in science literacy collections include *New England Journal of Medicine*, the *Lancet*, and the *British Medical Journal*. These are the primary sources for medical news that will later make it into the popular information resources.

Mayo Clinic Health Letter. Mayo Foundation, m. 0741–6245.

Subtitled "Reliable Information for a Healthier Life," this newsletter gives authoritative information on many subjects.

Medical World News. Miller Freeman Publications, s-m. 0025–763x.

"The news magazine of medicine." Reports on news broadly presented under several general sections of each issue. Covers all areas of the field.

World Health. World Health Organization, bi-m. 0043–8502.

Strong on public health from countries around the world. Most articles tend to cover news and breaking developments.

Audiovisual Resources

AIDS

AIDS: A Family Experience. Carle Medical Communications, 1989 (33 minutes, color).

The painful drama of a young gay man with AIDS and how he and his family have coped with the disease. Very personal, and very emotional—explores the uniquely difficult feelings that AIDS brings to the fore. (SBF v25 N 1989 p100)

AIDS in Africa. Filmmakers Library, 1991 (52 minutes, color).

As severe as AIDS has been in America, its impact in Africa, where the disease originated, has been so devastating that some say it threatens to become genocidal. (SBF v27 Mr 1991 p58)

Physicians and AIDS: The Ethical Response. Carle Medical Communications, 1991 (20 minutes, color).

Two films: the first portrays the moral ordeal of a physician who is called upon to perform surgery on an HIV positive patient, and the second is an expert panel discussion on relevant ethical issues. (SBF v29 D 1993 p278)

Risk. Landmark Films, 1990 (22 minutes, color).

A realistic and disturbing video that shows several young people with AIDS. They talk about how they got the disease, its complications, and their life prospects. Sober and eye-opening. (SBF v27 My 1991 p123)

Also Recommended

AIDS Babies. Cinema Guild, 1990 (58 minutes, color). A provocative and emotional look at how infants contract the disease and their prospects for life. (SBF v27 N 1991 p249)

Diseases and Disorders

Addictions. PBS Video, 1988 (60 minutes, color).
Studies the psychological and physiological mechanisms of addictions to drugs and alcohol. Discussions include some of the latest theories on genetic predispositions to addiction and addictive behavior. (SBF v25 My 1990 p287)

Alzheimer's Disease. Professional Research, 1989 (four parts, 10–12 minutes each, color).
The four parts in this series are "An Introduction for Caregivers," "Home Care and Problem Solving," "Alternative Care," and "Advocacy and Support Groups." Although designed to answer questions of those directly affected by Alzheimer's, these videos, especially the first two, can also give good general information on the destructive effects of the disease. (SBF v25 N 1989 p100)

Heart Attack: Every Minute Counts. Milner-Fenwick, 1991 (10 minutes, color).
How many people would know how to react if somebody near them had a sudden heart attack? When this occurs, time is of the essence, and action must be taken immediately to minimize complications and save lives. This video presents two possible outcomes based on how individuals respond to an emergency. (SBF v29 Mr 1993 p56)

Pain and Healing. PBS Video, 1988 (60 minutes, color).
This video shows victims who suffer from chronic pain and asserts strongly that their disease is real. Discusses various cures with an emphasis on the power of mental healing. (SBF v25 My 1990 p289)

Stroke: Frontiers of Hope. Films for the Humanities, 1989 (26 minutes, color).
An overview on strokes—their causes and prevention, as well as the prospects for rehabilitation of victims. (SBF v25 N 1989 p105)

Transplant. Pyramid Film and Video, 1989 (30 minutes, color).
This powerful video is made personal through showing the dramas of both organ donor and recipient. Shows the medical and ethical aspects of organ donation. (SBF v25 N 1989 p105)

Also Recommended

Blindness. Landmark Films, 1993 (30 minutes, color). How blindness is evaluated and treated, and the means by which sufferers adapt to life. (SBF N 1993 p248)

Coronary Heart Disease. Pyramid Film and Video, 1989 (12–17 minutes, color). Three parts: "Biology," "Prevention," and "Clinical Aspects." (SBF v25 N 1989 p101)

Living (series). Living Series, 1990– . Each of the many installments in this practical and informative series gives sound medical advice on how to live with various diseases and disorders, such as arthritis, diabetes, heart diseases, and so on. All come highly recommended. (SBF v27 Ap 1991 p91; SBF v27 My 1991 p122)

Medical Aspects of Alcohol. FMS Productions, 1991 (60 minutes, color). Lectures on how alcohol acts in the body and interacts with other systems. (SBF v27 Ag 1991 p185)

PUBLIC HEALTH TOPICS

Borderline Medicine. Carle Medical Communications, 1990 (57 minutes, color).
   Access to health care is a front page issue in America today. At the heart of the issue is universal health care, which has been available in Canada for some time. Walter Cronkite narrates this video, which compares and contrasts the U.S. and Canadian systems. (SBF v27 N 1991 p250)

A Case of Need: Media Coverage and Organ Transplants. Fanlight Productions, 1990 (16 minutes, color).
   Two cases are profiled: one in which the plight of a patient in need of a donor organ is publicized, and a second in which no media attention is forthcoming. Viewers are asked to consider the appropriate role of the media in cases such as these. (SBF v27 Mr 1991 p58)

Diet and Cancer. Professional Research, 1989 (18 minutes, color).
   Discusses the evidence for linking diet with cancer and explains why high-fat, high-calorie diets are so damaging. Includes interviews with several nutritionists. (SBF v25 N 1989 p101)

Fit or Fat for the Nineties. PBS Video, 1992 (77 minutes, color).
   Based on Covert Bailey's successful book on the same topic, this video looks at the health and physiological benefits of exercise. Uses some interesting computer graphics to show metabolic functions. Entertaining and humorous throughout. (SBF v28 D 1992 p277)

Our Nation's Health: A Question of Choice. Carle Medical Communications, 1990 (59 minutes, color).
   A good introductory look at many of the complex issues facing each of us individually as lifestyle choices for health, and collectively as matters of public health and costs of health care. (SBF v27 N 1991 p250)

Safer Sex. Films for the Humanities, 1989 (19 minutes, color).
   Practicing safer sex is not an issue of morality; it is a personal and public health imperative. This is a no-nonsense presentation for general audiences. (SBF v25 Mr 1990 p221)

PSYCHOLOGY AND PSYCHOTHERAPY

Facing Death. Filmmakers Library, 1993 (56 minutes, color).

In America, death has been treated like a taboo; hence, many of us have little knowledge about the experience or how to care for others who face death. This sympathetic but not saccharine Swedish documentary records the death of an elderly woman as seen through her son's eyes. (SBF v30 My 1994 p120)

Healing and the Mind. Ambrose Video, 1993 (87 minutes, color). Considers holistic approaches to medicine and health care, and the importance of mental health in overall mind/body well-being. From a series by Bill Moyers.

John's Not Mad. Filmmakers Library, 1992 (30 minutes, color).

Tourette's syndrome, a neurological disorder that leaves victims prone to loud public outbursts, has been the subject of current scientific research and popular curiosity. This sympathetic video portrays a sixteen-year-old sufferer of the disease and how he deals with everyday life. (SBF v29 D 1992 p24)

Personality Disorders. Annenberg–CPB Project, 1991 (60 minutes, color). From *World of Abnormal Psychology* series.

Looks at four general categories of disorders: (1) obsessive-compulsive, (2) narcissistic, (3) antisocial, and (4) borderline. Considers causes, symptoms, and treatments. (SBF v28 My 1992 p116)

CD-ROMs

Family Doctor. Creative Multimedia Corp., 1991 (one disc).

General information on over 2,000 commonly asked medical questions and over 1,600 prescription drugs. Also contains an anatomical module that shows the human body and systems.

Health for All. United Nations, Pan American Health Organization–Norman Ross Publishing, 1991 (one disc).

Full text information on health care access, management, and social aspects within international communities. Also strong on environmental and occupational health.

Heart, the Engine of Life. Updata Publications, 1992 (one disc).

Program allows users to interactively explore the workings of the human heart, in health and disease.

Lifesaver 2.0. Media Design Interactive, 1991 (one disc).

A striking multimedia first aid program with vivid graphics and relevant narration. Some of the video was shot during actual health emergencies.

Medicine and Health Care. Applied Optical Media Corp., 1991 (one disc).

With over 1,500 multimedia images, this program covers topics including anatomy, patient care, surgery, and many others. General interest.

Vital Signs: The Good Health Resource. Software Mart, 1991.
General personal and family health, at a practical level.

Chapter Eleven

Natural History

We are of the soil and the soil is of us. We love the birds and beasts that grew with us on this soil. They drank the same water as we did and breathed the same air. We are all one in nature. Believing so, there was in our hearts a great peace and a willing kindness for all living, growing things.

Luther Standing Bear, 1933

The more we know of other life forms, the more we enjoy and respect ourselves. Humanity is not exalted because we are so far above other living creatures, but because knowing them well elevates the very concept of life.

Edward O. Wilson

Nature, to be commanded, must be obeyed.

Francis Bacon, Novum Organum, 1620

Heightened public environmental awareness has no doubt encouraged the abundant publication of related information resources in the last decade. Still, this new environmental literature is part of a hearty tradition of naturalistic writing in America. Coming from the Old World, the original settlers saw this continent as a magnificent and pristine wilderness, its resources virtually limitless. Their attitude toward the land was paradoxical, at once awed and humbled, but also rapacious and exploitative. People thrust westward, driven largely by materialistic goals, but as they experienced the grandeur of the American wild areas, they wrote about them, and these reports, some fantastically exaggerated, encouraged further migrations. Some American nature writings of the nineteenth century, such as those of Audubon, Emerson, Muir, and Thoreau, have become scripture to contemporary environmentalists. Today, these works can be appreciated for their almost visionary preservationist ethic. Articulating that vision has been the task and privilege of several generations; never, however, has the message been more urgent.

Included in this chapter, which overlaps significantly with Chapter Seven, "Biological Sciences," are topics that relate to the overall natural history of planet Earth and the global life forms it has mothered. Examining some of the titles, it is clear that the environmental credo "Think globally, act locally" is taken to heart by many information producers. At the planetary level, unifying theories, such as James Lovelock's "Gaia" hypothesis, expound a holistic biology that sees all life as inextricably interconnected, every ecosystem a vital part of the whole "superorganism." The potency of these theories is that they expand one's perspective, making it possible to put individual ecological disturbances in proper context. Thus, studies of local ecological disruption—the destruction of rainforests, for example—can examine not only the immediate consequences, but also those of long term global significance. Works about very specific environments that might seen quite remote can still be of interest even to people who will never visit them.

Evolution is another subject that has consistently fascinated the lay audience since the time of Darwin. Annually, perhaps a dozen new books about evolution and human origins are published. New and refined theories appear with regularity, but that is a sign of the field's rapid development, and not, as some contend, of its dubiousness. In fact, basic evolutionary theory is the foundation of the study of all living things. For that reason, it is encouraging that so large and eager a market exists for popularizations of the theory.

Information resources on natural history and the environment are probably the most prevalent of all popular science genres. They can be read by anybody, requiring little or no technical background. Further, they appeal to us for the same reasons that previous generations were so enthralled with the writings of explorers—nature is powerful, awesome, and sacred. Today, it is also endangered, and many of these resources tell how individuals, by taking positive steps in their own households and communities, can contribute toward the solution of large-scale problems.

NATURAL HISTORY—GENERAL TOPICS

Reference

The Dictionary of Ecology and Environmental Science. Henry Art, ed. Holt, 1993. 632 p. 0–8050–2079–9.

Defines over 8,000 terms, many of which are unlikely to be found in most—if any—basic science reference works. The scope of this much-needed resource is broad enough to include all major disciplines that have contributed to modern environmental knowledge. (SciTech v17 S 1992 p2)

Milner, Richard. *The Encyclopedia of Evolution*. Facts on File, 1990. 482 p., illus., index. 0–8160–1472–8.

The liner notes state that "nothing in biology makes sense except in the light of evolution" (attributed to evolutionary geneticist Theodosius Dobzhansky). That is precisely why this one-of-a-kind resource is so valuable. Covers major topics and individuals, and also "rescues many unknown events from oblivion." With a foreword by Stephen Jay Gould. (BL v87 Ja 1, 1991 p950; Choice v28 Je 1991 p1619; LJ v15 D 1990 p120)

1994 Conservation Directory. Rue E. Gordon, ed. National Wildlife Federation, 1994. 477 p., index. 0–945051–7.

The thirty-ninth edition of this reference lists over 2,000 U.S. and Canadian organizations involved with environmental causes of all kinds. Alphabetical name and subject indexes make this practical and easy to use. (SBF v30 My 1994 p101)

World Resources Institute. *World Resources*. Oxford University Press, 1993. 383 p., illus., maps, index. 0–19–506231–0.

From a United Nations development committee, this book adopts global perspectives on sustainability from resource management, cultural, political, and economic points of view. Complex issues are well presented. (ARBA v24 1993 p739; SBF v28 Je 1992 p139)

Also Recommended

Allaby, Michael. *Dictionary of the Environment*. New York University Press, 1989. A standard source for quick reference. (Choice v27 S 1989 p75; SBF v25 S 1989 p15)

Lincoln, Roger J. *The Cambridge Illustrated Dictionary of Natural History*. Cambridge University Press, 1987. A layperson's dictionary with short definitions and drawings. (ARBA 1988 p609; BL v84 Mr 1, 1988 p1115; Choice v25 F 1988 p887)

The McGraw-Hill Encyclopedia of Environmental Science and Engineering. Sybil Parker and Robert Corbitt, eds. McGraw-Hill, 1993. In its third edition, the words "and Engineering" were added to the title, thus reflecting the inclusion of information on environmental protection technologies. (Choice v31 O 1993 p268; WLB v68 S 1993 p119)

(*Note*: See also reference sources listed under specific categories in this chapter.)

Autobiography and Biography

Desmond, Adrian, and James Moore. *Darwin: The Life and Times of a Tormented Evolutionist*. Warner, 1992. 808 p., photos, index. 0–446–51589–2.

Many biographies and studies of Darwin have been published. Of them, this is the most comprehensive, and it is one of those rare books that measures up to scholarly scrutiny at the same time it makes for engrossing reading. The authors draw upon voluminous primary sources to explore Darwin's ideas and convictions and to recreate their impact on Victorian society. Few figures in the history of science are as complex and influential as Darwin; his life continues to fascinate today. (KR v60 Ap 15, 1992 p510; LJ v117 My 15, 1992 p98; Sci v257 Jl 17, 1992 p419)

House, Adrian. *The Great Safari: The Lives of George and Joy Adamson*. Morrow, 1993. 0–688–10141–0.

The Adamsons of *Born Free* fame possessed a commitment to conservation that bound them to their adopted homeland of Kenya as well as to each other. In private they were two strong-willed people whose personalities sometimes clashed, but also energized and encouraged each other. (BL v909 S 15, 1993 p111; KR v61 S 1, 1993 p1119; PW v240 S 13, 1993 p108)

Richardson, Robert D. *Henry Thoreau: A Life of the Mind*. Illustrated by Barry Moser. University of California Press, 1986. 455 p., index. 0520054854.

Some people associate Thoreau with his nature writings and regard him as primarily a naturalist. Others more familiar with his poetry and other creative writings recognize him as a major American literary figure. This book examines these and other dimensions of this complex, multitalented, and uniquely American figure. (BL v83 S 1, 1986 p20; PW v230 Jl 18, 1986 p72)

Streshinsky, Shirley. *Audubon: Life and Art in the American Wilderness*. Villar, 1993. 0679408592.

Novelist Shirley Streshinsky conveys empathy and passion in her biography of this creative and energetic American legend. Audubon was filled with an insatiable curiosity about wildlife, which kept him wandering the continent most of his life. Still, his fascination was focused and productive, as is evidenced by his famous *Birds of America*. (BL v90 S 15, 1993 p110; KR v61 Ag 15, 1993 p1062; PW v240 S 20, 1993 p54)

Turner, Frederick. *Rediscovering America: John Muir in His Time and Ours*. Viking, 1986. 417 p., illus., notes, index. 0–670–80774–5.

This biography not only gives a complete and sympathetic account of the great American naturalist, but in doing so repeatedly reflects upon his work and assesses its enduring impact on wilderness and conservation movements today. (Choice v23 F 1986 p885; SciTech v 9 D 1985 p14; Sierra v71 Ja 1986 p147)

Also Recommended

Axelrod, Alan, and Charles Phillips. *The Environmentalists: A Biographical Dictionary from the Seventeenth Century to the Present.* Facts on File, 1993. A unique reference source, which defines "environmentalist" quite liberally, especially in the political arena. (LJ v118 Jl 1993 p66; SBF v30 Mr 1994 p40)

Bonner, John Tyler. *Life Cycles: Reflections of an Evolutionary Biologist.* Princeton University Press, 1993. "I have devoted my life to slime molds," the author begins. Bonner expounds his theory that the foundation of biology is rooted in the study of the life cycle. (LJ v118 My 15, 1993 p93; Nature v365 S 30, 1993 p400; SBF v30 Ja 1994 p8)

Hynes, Patricia H. *The Recurring Silent Spring.* Pergamon, 1989. A biography of Rachel Carson from a feminist point of view; controversial and provocative. (Choice v27 D 1989 p651)

Lurie, Edward. *Louis Agassiz: A Life in Science.* University of Chicago Press, 1960. Chronicles the life of the great nineteenth century biologist. (BRD 1961 p883; NYTBR D 11, 1960 p6)

Histories

Bowler, Peter. *The Norton History of the Environmental Sciences.* Norton, 1993. 633 p., index. 0–393–03535–2.
  A difficult history to write and conceptualize, this succeeds because it incorporates ideas and developments from all sciences that deal with our physical environment and connects them through common philosophies. To date, this is the only comprehensive history of the environmental sciences available, and its accessibility to general readers is a great bonus. (BL v89 Jl 1993 p1930; LJ v18 Jl 1993 p112; NYTBR v98 O 17, 1993 p34)

Also Recommended

Worster, Donald. *The Wealth of Nature: Environmental History and the Ecological Imagination.* Oxford University Press, 1993. A study of attitudes and experiences related to the environment. (Aud v95 S 1993 p122; KR v61 Ap 1, 1993 p448; LJ v118 Je 15, 1993 p93)

General Topics

Commoner, Barry. *Making Peace with the Planet.* Pantheon, 1990. 292 p., index. 0–394–56598–3.
  Noted ecologist Commoner gives a thoughtful analysis of the environmental sciences, human nature, and social realities. He contends that a conflict has arisen between the "eco-sphere" and the "techno-sphere," but that this is a relatively new situation in human existence, and it can be resolved by simply

returning to the sustainable practices of the past. (LJ v116 Mr 1, 1991 p62; Nature v348 N 29, 1990 p402)

DiSilvestro, Roger. *Reclaiming the Last Wild Places: A New Agenda for Biodiversity.* Wiley, 1993. 266 p., biblio., index. 0–471–57244–6.

This is a book about boundaries. In the wild, ecosystem and species boundaries are natural. Unfortunately, too often conservation-minded humans impose unnatural boundaries on species in well-intentioned but unscientific efforts to preserve them. Instead, DiSilvestro calls for an "enlightened biodiversity" philosophy. (KR v61 My 15, 1993 p640; LJ v118 Je 15, 1993 p93; New Sci v139 S 11, 1993 p48)

Durrell, Gerald, with Lee Durrell. *The Amateur Naturalist.* Knopf, 1983. 320 p., illus., index. 0–394–53390–9.

Gerald Durrell says, "In writing this book, Lee and I have tried to produce the sort of work we ourselves, as young aspiring naturalists, would have liked to possess." Accordingly, they discuss nature study techniques for a variety of species and ecosystems. As the Durrells would have wished, this is a good book for anybody with a yen to enter the field or learn more about it. (LJ v108 O 1, 1983 p1884; TES D 17, 1982 p28)

Ehrlich, Paul. *The Population Bomb.* Ballantine, 1968. 223 p., biblio.; Paul Ehrlich and Anne Ehrlich. *The Population Explosion.* Simon and Schuster, 1990. 320 p., index. 0–671–68984–3.

The Ehrlichs' central thesis described in these two books is that today's most serious ecological problems result directly from the pressures placed on the planet by the sheer size of the world's population. Thus, solving environmental problems requires, first, dealing with the human ones. (LJ v116 Mr 1, 1991 p62; Sierra v76 Jl 1991 p52; SLJ v36 N 1990 p154)

Gallagher, Winifred. *The Power of Place: How Our Surroundings Shape Our Thoughts, Emotions, and Actions.* Poseidon Press, 1993. 240 p., index. 0–671–72410–x.

According to a growing body of research, the physical environment in which we live has a profound effect on how we act and think. Gallagher draws on diverse literature in arguing that by cultivating a deeper sense of place, we can develop a greater sensitivity to the environment. (BL v89 F 15, 1993 p1014; Choice v31 O 1993 p368; Sierra v78 S 1, 1993 p89)

Gibbons, Whit. *Keeping All the Pieces: Perspectives on Natural History and the Environment.* Smithsonian, 1993. 182 p., index. 1–56098–224–1.

At the beginning of this book, the author envisions a worst case scenario for environmental despoilment. From there, he identifies his goals for the book as being to enhance awareness of the natural world, to foster a basic understanding of the interconnectedness of life, and to stimulate responsible human action to preserve the environment. (SBF v30 Ap 1994 p71; New Sci v140 D 11, 1993 p45)

Gould, Stephen Jay. *Ever Since Darwin.* Norton, 1977; *The Panda's Thumb.* Norton, 1980; *Hen's Teeth and Horses' Toes.* Norton, 1983; *The Flamingo's Smile.* Norton, 1985; *Bully for Brontosaurus.* Norton, 1991; *Eight Little Piggies.* Norton, 1993.

Gould is perhaps the leading scientific essayist of today. As a series, these books build upon several currents of thought and reflect issues that have changed over time. At the same time, each stands on its own as a whole statement. (BL v87 Mr 15, 1991 p1433; KR v59 Ap 1, 1991 p452; LJ v117 N 15, 1992 p96)

Lovelock, J. E. *Gaia: A New Look at Life on Earth.* Oxford University Press, 1979. 157 p., illus. 0–19–217665–x.

Lovelock drew upon Greek mythology in naming "Gaia," which is his conception of life on Earth existing within a planetary biosphere that functions as a single cybernetic system. This is the author's personal theory presented to a general audience. While many scientists have criticized it as unscientific, it has been extraordinarily influential in providing a model for contemporary environmentalism. (Choice v17 Ap 1980 p237; SA v242 Mr 1980 p44)

Also Recommended

Eiseley, Loren. *The Immense Journey.* Vintage, 1957. Some of the very finest nature essays ever written. (BL v54 S 15, 1957 p40; BRD 1957 p284; KR v25 Je 15, 1957 p433)

Merideth, Robert. *The Environmentalist's Bookshelf.* G. K. Hall, 1993. Annotated bibliographies of essential environmental books, as selected by scientists and activists.

Nash, Roderick. *Wilderness and the American Mind.* Yale University Press, 1982. The American view of wilderness as both a source of wonderment and a challenge to conquer. (Sierra v68 Jl 1983 p73)

Thomson, Keith Stewart. *The Common but Less Frequent Loon and Other Essays.* Yale University Press, 1993. Essays by the popular biology writer and frequent contributor to *American Scientist.* (KR v61 O 1, 1993 p1259; LJ v118 O 15, 1993 p85)

NATURAL HISTORY—SPECIAL TOPICS

Archaeology

McMillon, Bill. *The Archaeology Handbook.* Wiley, 1991. 259 p. 0–471–53051–4.

Much more than just a field guide. Even the merely curious who have no intention of hitching onto an archaeological dig will find this an interesting source for introductory information on the methods and principles of archaeology. Nearly half of the book is devoted to listing resources, so it could serve for ready reference, too. This book would be a good one to give a student thinking about archaeology as a career or major. (LJ v116 S 15, 1991 p96)

Stiebing, William. *Uncovering the Past: A History of Archaeology.* Prometheus, 1993. 315 p., biblio., index. 0–87975–764–7.
In the first part of this book, the author describes the field exploits of excavators and archaeologists from the discipline's Heroic Age (1450–1925). The second part focuses on modern archaeology, which uses methods from the physical sciences and mathematics for much more rigorous research. (BL v89 Mr 1, 1993 p1154; LJ v118 Mr 15, 1993 p92)

Also Recommended

Lamberg-Karlovsky, C. C. *Archaeological Thought in America.* Cambridge University Press, 1989. Eighteen scholars from the field contribute essays on methods, theories, and regional case studies. (Choice v27 Ap 1990 p1355)

Biogeography and Population Genetics

Mielke, H. W. *Patterns of Life: Biogeography of a Changing World.* Unwin Hyman, 1989. 370 p., illus., index. 0–04–574033–x.
Biogeography is the scientific study of the Earth's global pattern of plants and animals. This text borrows from various other fields in the life and physical sciences in defining a relatively young discipline. (GJ v156 Jl 1990 p220)

Also Recommended

Browne, Janet. *The Secular Ark: Studies in the History of Biogeography.* Yale University Press, 1983. Biogeographers treat the Earth as a "chessboard." This semi-technical book relates the history of the field. (Choice v21 S 1983 p124; LJ v108 Ag 1983 p1491)

Maynard-Smith, John. *Evolutionary Genetics.* Oxford University Press, 1989. A primer of population genetics and population biology.

Ecology

Botkin, Daniel B. *Discordant Harmonies: A New Ecology for the Twenty-First Century.* Oxford University Press, 1990. 241 p., illus., index. 0–201–10639–6.
Entrenched ideas about nature and the role of humans as natural agents have impeded our development of sound ecological practices. In a somewhat technical presentation, the author draws upon analogies between the complexities of nature and those of computer systems. For more serious readers. (BL v86 Ap 15, 1990 p1589; NYTBR v95 Ap 29, 1990 p43; SBF v27 Ja 1991 p4)

Goldsmith, Edward. *Imperiled Planet: Restoring Our Endangered Ecosystems.* MIT Press, 1990. 288 p., index. 0–262–07132–0.

This broad ecological survey shows that, to varying degrees, every ecosystem on Earth has been affected by human activities. Emphasizes the need to develop renewable and sustainable alternatives to destructive practices. (SBF v27 Ap 1991 p69)

Odum, Eugene P. *Ecology and Our Endangered Life Support Systems.* 2nd ed. Sinauer, 1993. 301 p., biblio., index. 0–87893–634–3.

This book is a hybrid of a textbook and a citizen's guide. It begins with a discussion of Apollo 13, which produced the widely distributed photographs of "spaceship Earth" that have illustrated the concept of planet ecology in a powerful way. All in all, this is a fine introduction to the science of ecology for students and generalists. (SBF v30 Ag 1993 p169; SciTech v17 Je 1993 p17)

Pimm, Stuart L. *The Balance of Nature? Ecological Issues in the Conservation of Species and Communities.* University of Chicago Press, 1992. 426 p., index. 0–226–66829–0.

A thorough, balanced overview of what we know about species conservation. As the author notes, actual data are scarce, while theories seem to abound. The two must be joined in order to achieve any real understanding. (BioSci v42 S 1992 p628; Nature v357 My 14, 1992 p124; SBF v28 My 1992 p103)

Tudge, Colin. *Global Ecology.* Oxford University Press, 1991. 173 p., gloss., index. 0–19–520904–4.

A fine introduction for any student or layperson. This book was written for an ecology exhibit at the London Museum of Natural History. (Choice v29 My 1992 p1420; SBF v28 Mr 1992 p42; SLJ v38 Je 1992 p152)

Also Recommended

Hagan, Joel B. *An Entangled Bank: The Origins of Ecosystem Ecology.* Rutgers University Press, 1992. A history of ecology in the United States, from Darwin's time to modern "Big Ecology." (Choice v30 My 1993 p1489; SBF v29 Mr 1993 p40; Sci v261 Jl 23, 1993 p497)

Lanier-Graham, Susan. *The Ecology of War: Environmental Impacts of Weaponry and Warfare.* Walker, 1993. The toll of war can be measured not only in human loss, but also in environmental degradation. (BL v89 My 1, 1993 p1553; KR v61 My 1, 1993 p573)

Environmental Topics—General

Brown, Lester, et al. *State of the World.* Worldwatch Institute (annual).

Each year the respected Worldwatch Institute issues its report on progress toward a sustainable society. This provides a good summary of broad environmental topics and initiatives throughout the world. It should be purchased annually by all libraries.

Carson, Rachel. *Silent Spring*. Houghton Mifflin, 1962. 368 p., illus.

Publication of Carson's seminal book is often cited as a defining event in environmental science history. Her harsh criticism of the indiscriminate use of pesticides raised the ecological consciousness of an entire generation. Will remain worth reading in the twenty-first century. (BRD 1962 p202; LJ v87 S 15, 1962 p3059; NYTBR S 23, 1962 p1)

Ehrenfeld, David. *Beginning Again: People and Nature in the New Millennium*. Oxford University Press, 1993. 205 p., index. 0–19–507812–8.

Hindsight and reflection on past mistakes are important factors to consider in formulating any future vision of the environment. The author outlines some of the foibles of the past and suggests what we can learn from them as we look forward to the next millennium. (Choice v30 Jl 1993 p1789; KR v60 N 1, 1992 p1346; SBF v29 Ag 1993 p165)

Ehrlich, Paul, and Anne Ehrlich. *Healing the Planet: Strategies for Resolving the Environmental Crisis*. Addison-Wesley, 1991. 366 p., index. 0–201–55046–6.

The Ehrlichs take a population biology approach to solving environmental problems, arguing that the Earth has reached the carrying capacity of its life support systems. The key is to reduce needs and demands upon the planet's resources. Any viable solution must cut across political and geographical boundaries. (KR v59 Ag 15, 1991 p1058; LJ v116 O 1, 1991 p132)

Global Tomorrow Coalition. *The Global Ecology Handbook: What You Can Do about the Environment*. Beacon Press, 1990. 414 p., biblio., illus., index. 0–8070–8500–6.

Organized into chapters on over a dozen separate categories of environmental concerns, this book thoroughly documents the nature, history, and extent of these threats and ends with a "What You Can Do" section. Gives copious references and resources. (BL v86 My 15, 1990 p1760; Choice v27 Jl 1990 p1805; LJ v115 Ap 15, 1990 p118)

Meadows, Donella, Dennis Meadows, and Jorgen Randers. *Beyond the Limits: Confronting Global Collapse, Envisioning a Sustainable Future*. Chelsea Green, 1992. 300 p. 090031555.

Using high-tech computer modeling, the authors project the state of the environment in the next century if today's trends are not halted or reversed. Contains excellent graphs and charts. The title refers to a classic 1972 work, *The Limits to Growth* by William Behrens III. (BL v88 My 15, 1992 p1646; LJ v117 Ap 15, 1992 p118; SBF v28 D 1992 p262)

Also Recommended

Dixon, John A. *Economics of Protected Areas: A New Look at the Benefits and Costs.* Island Press, 1990. The attempt to apply market economics to environmental issues is philosophically distasteful to some, but a cost/benefit argument is the most likely to succeed in certain political arenas. Scholarly and dispassionate, but accessible. (Choice v28 Jl 1991 p1799; SBF v27 Mr 1991 p36; TLS S 13, 1991 p6)

Gore, Al. *Earth in the Balance.* Houghton Mifflin, 1992. Not the partisan political statement its detractors have made it out to be, this book is thoughtful and, while it sets a political agenda, it is based upon science. (BL v88 F 1, 1992 p993; Bus W Ag 17, 1992 p16; New Sci v135 Ag 1, 1992 p38)

Environmental Topics—Atmospheric Pollution

Nance, John J. *What Goes Up: The Global Assault on Our Atmosphere.* Morrow, 1991. 324 p., notes, index. 0–688–08952–6.

In his preface, Nance tells readers that he wrote this book out of a sense of need. Human beings must know about the millions of tons of waste gases that are being passed into the atmosphere and the potentially deadly effects that they could have. This book sounds an alarm and personalizes the crisis. (BL v87 Je 1, 1991 p1847; Bus W Jl 22, 1991 p10; NYTBR S 8, 1991 p30)

Oppenheimer, Michael, and Robert H. Boyle. *Dead Heat: The Race Against the Greenhouse Effect.* Basic, 1990. 268 p., index. 0–465–09804–5.

This book begins with ''The End,'' which envisions the Earth in 2050 if greenhouse warming is not corrected, and it ends with ''The Beginning,'' a more cheerful scenario that can come to be if measures are taken now. In between, the authors, a scientist and a journalist, give their prescriptions for making the latter happen. (Choice v28 O 1990 p330; Sierra v75 S 1990 p93; SBF v27 Ja 1991 p4)

Roan, Sharon L. *Ozone Crisis: The Fifteen Year Evolution of a Sudden Global Emergency.* Wiley, 1989. 270 p., biblio., index. 0–471–52823–4.

This book chronicles the ''discovery'' of ozone depletion, the fight to ban chlorofluorocarbons, and prospects for the future. The central theme is that fifteen years have already passed since ozonic depletion was identified, and nothing has happened; we can afford to wait no longer. (BL v85 Jl 1989 p1855; LJ v114 Jl 1989 p103)

Environmental Topics—Waste Disposal

Alexander, Judd H. *In Defense of Garbage.* Praeger, 1993. 239 p., biblio., index. 0–275–93627–9.

Styled as ''the other side of the story,'' this is an industrialist's thoughtful response to the waste management crisis. The author contends that industry is not at fault, but rather problems arise because of politics and the propagation of misinformation. (BL v89 Mr 15, 1993 p1279; Choice v31 O 1993 p310; LJ v118 Ap 1, 1993 p124)

Moyers, Bill. *Global Dumping Ground: The International Traffic in Hazardous Waste.* Seven Locks Press, 1990. 152 p., index. 0–932020–95–x.
    Every day it seems the news brings some new report on an ecological disaster that occurred because of inadequate disposal of dangerous wastes. Shortsighted cost considerations and shoddy research are generally to blame. Sound, scientific waste management is sorely needed—the United States today is the world's leading exporter of toxic wastes, and even greater problems are likely to arise when these materials pass national borders. (LJ v115 D 1990 p156; NYTBR v95 N 25, 1990 p14; SBF v27 Ap 1991 p72)

Rathje, William, and Cullen Murphy. *Rubbish! The Archaeology of Garbage.* HarperCollins, 1992. 250 p., index. 0–06–016603–7.
    You can tell a lot about people from their garbage. In the University of Arizona's Garbage Project, researchers sorted through, classified, and analyzed tons of garbage. What they discovered reveals much about human nature, recycling practices and realities, and solid waste disposal in general. (BL v88 Jl 1992 p1905; LJ v117 Jl 1992 p117; PW v239 Je 1, 1992 p45)

Also Recommended

Schweitzer, Glenn E. *Borrowed Earth, Borrowed Time: Healing America's Chemical Wounds.* Plenum, 1991. Environmental protection measures that can be adopted to clean up and control chemical wastes in the United States. (NYTBR v96 Ap 21, 1991 p18; PW v238 Mr 22, 1991 p68; SBF v27 O 1991 p205)

Wallace, Deborah. *In the Mouth of Dragons: Toxic Fires in the Age of Plastics.* Avery, 1990. Plastics, which are highly flammable, can also be very dangerous in landfills. (LJ v115 S 1, 1990 p252; SBF v27 Ap 1991 p73)

Environmentalism and Environmentalists

Grove, Noel. *Preserving Eden: The Nature Conservancy.* Abrams, 1992. 176 p., index. 0–8109–3663–1.
    For nearly forty years, the Nature Conservancy has actively worked to preserve endangered species and habitats. Part of this organization's strategy has been to buy large tracts of land, which are then left wild in perpetuity. This oversize book describes and shows photographs of some of the conservancy's successful projects. (BL v88 Je 15, 1992 p1790; SBF v28 O 1992 p197)

Schwab, Jim. *Deeper Shades of Green: The Rise of Blue-Collar and Minority Environmentalism in America.* Sierra Club, 1994. 480 p., biblio. 0–87156–462–9.

Many politicians facilely decry conservationist measures as representing the interests of "environmental extremists." Increasingly, however, environmental awareness penetrates to all levels of society, so that, far from being limited to activists and crusaders, it has become a daily, working mentality among ordinary people who care about their communities. This book chronicles that shift of attitudes.

Zakin, Susan. *Coyotes and Town Dogs: Earth First! and the Environmental Movement.* Viking, 1993. 483 p., index. 0–670–83618–4.
    Earth First! is considered extreme by many in the mainstream environmental movement. Members, however, regard their no-compromise philosophy as a matter of absolute necessity. This history of the movement considers its current actions and doctrines within a context of American environmental philosophy, which is in fact deeply rooted in our culture. (BL v90 S 1, 1993 p16; KR v61 Je 15, 1993 p777; LJ v118 Jl 1993 p113)

Also Recommended

Lewis, Martin. *Green Delusions: An Environmentalist's Critique of Radical Environmentalism.* Duke University Press, 1992. The author contends that not only are extreme environmentalists on the political fringe, but their ideologies are not scientific and, on the whole, do more harm than good. (PW v239 S 7, 1992 p89)

The Wilderness Condition: Essays on Environment and Civilization. Max Oelschlaeger, ed. Sierra Club, 1992. Intelligent essays on the environmental movement, its philosophies and thought. (Choice v30 N 1992 p489; LJ v118 Mr 1, 1993 p43)

Evolution—General

Berra, Tim M. *Evolution and the Myth of Creationism: A Basic Guide to the Facts in the Evolution Debate.* Stanford University Press, 1990. 198 p., gloss., index. 0–8047–1548–3.
    This book first explains evolution for general readers, then distinguishes between scientific evolutionary theory and creationism. Unlike other books, it does not focus on the narrow debate, but invokes a context that includes all of the biological sciences. In addition to being provocative reading on its own, this could also supplement other readings and classroom activities. (SBF v27 Mr 1991 p37)

Darwin, Charles. *The Origin of Species by Means of Natural Selection* (several editions exist).
    One of the greatest works in the history of science. Its prose might be somewhat labored, but Darwin wrote it so that its basic ideas would be accessible to all. Many people today still read it.

Gould, Stephen Jay. *Wonderful Life: The Burgess Shale and the Nature of History*. Norton, 1989. 347 p., illus., biblio., index. 0–393–02705–8.

The Burgess Shale formation in British Columbia is one of the richest fossil localities in the world. The story of its discovery, study, and interpretation spans over eighty years, during which findings have contributed significantly to knowledge in evolutionary biology. Gould looks at the formation and the odd life forms whose fossils it preserves. (BL v86 O 1, 1989 p247; LJ v114 S 1, 1989 p214; NYTBR v94 O 22, 1989 p1)

Ward, Peter Douglas. *On Methuselah's Trail: Living Fossils and the Great Extinctions*. Freeman, 1991. 212 p., index. 0–7167–2203–8.

"Living fossils" (the term was coined by Darwin) are creatures that have existed virtually unchanged since the age of the dinosaurs. The study of these extremely ancient species can shed considerable light on evolution and the cycles of extinction. (LJ v116 S 15, 1991 p108)

Whitfield, Philip. *From So Simple a Beginning: The Book of Evolution*. Macmillan, 1993. Gloss., index. 0–02–627115–x.

An overview of the basic doctrines and fundamental assumptions of evolutionary theory. Well organized and clearly presented, this is a good book for beginners. (SBF v30 My 1994 p108)

Also Recommended

Environmental Evolution: Effects of the Origin and Evolution of Life on Planet Earth. Lynn Margulis and Lorraine Olendzenski, eds. MIT Press, 1992. This text provides an accessible introduction to changes in the biosphere since life's origin. (SBF v28 Ag 1992 p166; SciTech v16 My 1992 p14)

Mayr, Ernst. *One Long Argument: Charles Darwin and the Genesis of Modern Evolutionary Thought*. Harvard University Press, 1991. A distillation of Darwinian thought and its legacy. (SBF v27 N 1991 p232)

Miller, Jonathan. *Darwin for Beginners*. Pantheon, 1985. A simple introduction to Darwinian theory. (Brit Bk N Ap 1985 p199; New R v192 F 25, 1985 p29)

Whitfield, Philip. *The Natural History of Evolution*. Doubleday, 1993. A coffee table book on the riches and oddities of evolution. (New Sci v140 N 20, 1993 p43)

Williams, George C. *Natural Selection: Domains, Levels, and Challenges*. Oxford University Press, 1992. Covers ideas in natural selection at broad, hierarchical levels. The author strives to present theoretical unities in natural selection. This is a semitechnical treatment for readers who have a basic background in the subject. (BioSci v43 O 1993 p641; Choice v30 Je 1993 p1652; SBF v29 O 1993 p198)

Evolution—Human

Gribbin, John, and Mary Gribbin. *Children of the Ice: Climate and Human Origins*. Blackwell, 1990. 199 p., index. 0–631–16817–6.

According to the theory promulgated in this book, when glaciers encroached upon the lands occupied by early hominids, the associated climate and habitat changes engendered a succession of evolutionary adaptations that led to the rise of Homo Sapiens. (LJ v115 Mr 1, 1990 p110; SBF v27 Mr 1991 p39)

Leakey, Richard, and Roger Lewin. *Origins Reconsidered: In Search of What Makes Us Human*. Doubleday, 1992. 375 p., illus., index. 0–385–41264–9.

An update of Leakey's 1977 book *Origins*, this account begins with the 1984 discovery of "Turkana Boy," a nearly intact Homo Erectus skeleton. Leakey's chief interest, however, is not physical evolution, but rather the development of those traits that we associate with being human—our intellects and emotions. (BL v89 S 1, 1992 p3; KR v60 Ag 1, 1992 p967; PW v239 Ag 31, 1992 p58)

Margulis, Lynn, and Dorion Sagan. *Mystery Dance: On the Evolution of Human Sexuality*. Summit Books, 1991. 215 p., index. 0–671–63341–4.

The "mystery dance" is depicted as an "evolutionary striptease," in which our basic sexuality is denuded to reveal its primate, reptilian, and, ultimately, even protist origins. Through the dance, the authors suggest possible anatomical and behavioral significance of male and female sexualities. (BL v87 Jl 1, 1991 p2014; LJ v116 Jl 1991 p128; NYTBR v96 Ag 25, 1991 p11)

Prochiantz, Alan. *How the Brain Evolved*. McGraw-Hill, 1992. 118 p. 0–07–050929–8.

A brief but highly informative treatment of human intelligence and the physical evolution of the brain. Covers topics directly related to higher brain capabilities, like language development, but also those that are indirectly linked, such as social behaviors and cultural institutions. The author also speculates on the role of nature versus nurture and whether a human being can ever be cloned. (PW v239 F 17, 1992 p61; SBF v28 Ag 1992 p167)

Ridley, Matt. *The Red Queen: Sex and the Evolution of Human Nature*. Macmillan, 1994. 405 p., biblio., index. 0–02–603340–2.

In "Alice in Wonderland," the Red Queen runs faster and faster just to stay in place. So-called Red Queen evolutionary theory suggests that human evolution is a bit like running in place—especially where sex is involved. Ridley describes this theory which holds that the purpose of sexual reproduction is simply to maintain a healthy genetic mix, which in turn gives us and our offspring better abilities to adapt to our changing environments. (New Sci v142 My 28, 1994 p39; NYTBR v99 My 22, 1994 p15)

Russell, Robert Jay. *The Lemur's Legacy: The Evolution of Power, Sex, and Love*. Tarcher, 1993. 274 p., notes, index. 0–87477–714–3.

Most studies of human evolution start with our kinship to higher apes; this controversial book looks at commonalities with another type of primate, the lemur. Russell takes a bold approach and suggests possible behavioral linkages between the species in areas such as courtship, socialization, and power seeking. (BL v89 Ag 15, 1993 p1478; KR v61 Ap 1, 1993 p439; Nature v364 Ag 12, 1993 p585)

Sagan, Carl, and Ann Druyan. *Shadows of Forgotten Ancestors.* Random House, 1992. 0–394–53481–6.

This book's currency, ease of reading, comprehensive approach, and, not least of all, the name appeal of its authors make it a first choice of many library patrons. (Choice v30 Je 1993 p1647; KR v60 Ag 15, 1992 p1048)

Tattersall, Ian. *The Human Odyssey: Four Million Years of Human Evolution.* Prentice-Hall, 1993. 191 p., gloss., index. 0–671–85005–9.

Based on the Hall of Human Biology and Evolution exhibit at the American Museum of Natural History, this book tells how scientists have traced the path of human evolution. A good first book on the subject. Well illustrated. (Choice v31 S 1993 p174; LJ v118 Jl 1993 p88)

Trinkaus, Erik, and Pat Shipman. *The Neanderthals: Changing the Image of Mankind.* Knopf, 1993. 454 p., illus., index. 0–394–58900–9.

The popular image of Neanderthals is of plodding, stocky, and dumb creatures, barely different than apes. Today, however, scientists see them as representing a species with qualities quite unlike those portrayed by popular images. This is the definitive examination of what we know about Neanderthals, and what we think we know. (BL v89 Ja 1, 1993 p779; LJ v117 D 1992 p148; New Sci v139 Jl 3, 1993 p38)

Also Recommended

Brown, Michael. *The Search for Eve.* Harper, 1990. In 1987 a team of American geneticists propounded the "Eve" theory, which posits that all humans alive today descended from a single hominid female. (BL v86 Ap 15, 1990 p1593; Choice v28 S 1990 p167; LJ v115 F 1, 1990 p90)

Diamond, Jared. *The Third Chimpanzee.* HarperCollins, 1992. Some jeer at this book, while others cheer it. Diamond's informed but controversial opinions on human evolution leave no middle ground. (BL v88 Mr 15, 1992 p1323; LJ v117 Mr 15, 1992 p120; Nature v357 My 28, 1992 p290)

Donald, Merlin. *Origins of the Modern Mind.* Harvard University Press, 1991. A somewhat technical treatment, for those who would like something more challenging on the subject. (KR v59 Ag 15, 1991 p1057)

Howells, William. *Getting Here: The Story of Human Evolution.* Compass Press, 1993. An anthropological review of the stages of human evolution. Good introduction. (LJ v118 Ja 1993 p126; Nature v365 S 2, 1993 p25)

Sagan, Carl. *The Dragons of Eden: Speculations on the Evolution of Human Intelligence.* Random House, 1977. Although dated, this 1978 Pulitzer Prize winning book is still a favorite. (BRD 1977 p1164; Choice v14 D 1977 p1386; LJ v102 My 15, 1977 p1198)

Extinctions

DeBlieu, Jan. *Meant to Be Wild: The Struggle to Save Endangered Species Through Captive Breeding.* Fulcrum, 1991. 302 p., photos, index. 1–55591–074–2.

Breeding animals in captivity and later releasing them into their natural environments is a last resort for saving some species facing extinction. These efforts are fraught with difficulty, however, and although they may be a last hope, little is known about how effective they will be in the long run. (KR v59 Ag 15, 1991 p1056; LJ v116 O 1, 1991 p132; PW v238 Ag 16, 1991 p43)

Eldredge, Niles. *The Miner's Canary: A Paleontologist Unravels the Mysteries of Extinction.* Prentice-Hall, 1991. 256 p. 0–13–583659–x.

The title derives from the practice whereby miners take canaries underground with them; if the bird, which is more sensitive than humans to certain odorless, poisonous gases, dies or becomes inactive, it serves as an early warning to get out quickly. Eldredge sees in this an analogy for the contemporary phenomenon of species extinction—the warning must be heeded; we must take steps to rehabilitate the environment, or our very existence could be threatened. (BL v88 S 15, 1991 p104; KR v59 S 1, 1991 p1131; PW v238 Ag 23, 1991 p50)

La Bastille, Anne. *Mama Poc: An Ecologist's Account of the Extinction of a Species.* Norton, 1990. 0–393–02830–5.

A rare species of giant grebe once flourished along the coasts of Lake Atitlan of Guatemala. That species is now extinct, fallen to a host of natural and human-made pressures. This book is an ecologist's report on the dynamics of extinction, a case study of desperate preservation efforts, and, ultimately, a moving obituary for the species. (BL v86 Je 1, 1990 p1860; KR v58 My 1, 1990 p630; LJ v115 Je 15, 1990 p130)

Walters, Mark Jerome. *A Shadow and a Song: The Struggle to Save an Endangered Species.* Chelsea Green, 1992. 238 p., index. 0–930031–58–x.

The dusky seaside sparrow, which once thrived in its habitat near Cape Canaveral, Florida, fell victim to development that began with the establishment of the Kennedy Space Center. Well-intentioned but misguided efforts by the U.S. Fish and Wildlife Service and even Disney World failed to stave off its extinction. A literate and compassionate account. (KR v60 Ag 15, 1992 p1051; PW v23 S 14, 1992 p92)

Also Recommended

Raup, David. *Extinction: Bad Genes or Bad Luck?* Norton, 1991. This book begins by noting that over 90 percent of all species that have ever lived are now extinct, then looks at some of the reasons. (BL v87 Ag 1991 p2085; KR v59 Jl 1, 1991 p849; LJ v116 Ag 1991 p139)

Ward, Peter. *The End of Evolution: On Mass Extinction and the Preservation of Biodiversity.* Bantam, 1994. A personal narrative on the phenomenon of extinctions, past and present. (KR v62 Ap 15, 1994 p545; LJ v119 My 15, 1994 p95)

Geology—General

Allegre, Claude. *From Stone to Star: A View of Modern Geology.* Harvard University Press, 1992. 287 p., index. 0–674–83866–1.

In the latter part of this century, increasing knowledge of the chemistry and geology of the solar system has furthered knowledge of our own planet. Allegre, who was a consulting scientist on several interplanetary space missions, incorporates that knowledge into this popular presentation of developing theories in earth sciences. (Nature v358 Ag 27, 1992 p719; SBF v28 Ag 1992 p166)

Dixon, Dougal. *The Practical Geologist.* Fireside Books, 1992. 160 p., photos, gloss., index. 0–671–74897–9.

A high school to lower division undergraduate text that would also appeal to general readers. There are over 200 color photographs. (SBF v28 D 1992 p269)

Gould, Stephen Jay. *Time's Arrow, Time's Cycle.* Harvard University Press, 1987. 222 p. 0674891988.

''Deep time'' is required for the geological processes as we understand them to have created the world we live in today. Historically, the concept ran counter to religious beliefs, and thus it was difficult for many people, even many scientists, to accept. Gould looks at some of the basic texts that were persuasive in leading to the adoption of this concept in the earth sciences. (BL v83 Ap 1, 1987 p1160; LJ v112 Mr 15, 1987 p84; Nature v326 Ap 30, 1987 p898)

Lambert, David, and the Diagram Group. *The Field Guide to Geology.* Facts on File, 1989. 256 p., index. 0–8160–2032–9.

Twelve chapters cover general topics such as the Earth's origin, plate tectonics, volcanology, weathering, and so on. Most of all, however, this book is meant to serve as a field guide for classroom activities at the high school to beginning college level. Maps, charts, and photographs serve well in this regard. (SBF v25 N 1989 p71)

McPhee, John. *Basin and Range.* Farrar Straus Giroux, 1981. 215 p. 0374109141; *In Suspect Terrain.* Farrar Straus Giroux, 1983. 209 p.; *Rising from the Plains.* Farrar Straus Giroux, 1986. 213 p.; *Assembling California.* Farrar Straus Giroux, 1993. 303 p. 0374106452.

McPhee's geological histories of the American West have garnered an avid following and, while perhaps best read in sequence, can be appreciated individually as well. (BL v89 D 1, 1992 p633; KR v60 D 1, 1992 p1485)

Officer, Charles, and Jake Page. *Tales of the Earth: Paroxysms and Perturbations of the Blue Planet.* Oxford University Press, 1993. 224 p., refs., index. 0–19–507785–7.

Castastrophes of every sort are part of the Earth's natural history. Floods, earthquakes, and volcanic eruptions are inevitable dangers with which we must cope on this planet. Other disasters, however, are created or at least exacerbated by human beings. This book by an earth scientist and a journalist uses many first person accounts when describing these events. (BL v89 My 1, 1993 p1556; KR v61 Ap 1, 1993 p437; LJ v118 My 1, 1993 p110)

Redfern, Ron. *The Making of a Continent.* Times Books, 1983. 242 p. 0–8129–1079–6.

A tie-in with a PBS series, this book looks at the forces that shaped North America. In addition to describing how the physical environment came to be, the author also ventures his opinions on how local and regional geologies have affected human culture and social institutions. (LJ v111 N 1, 1986 p104; PW v230 Ag 8, 1986 p66)

Also Recommended

Dietrich, R. V., and E. Reed Wicander. *Minerals, Rocks and Fossils.* Wiley, 1983. Everything you need to know to get started collecting. (BL v79 Je 15, 1983 p1309)

Erickson, Jon. *Craters, Caverns and Canyons: Delving Beneath the Earth's Surface.* Facts on File, 1993; *Rock Formation and Unusual Geologic Structures.* Facts on File, 1993. Two unique books from a prolific popularizer of earth sciences. (BL v89 My 1, 1993 p1556; SBF v29 N 1993 p235)

Hambrey, Michael, and Jurg Alean. *Glaciers.* Cambridge University Press, 1992. With vivid photographs of glacial landscapes from across the world, this book delivers a good overview of glaciology and the power of ice. (Nature v358 Jl 9, 1992 p1115)

Parker, Ronald B. *The Tenth Muse: The Pursuit of Earth Science.* Scribner's, 1986. Miscellaneous observations expressed in lively, literate essays. (BL v82 Jl 1986 p1569; LJ v111 Jl 1986 p96)

Schumann, Walter. *Minerals of the World.* Sterling, 1992. A clear, well-illustrated guide to mineralogy. (SBF v29 Mr 1993 p45)

Skinner, Brian J., and Stephen C. Porter. *The Dynamic Earth: An Introduction to Physical Geology.* 2nd ed. Wiley, 1992. A standard, up-to-date text. (SBF v25 N 1989 p72)

Geology—Earthquakes and Volcanoes

Bolt, Bruce A. *Earthquakes: A Primer.* Rev. ed. Freeman, 1993. 320 p., gloss., index. 0–7167–2236–4.

Recognizing earthquakes as natural and inevitable phenomena, the author describes what causes them, how they are measured, what they tell us about our planet, and, in conclusion, some steps that can be taken to minimize their impact on human beings. (SBF v29 Ag 1993 p174)

Decker, Robert, and Barbara Decker. *Volcanoes.* Rev. ed. Freeman, 1989. 244 p., gloss., photos, index. 0–7167–1851–0.

"Volcanoes assail the senses. They are beautiful in repose and awesome in eruption; they hiss and roar, they smell of brimstone." This book starts out on that very literary note and proceeds to give a complete introduction to volcanology. Spectacular photographs. (LATBR S 24, 1989 p14)

Also Recommended

Simkin, Tom, and Richard S. Fiske. *Krakatau 1883.* Smithsonian, 1983. The terrible eruption of Krakatau and its geologic, biological, and meterological consequences. (Atl v252 N 1983 p149)

Volcanoes of North America: United States and Canada. Charles A. Wood and Jurgen Kienle, eds. Cambridge University Press, 1990. A definitive reference on 262 active and potentially active volcanoes. (LJ v118 Mr 1, 1991 p62)

Hydrology

Reisner, Marc. *Cadillac Desert: The American West and Its Disappearing Water.* Douglas and McIntyre, 1993. 582 p., illus. 1–550–5408–07; Marc Reisner and Sarah Bates. *Overtapped Oasis.* Island Press, 1990.

Water is a most precious resource in the American West. Reisner charges that well-intentioned but misguided efforts by such federal agencies as the Bureau of Reclamation and the Army Corps of Engineers have resulted in the diminution and unfair management of the West's water supplies. These books examine the problem and offer prognoses of its severity and possible solutions. (BRD 1987 p1557; Choice v24 D 1986 p645; NYTBR v95 S 2, 1990 p17)

Owens, Owen. *Living Waters: How to Save Your Local Stream.* Rutgers University Press, 1993. 220 p., index. 0–8135–1997–7.

Protecting inland waterways is vital to all conservation efforts, and it is an area where grassroots efforts can be particularly effective. The author suggests many ways to accomplish this and, drawing in part from Native American belief systems, ties them together philosophically. The resource guide is useful. (SBF v30 Ap 1994 p75)

Also Recommended

National Research Council. *Restoration of Aquatic Ecosystems.* National Academy Press, 1992. This treatise examines wetland and aquatic ecologies and suggests public policy measures for their preservation. (SBF v28 D 1992 p262)

Opie, John. *Ogallala: Water for a Dry Land.* University of Nebraska Press, 1993. The Ogallala aquifer, which provides water to much of the central part of the United States, is desperately threatened. (Choice v31 D 1993 p624)

Meteorology and Climatology

Ludlum, David McWilliams. *The Audubon Society Field Guide to North American Weather.* Knopf, 1991. 636 p. 0679733531.
The first section of this book gives information on common meteorological events and their causes. Also included are what might be called natural histories of weather phenomena in which their geographic, seasonal, and cyclic occurrences are described. Good illustrations throughout. (SLJ v38 Mr 1992 p270)

Schotterer, Ulrich. *Climate—Our Future?* University of Minnesota Press, 1992. 175 p., illus. 0–8166–2130–6.
Begins with a primer on the Earth's atmosphere and a history of climate changes, then proceeds to analyze global climatic trends and current research. Concludes with alternative visions of future climatic changes. Beautifully illustrated. (SBF v29 Mr 1993 p45; SciTech v16 N 1992 p13)

Also Recommended

Lewis, R.P.W. *Meteorological Glossary.* 6th ed. Unipub, 1991. A very inclusive, encyclopedic reference for students, teachers, and professionals in the field. (SBF v28 Mr 1992 p39)

Trefil, James. *Meditations at Sunset: A Scientist Looks at the Sky.* Scribner's, 1987. The ruminations of a scientist as he ponders the sunset. (BL v83 Je 15, 1987 p1545; LJ v112 Je 15, 1987 p78; NYTBR v92 Jl 12, 1987 p7)

Wagner, Ronald, and Bill Adler, Jr. *The Weather Sourcebook.* Globe Pequot, 1994. A one-stop guide to all types of weather.

Williams, Jack. *The Weather Book.* Vintage, 1992. With its bright illustrations and vivid photographs, *USA Today*'s weather page is one of its favorite features. This book—surprisingly detailed—is by one of that column's founders. (LATBR v117 S 15, 1992 p90; LJ v117 S 15, 1992 p90; PW v239 Mr 2, 1992 p62)

Natural Histories of Specific Environments

(*Note*: Natural histories are extremely numerous. Because of the great abundance of nature writings, and because individual books and collections of essays span the entire globe in terms of coverage, some at very specific geographic levels, it is very difficult to represent this literature in a selective bibliography. For collection development purposes, there should be two general goals: (1) to collect a handful of the most popular writings that have long shelf lives, and (2) to select books that cover geographic areas of particular interest, usually those of a regional nature. The following citations and brief annotations are directed more toward the first goal, but also list books on area ecosystems that are of current general interest. Finally, see also citations in other categories throughout this section.)

Abbey, Edward. *Desert Solitaire: A Season in the Wilderness*. McGraw-Hill, 1968. 269 p.
  Abbey, an original character and a favorite among environmentalists, writes of his two seasons as a park ranger at Arches National Park in Utah. (BRD 1968 p1; LJ v93 Ja 1, 1968 p67)

Alcock, John. *Sonoran Desert Summer*. University of Arizona Press, 1990. 187 p., illus., index. 0–8165–1150–0.
  Alcock, a biologist, is intimately familiar with this harsh desert environment in southern Arizona and Mexico. Despite its intense heat and scarce rainfall, the Sonoran harbors a tremendous variety of plant and animal life. (BL v86 Mr 1, 1990 p1250; LJ v15 Mr 15, 1990 p109; New Sci v126 Je 2, 1990 p60)

Bass, Rick. *Winter: Notes from Montana*. Houghton Mifflin, 1991. 162 p. 0395517419.
  A winter spent in a tiny, scarcely populated valley in extreme northwestern Montana. Poetically written. (BL v87 F 1, 1991 p1111; LJ v116 F 15, 1991 p218)

Bonta, Marcia. *Appalachian Spring*. University of Pittsburgh Press, 1991. 187 p., biblio., illus. 0822936585.
  A personal and sensitive ode to the Appalachian woodland in central Pennsylvania. While reveling in the land's beauty and its wildlife, the author is also alert to the factors that threaten it. (BL v87 Ap 1, 1991 p1534; KR v59 F 1, 1991 p149; LJ v116 F 15, 1991 p218)

Brower, Kenneth, and Eliot Porter. 2 vols. *Galapagos: The Flow of Wildness*. Sierra Club, 1970.
  The stunning biodiversity of the Galapagos Islands started Darwin thinking about natural selection and evolution. To many scientists today, they remain a laboratory for life. The two volumes in this illustrated set are *Discovery* and *Prospect*.

Campbell, David G. *The Crystal Desert: Summers in Antarctica.* Houghton Mifflin, 1992. 308 p., index. 0–395–58969–x.

It has been said that a person's best day in Antarctica is the first day there, and the second best day is the last. However remote, Antarctica is a vital part of the global ecosystem, and a threatened environment. Campbell spent three summers there and conveys the majesty and harshness of the land in this book. (KR v60 S 15, 1992 p1159)

Chase, Alston. *Playing God in Yellowstone: The Destruction of America's First National Park.* Atlantic Monthly, 1986. 446 p. 0871130254.

Chase, a biologist intimately familiar with Yellowstone, issues a searing indictment of policy and management practices at the nation's oldest park. The controversial Chase makes some bold accusations, but throughout he insists upon science as the basis for his claims. (KR v54 Mr 1, 1986 p356; LJ v111 Ap 15, 1986 p70; NYTBR v91 Je 1, 1986 p15)

Cowell, Adrian. *The Decade of Destruction: The Crusade to Save the Amazon Rain Forest.* Holt, 1990. 217 p. 0–8050–1494–2.

Cowell, a filmmaker and naturalist who has lived and worked extensively in Amazonia, documents the devastation that has taken place in the rainforest and shows it to have resulted from human greed, duplicity, and shortsightedness. This is a personal plea for conservation of this region while time still remains. (BL v87 S 15, 1990 p97; KR v58 Ag 1, 1990 p1058; LJ v115 S 15, 1990 p96)

Deserts: The Encroaching Wilderness. Tony Allan and Andrew Warren, eds. Oxford University Press, 1993. 176 p., maps, photos, index. 0–19–520941–9.

Published under the auspices of the International Union for the Conservation of Nature and Natural Resources, this expansive reference examines the full range of human and ecological factors affecting world desert climates. (LJ v118 Je 15, 1993 p62; Nature v363 My 27, 1993 p313)

Dillard, Annie. *Pilgrim at Tinker Creek.* Harper's, 1974. 271 p. 0–06–121980–0.

Dillard reflects on her experiences in a pastoral part of Virginia near the Blue Ridge Mountains. In her writings she conveys a sense of serenity, spirituality, and love of the land. (BRD 1974 p293; LJ v99 My 1, 1975 p1293)

Fitzpatrick, Tony. *Signals from the Heartland.* Walker, 1993. 230 p. 0–8027–1260–6.

The devastating 1993 floods along the Missouri underscored the environmental problems that exist in America's heartland. Fitzpatrick, who hails from the area about which he writes, brings insight into the natural history of a part of America often taken for granted. (BL v89 D 15, 1992 p701; LJ v118 Ja 1993 p160)

Leopold, Aldo. *A Sand County Almanac* and *Sketches Here and There.* Oxford University Press, 1949.

Wildlife ecologist Aldo Leopold pioneered many of the modern principles of conservation and wildlife management. This beautifully written collection of essays about the natural milieu of his Wisconsin farm is part of the canon of American environmental literature. (BRD 1949 p548)

Lopez, Barry. *Arctic Dreams*. Scribner's, 1986. 0684185784.
Lopez, who lived among the native peoples of the Arctic for several seasons, expresses deep respect for their culture and the brutal land they inhabit. He writes in beautiful prose on the Arctic landscape and the animals that live there. (BL v82 Ja 1, 1986 p643; Choice v23 My 1986 p1390; LJ v111 Mr 1, 1986 p102)

Muir, John. *The Yosemite*. Sierra Club, 1988. 215 p. 0871567822.
Muir's classic "guidebook" makes it evident why he called Yosemite his favorite place on Earth. Some of the most moving nature writing in existence. (PW v236 S 1, 1989 p71; WSJ v 214 O 3, 1989 pA20)

Sax, Joseph L. *Mountains Without Handrails: Reflections on the National Parks*. University of Michigan Press, 1980. 152 p. 047206324x.
Sax summarizes the salient issues regarding the use and protection of our national parks. He sees two sides in the debate, the "preservationists," who would lock up the land, and the "automobile tourists," whose attitudes and experiences at parks tend to be destructive. Sax's approach is to encourage a new breed of "contemplative tourists." (BL v77 F 15, 1981 p791; Choice v18 Mr 1981 p986; LJ v106 My 1, 1981 p942)

Thoreau, Henry David. *Walden* (several editions exist); *Faith in a Seed*. Island Press, 1993. 283 p. 1559631813.
Every library in America should have a sampling of Thoreau's writings. *Faith in a Seed* is a recent, beautifully produced collection of some of his later natural histories. (BioSci v44 Mr 1994 p184; SciTech v17 D 1993 p18)

Wetlands. Max Finlayson and Michael Moser, eds. Facts on File, 1993. 224 p., photos, gloss., index. 0–8160–2556–8.
Historically, some have equated wetlands with wastelands. Today, they are recognized as very biodiverse and ecologically vital areas, but they are drying up rapidly. This book looks at world wetlands in words and stunning photographs. (SciTech v16 S 1992 p3)

Wheelwright, Jeff. *Degrees of Disaster: Prince William Sound—How Nature Reels and Rebounds*. Simon and Schuster, 1994. 348 p., index. 0–671–70241–6.
Twice in twenty-five years, Alaska's magnificent Prince William Sound was rocked by catastrophe. One was natural—the huge 1964 earthquake; the other was man-made—the Exxon *Valdez* disaster. In both cases, the natural systems of the sound responded vigorously, aided somewhat by human activity, but

spurred mostly by nature's own restorative powers. (KR v62 Je 1, 1994 p765; LJ v119 J 1994 p123)

Wild Africa: Three Centuries of Nature Writing from Africa. John Murray, ed. Oxford University Press, 1993. 318 p., biblio., index. 0–19–507377–0.

As Europeans explored the vast African continent, they produced a spellbinding literature of adventures and natural histories. Today, conservationists and naturalists maintain that tradition, incorporating a new environmental ethic into their writings. This anthology spans three periods: the Age of Exploration, the Age of Exploitation, and the Age of Environmentalism. (BL v89 My 1, 1993 p1567; LJ v118 My 1, 1993 p112)

Ocean Sciences

Brower, Kenneth. *Realms of the Sea.* National Geographic, 1992. 274 p., photos, index. 0–87044–855–2.

This book has six general parts and twenty chapters, but its most conspicuous attribute is the over one hundred photographs and detailed drawings. Good balance overall. (SBF v29 D 1993 p9)

Cone, Joseph. *Fire under the Sea: The Discovery of Hot Springs on the Ocean Floor and the Origin of Life.* Morrow, 1991. 320 p., illus., biblio., index. 0–688–09384–7.

Deep beneath the ocean, superheated seawater vents spew hot water through faults on the ocean floor. Along these vents, some of the most exotic life forms on Earth flourish. The study of ocean hot springs and the biota they nurture is important for several fields in the natural sciences. (BL v87 Jl 1991 p2017; KR v59 Je 15, 1991 p767; LJ v116 Jl 1991 p128)

Thorne-Miller, Boyce, and John Catena. *The Living Ocean: Understanding and Protecting Marine Biodiversity.* Island Press, 1991. Index. 1–55963–064–7.

The authors discuss various oceanic and wetland ecosystems, the life that inhabits them, and conservation strategies. As formidable as the problems of clean-up are on land, those of the ocean might be even more daunting. Unfortunately, it is easier for human beings to overlook oceanic pollution, so the need to take action is even more critical. (Choice v29 S 1991 p130; LJ v116 F 15, 1991 p218)

Also Recommended

Borgese, Elisabeth Mann. *Ocean Frontiers: Explorations by Oceanographers on Five Continents.* Abrams, 1992. The history and accomplishments of twelve significant ocean research centers. An oversize book format, lavishly illustrated. (LJ v118 Mr 1, 1993 p44)

Ross, David. *Introduction to Oceanography.* 4th ed. Prentice-Hall, 1988. A thorough text for a beginning course. (SciTech v12 Jl 1, 1988 p1)

Sammon, Rick. *Seven Underwater Wonders of the World.* Thomason-Grant, 1992. Reefs, ocean vents, archipelagos, and other striking submarine wonders, in words and photographs.

Paleontology

Dixon, Dougal, et al. *The Macmillan Illustrated Encyclopedia of Dinosaurs and Prehistoric Animals.* Macmillan, 1988. 312 p. 0–02–580191–0.
A one-volume encyclopedia of dinosaurs and other extinct prehistoric animals; for high school students up. In all, over 500 species are included. (SBF v25 S 1989 p14)

Eldredge, Niles. *Fossils: The Evolution and Extinction of Species.* Abrams, 1991. 220 p., illus., photos, index. 0–8109–3305–5.
A leading theorist of evolution provides an intelligently written and superbly illustrated volume on fossils as they relate to Earth history and paleontology. (Choice v30 S 1992 p154; Nature v356 Ap 2, 1992 p393; SBF v28 My 1992 p108)

Fortey, Richard. *Fossils: The Key to the Past.* New ed. Harvard University Press, 1991. 187 p., gloss., index. 0–674–31135–3.
In nine chapters, the author provides summary discussions of how fossils are created, what they are, and why they are important to researchers today. Covers both vertebrate and invertebrate paleontology. For all readers. (SBF v27 N 1991 p233)

Horner, John R., and James Gorman. *Digging Dinosaurs.* Workman, 1988. 210 p., illus., photos, index. 0–89480–220–8.
In his preface, David Attenborough asks, ''How can one account for the thrill of finding a fossil?'' Horner, who has made some major discoveries, gives an in-the-trenches account of the difficult but eminently rewarding work of paleontologists. (Am Sci v88 Mr 1989 p208; BL v85 N 1, 1988 p439; BL v85 N 1, 1988 p439)

Horner, John R., and Don Lessem. *The Complete T Rex.* Simon and Schuster, 1993. 239 p., index. 0–671–74185–3.
In popular lore, T. Rex is the quintessential predator. While its rapacious reputation is well deserved, the authors draw upon what has been learned from recent excavations to show that the beast was more complex than commonly believed. (SBF v29 D 1993 p269)

Lessem, Don. *Kings of Creation: How a New Breed of Scientists Is Revolutionizing Our Understanding of Dinosaurs.* Simon and Schuster, 1992. 353 p., illus., index. 0–671–73491–1.
This is a good general summary of the state of our knowledge about dinosaurs. Lessem covers the revolutionary discoveries that paleontologists have

made over the last thirty years, during which time more has been learned about the "terrible lizards" than all that was known previously. (BL v88 Ap 1, 1992 p1419; KR v60 Mr 1, 1992 p1435; LJ v117 Ap 15, 1992 p118)

McGowan, Christopher. *Dinosaurs, Spitfires and Sea Dragons.* Harvard University Press, 1991. 365 p., index. 0–674–20769–6.

The Mesozoic era, which ended 65 million years ago, is called the Age of the Dinosaurs. Although the extant relics from that era are piecemeal and scant, McGowan takes the position that we can use these fragmentary clues, look at some of the basic concepts of biology and other sciences that we have at our disposal today, and draw inferences that can in fact tell us quite a lot. (Choice v28 Jl 1991 p1803; Nature v351 My 30, 1991 p361; SBF v27 Je 1991 p134)

Norman, David. *Dinosaur!* Prentice-Hall, 1991. 192 p., illus., biblio., index. 0–13–218140–1.

A tie-in with the Arts and Entertainment network series of the same name, *Dinosaur!* is a general overview of the lives and times of these creatures, and of the contemporary scientists who study them. (BL v87 D 15, 1990 p849; LJ v116 N 1, 1991 p129; SLJ v38 Mr 1992 p269)

Spalding, David. *Dinosaur Hunters: Eccentric Amateurs and Obsessed Professionals.* Prima, 1993. 309 p., index. 1–55958–x.

Paleontologists are a rare breed. Possessed of equal measures of a child's fascination with dinosaurs and a scientist's fastidiousness for detail, they fit no stereotype. The author emphasizes the human element in this field and the individuals it attracts. (KR v61 Ag 1, 1993 p990; LJ v118 S 1, 1993 p217)

Also Recommended

The Dinosauria. David Weishampel et al., eds. University of California Press, 1990. 733 p., biblio., index. A comprehensive, technical reference for college and larger public libraries. (Choice v28 Mr 1991 p1166; LJ v116 Mr 15, 1991 p112; Nature v348 D 20 p686)

Fenton, Carroll Lane, and Mildred Adams Fenton. *The Fossil Book: A Record of Prehistoric Life.* Rev. ed. Doubleday, 1989. Vertebrate fossils and their geographic distribution. (SBF v25 My 1990 p245)

Guthrie, R. Dale. *Frozen Fauna of the Mammoth Steppe: The Story of Blue Babe.* University of Chicago Press, 1990. Discovery and study of "Blue Babe," a Pleistocene bison found frozen near Fairbanks. (SBF v26 N 1990 p110)

Wilford, John Noble. *The Riddle of the Dinosaur.* Knopf, 1985. Although slightly dated, this eminently readable book by a Pulitzer Prize winning journalist is still a first choice of many. (BL v82 Ja 15, 1986 p714; Choice v23 My 1986 p1407; LJ v111 Ja 1986 p93)

Philosophical Topics

Oelschlaeger, Max. *The Idea of Wilderness: From Prehistory to the Age of Ecology.* Yale University Press, 1991. 477 p. 0300048513.

Oelschlaeger describes the implicit and explicit philosophies of nature as contained in the works of the likes of Descartes, Bacon, Spinoza, and Rousseau, as well as many contemporary environmental writers like Aldo Leopold and James Lovelock. (A Lib v63 D 1991 p784; Choice v29 Ja 1992 p765)

Sheldrake, Rupert. *The Rebirth of Nature: The Greening of Science and God.* Bantam, 1991. 272 p., index. 0–553–07105–x.

All cultures have developed ideologies that establish a place and role for human beings in nature. The dominant Judeo-Christian belief has been that humans are masters of nature. In this book, the author assimilates a vast amount of cross-cultural information in developing a framework for a philosophical partnership between human beings and the natural world which, in part, requires an acceptance of the limitations of science. (SBF v27 Ag 1991 p170)

OTHER RESOURCES

Periodicals for General Readers

Audubon. National Audubon Society, bi-m. 0004–7694.

The popular vehicle of the National Audubon Society reflects the group's interests and agendas. Articles are founded in science, but written for generalists, and the photographs alone are worth the subscription price.

Buzzworm: The Environmental Journal. Buzzworm, Inc., bi-m., revs. 0898–2996.

Articles on a broad range of environmental issues from a nonpartisan standpoint. Will appeal to activists and dilettantes as well. Conveys news and information on events, groups, and so on.

National Geographic. National Geographic Society, m. 0027–9358.

With a circulation of over 10 million, this magazine is undeniably popular and has a time-tested formula for taking field research and presenting it in a readable, almost adventurous way.

National Parks. National Parks and Conservation Association, bi-m. 0276–8168.

Publishing by a nonprofit association that focuses solely on defending, promoting, and improving America's national park system, this magazine is the chief vehicle by which the group realizes its public education goals. Includes news, regional reports, and articles on current and historical topics.

Natural History. American Museum of Natural History, m., revs. 0028–9358.

Stephen Jay Gould's essays are probably the most popular feature in this magazine and worth the subscription price alone. Additionally, the articles contained here present appealing surveys of topics from all realms of the field, including archaeology, paleontology, ecology, and others.

Sierra. Sierra Club, bi-m., revs. 0161–7362.

A good source of information representative of mainstream environmental issues and philosophies. Includes factual articles as well as occasional literary offerings.

Audiovisual Resources

EARTH SCIENCES

Earthquake Country. Filmmakers Library, 1991 (52 minutes color).

A history and assessment of present dangers along the San Andreas fault in southern California. Includes information on earthquake preparedness in affected regions and how scientists are monitoring seismic activity. (SBF v27 Mr 1991 p56)

Earthquakes: Seismology at Work. Media Guild, 1989 (25 minutes, color).

Vivid and sometimes chilling footage from the 1964 Alaskan earthquake makes this video unforgettable. Also visits a seismology lab for a look at some high-tech earthquake monitoring equipment. (SBF v26 S 1990 p71)

The Hawaiian Volcano: A Force for Creation. Harada Productions, 1989 (33 minutes, color).

Hawaii Volcanoes National Park, which attracts tourists from around the world, is the center of intense study. The causes and consequences of Hawaiian volcanism are discussed, as well as how scientists keep track of geothermal activity. (SBF v25 N 1989 p92)

In the Path of a Killer Volcano. Films for the Humanities, 1993 (60 minutes, color).

Actor Hal Holbrook narrates this outstanding look at the 1991 eruption of Mount Pinatubo in the Philippines, the scientists who tried to predict it, and what has been learned from the experience. The real-life documentary approach adds drama and insight. (SBF v30 My 1994 p120)

Physical Geography of North America Series (series). National Geographic Society, 1986 (20 minutes each, color).

An excellent six-part series from the National Geographic Society. Each installment gives information on the topography, climate, habitats, life forms, and other characteristics of a particular region. Segments are (1) "The Central Lowlands," (2) "The East," (3) "The Northlands," (4) "The Pacific Edge," (5) "The Rocky Mountains," and (6) "Western Dry Lands." (SBF v25 N 1989 p107)

Also Recommended

Drifting Continents. Media Guild (24 minutes, color). The movement of continental plates and its climatological effects. A British production. (SBF v25 My 1990 p283)

Return to Mount Saint Helens. British Broadcasting Co., 1991 (58 minutes, color). Examines the geological information gathered from the mountain's eruption and the ways plants and animals have reclaimed the land.

A World of Water. National Geographic Society, 1989 (16 minutes, color). Two-thirds of the Earth is covered with water. Even the most landlocked places on Earth are affected meteorologically by the world's oceans. (SBF v25 My 1990 p283)

ENVIRONMENTAL TOPICS

Danger at the Beach. PBS Video, 1991 (58 minutes, color).
Recently, when solid wastes began washing onto beaches along the American eastern seaboard, public concern and fear about the health of the oceans were greatly heightened. Actor Ted Danson narrates this satisfying documentary on the relationships between human activities in a variety of marine locales and the cleanliness of these waters. (SBF v28 My 1992 p117)

The Fateful Balance. Landmark Films, 1992 (52 minutes, color).
A general overview of human impacts upon the environment and the atmosphere, and possible climatological consequences. This Canadian production shows interviews with scientists who offer data, observations, and conclusions based on their own research. (SBF v28 O 1992 p212)

In Partnership with the Earth. Versar Inc., 1990 (57 minutes, color).
John Denver narrates this video, which looks at individuals from various environmental groups and industries that endorse environmentally sound practices. The theme is that we all share responsibility for maintaining the planet's environment, and the interviews and footage shown reinforce that point. (SBF v27 Ap 1991 p86)

Man and the Biosphere (series). Films for the Humanities, 1990 (28 minutes each, color).
This series explores how human beings have altered several specific environments, such as coastlines, coral reefs, and others. Shows both adverse impacts and positive steps that can be taken to remedy them. (SBF v27 N 1991 p246; SBF v28 Mr 1992 p56)

Protecting the Global Environment (series). Cinema Guild–Better World Society, 1990.
A production of the Better World Society and the Television Trust for the Environment (Great Britain), this series looks at a number of general environ-

mental topics, especially those related to plant and animal biodiversity. Most segments include input from researchers and experts in the field. (SBF v27 Mr 1991 p55)

Also Recommended

Alterations in the Atmosphere. Films for the Humanities, 1990 (18 minutes). From the *Fragile Planet* series, this looks at how the Earth's atmosphere makes life possible and how pollutants deplete and contaminate it. (SBF v27 Ja 1991 p22)

Our Restless Atmosphere. United Learning (12 minutes). A brief production on ozone depletion, the greenhouse effect, and acid rain. (SBF v30 Ja 1994 p24)

EVOLUTION, ARCHAEOLOGY, AND PALEONTOLOGY

Mysteries of Mankind. National Geographic Society, 1988 (59 minutes, color).
An interdisciplinary approach to the study of evolution and human ancestry. Combines knowledge from paleontology, primatology, zoology, molecular genetics, and other fields for a holistic, intelligent overview. Also shows scenes from field excavations. (SBF v25 S 1989 p52)

Patterns of Evolution. Coronet Films and Video, 1988 (25 minutes, color).
From a series entitled *Genetics*, this video focuses on breeding, genetics, and natural selection as agents of evolution. Also demonstrates techniques that scientists use for assessing genetic differences and similarities among species. (SBF v25 S 1989 p52)

Also Recommended

A Conversation with Stephen Jay Gould. Carolina Biological Supply Co., 1988 (28 minutes). Gould's ideas and works should provide motivation for students interested in the biological sciences.

Evolution and Human Equality. Insight Video, 1987 (42 minutes). Thoughts on human evolution and intelligence.

LIFE ON EARTH

Living Planet. Ambrose Video, 1992 (55 minutes each).
Originally broadcast in the mid-1980s, this excellent series produced by David Attenborough was finally brought to VHS by popular demand in the 1990s. In all, there are twelve episodes, from "A Portrait of the Earth," which begins with the building of the planet, to "New Worlds," a discussion of the Earth as a global biosphere. In between, segments study various ecosystems and environments, and the life forms they nurture. (SBF v29 Ap 1993 p90; SBF v29 Ja 1993 p24)

Patterns of Diversity. Coronet Film and Video, 1988 (25 minutes).
Visits some familiar and exotic habitats. A large portion of the video con-

centrates on the minute differences found within snail species and how these
can be associated with their habitats. Also looks at how continental drift has
affected the distribution of species. (SBF v25 S 1989 p52)

NATURAL HISTORIES

Amazonia: A Celebration of Life. Landmark Films, 1993 (23 minutes, color).
  The goal of this video is to give viewers an appreciation and understanding
of the biological wealth of Amazonia. Shows how species have adapted to very
specialized niches and how delicate those environments are. The message is
clear: human beings can undo in short order what it has taken evolution millions
of years to achieve. From the *Wildlife Environment* series. (SBF v30 Mr 1994
p54)

Arctic Refuge: A Vanishing Wilderness. PBS Video–National Audubon Society,
1990 (60 minutes, color).
  In extreme northeast Alaska there is a 19 million acre wildlife refuge which
is threatened because of proposed extraction of its petroleum resources. This
video highlights the clashing ideologies of conservation and economics, thus
giving a view into the pressures that can be brought to bear in the conflict. It
also gives a vista of this remote, magnificent land. (SBF v27 Ag 1991 p182)

Spirit of the Rainforest. F. P. Video–Discovery Channel, 1990 (90 minutes,
color).
  Peru's Manu Biosphere Reserve is a 7,000 square mile nature reserve on the
eastern slopes of the Andes that contains mountain, grassland, and jungle ecosystems. Thousands of plant and animal species have found sanctuary there. In
all, it is the world's most biologically diverse protected environment. This video
explores its riches. (SBF v29 N 1993 p247)

*The Wilderness Idea: John Muir, Gifford Pinchot and the First Great Battle for
Wilderness.* Direct Cinema Ltd., 1989 (58 minutes, color).
  An inspired look at the careers and conservation efforts of Muir and Pinchot,
who had different, yet complementary philosophies of nature. Uses some rare
black and white films to capture the times. Also interviews historians and biographers. (SBF v26 N 1990 p137)

Also Recommended

Antarctica. Landmark Films, 1992 (58 minutes, color). A natural history of this
forbidding but surprisingly fragile land at the bottom of the Earth. (SBF v28 O
1992 p216)

Baikal: Blue Eye of Siberia. Films for the Humanities, 1991 (107 minutes,
color). Baikal is the largest freshwater lake in the world, but it is largely unknown to Westerners. This video is a good introduction to its natural history.
(SBF Ag 1992 p185)

Nature of Things (series). Filmmakers Library–Canadian Broadcasting Corporation, 1991. This Canadian series includes segments on a variety of topics. Some relevant here include "Grizzly Bear," "And God Created the Great Whales," "Dinosaurs, Remains to Be Seen," and "The Restless Sky." (SBF v27 Mr 1991 p55; SBF v25 S 1989 p50)

Serengeti Diary. National Geographic Society, 1989 (59 minutes, color). The Serengeti ecosystems and the zoologists who have done research there. (SBF v25 Mr 1990 p217)

Wilderness Ecology. Cornonet Films and Video, 1988 (20 minutes each, color). Three sections: "Deciduous Trees," "Coniferous Trees," and "Lakes and Streams." Produced in Great Britain. (SBF v25 S 1989 p51)

CD-ROMs

Coral Kingdom. Wings for Learning–Sunburst, 1992.
   Principles of ecology and marine science are explored through "missions" to coral reef environments.

Dinosaurs: The Multimedia Encyclopedia. Media Design Interactive, 1992 (one disc).
   A comprehensive encyclopedia, with index, of information on dinosaur life. Includes color pictures, animations, and interviews with paleontologists.

ECODISC CD-ROM. BBC Enterprises, 1990 (one disc).
   Users visit wildlife preserves season by season and spot animals in their habitats. Simulation also allows the user to assume the role of wildlife manager.

Encyclopedia of Dinosaurs. Applied Optical Media Association, 1992.
   Covers topics in paleontology for students at the secondary level and up.

Environmental Data. PEMD Corp., 1991.
   Data on many areas of environmental concern from the book *World Resources, 1990–91.* Topics include greenhouse warming, ozone depletion, agricultural supply and demand, and others.

Guide to the Earth. Edge Interactive Media, 1991.
   Ecosystem management presented interactively, with special emphasis on practical steps that can be taken to cultivate a green lifestyle.

Last Chance to See. Voyager Co.
   Fashioned to be an excursion into unknown lands, where users encounter endangered animals and environments, as well as some that are familiar but no less fascinating. Program contains over 700 full screen photographs.

Chapter Twelve

Physics

Once upon a time, way back in the infinitesimal
First fraction of a second attending our creation,
A tiny drop containing all of it, all energy
And all its guises, burst upon the scene,
Exploding out of nothing into everything
Virtually instantaneously, the way our thoughts
Leap eagerly to occupy the abhorrent void.

George Bradley, About Planck Time

As I write this I happen to be in an airplane at 30,000 feet, flying over Wyoming en route home from San Francisco to Boston. Below, the earth looks very soft and comfortable. . . . It is very hard to realize that this all is just a tiny part of an overwhelmingly hostile universe. It is even harder to realize that this present universe evolved from an unspeakably unfamiliar early condition, and faces a future extinction of endless cold or intolerable heat. The more the universe seems comprehensible, the more it also seems pointless.

Steven Weinberg, The First Thee Minutes

During the last fifty years or so, more popular works have appeared on subjects in physics than in almost any other field of science. As researchers learn more about matter and the forces of the universe, their discoveries lend scientific credibility to some mind-boggling possibilities. In the literature of contemporary physics, such chimeras as time travel and parallel universes, long the stuff of science fiction, are written about in sober, technical detail by scholars who consider them entirely plausible.

Ironically, particle physics, the study of the infinitely small, has become wed to astrophysics, the science that explores the infinitely large. By understanding how subatomic particles interact, physicists theorize as to how the universe was fabricated in the Big Bang and speculate about its ultimate fate. This, however, is not the mechanistic universe of Isaac Newton, where a clockwork order governs all interactions and only an occasional nudge from the Creator is needed to keep things running properly. Rather, the cosmos described by contemporary physics is an exotic, ever-changing place, where space is curved, time is relative, and things that have been called voids, bubbles, wormholes, baby universes, and magnetic monopoles may exist in abundance. There is hope that eventually a Grand Unified Theory will be discovered, which will meld the fundamental forces of nature and reveal them to be aspects of the same "superforce." Such a theory, if achieved, would be perhaps the definitive masterpiece of human intellectual history.

The conjectures of theoretical physicists sometimes stray into the realms of philosophy and religion. They address questions of existence and ultimate meanings. Some of the concepts are decidedly counterintuitive. For example, Einstein himself dismissed quantum theory because of what he saw as its unappealingly bizarre consequences. While these ideas might seem esoteric to the layperson, they are also irresistibly tantalizing for the same reason. The science of physics exercises the imagination in some of the most fantastic ways.

Knowledge is moving swiftly in physics. By the time this book is published, new discoveries will no doubt have been made. Within the last two months, scientists have announced what they believe is the discovery of a massive galactic black hole and identified, through experimentation, the predicted "top" quark. When building science literacy collections for libraries, physics is one field that requires frequent monitoring.

PHYSICS—GENERAL

Reference

Encyclopedia of Modern Physics. Robert A. Meyers, ed. Academic Press, 1990. 733 p., illus., index. 0–12–226692–7.

This well-organized reference work contains thirty-four articles that collectively span the entire discipline at a broad, not deep level. Serves the needs of generalists and beginning to intermediate students. (ARBA v23 1992 p708; Choice Je 1990 p1650)

The Facts on File Dictionary of Physics. Rev. ed. John Daintith, ed. Facts on File, 1988. 235 p. 0–8160–1868–5.

Libraries that can afford just one physics reference book ought to consider this. It is pithy, readable, and relevant. It won't do everything that an encyclo-pedia can, but it covers the field well. (ARBA 1989 p658; SBF v24 My 1989 p288)

Also Recommended

Encyclopedia of Physics. 2nd ed. Rita Lerner and George L. Trigg, eds. VCH, 1991. A voluminous work with an impressive roster of contributors. Fairly tech-nical; for science majors and up. (ARBA v23 1992 p708)

The McGraw-Hill Encyclopedia of Physics. 2nd ed. Sybil Parker, ed. McGraw-Hill, 1993. The 828 entries here are taken from the parent encyclopedia. (ARBA v25 1994 p791; Choice O 1993 p268; JAL S 1993 p275)

Autobiography and Biography

Alvarez, Louis. *Alvarez: The Adventures of a Physicist.* Basic, 1987. 292 p. 0–4650–0115–7.

Alvarez wrote this book to tell his children of his life, so it is personal and thoughtful. He recounts his work at many major international research centers for physics, and, in doing so, reflects on the ways that work has impacted upon society, for better or worse. (BL v83 Ap 15, 1987 p1237; LJ v112 Ap 1, 1987 p155; Nature v326 Ap 30, 1987 p916)

Bernstein, Jeremy. *The Life It Brings: One Physicist Remembers.* Ticknor and Fields, 1987. 171 p. 0–899–1947–02.

How one young man was attracted to physics. Includes his reminiscences about friendships with colleagues like Oppenheimer, Dyson, and Gell-Mann, as well as others, such as musicians in the Duke Ellington orchestra. Anecdotal, insightful, and introspective. (BL v83 Mr 15, 1987 p1082; Choice v24 Jl 1987 p1715; SBF v23 S 1987 p15)

Blaedel, Niels. *Harmony and Unity: The Life of Niels Bohr.* Science Books, 1988. 323 p., index. 0–910239–2.

Some contend that Bohr was every bit as influential in the development of physics in the early part of this century as Einstein. His quantum theory, for all of the violence it does to common sense, is the bedrock in theoretical physics today. This biography highlights the human side of the scientist; it includes passages from original correspondence. (SBF v24 My 1989 p277; LJ v113 Ag 1988 p153; PW v233 Ap 22, 1988 p73)

Casidy, David. *Uncertainty: The Life and Science of Werner Heisenberg.* Free-man, 1991. 544 p., illus., photos, index. 0–7167–2243–7.

Heisenberg's famous "uncertainty principle" has become a potent metaphor for modern science. Unlike many others, this genius of modern physics remained in Germany during the World War II. His work on the German Bomb project put him at the center of political intrigue and forced him to face an impossible moral quandary. Heisenberg was not only a major figure in science, but he lived and thrived in a tense political climate as well. (LJ v117 Mr 1, 1993 p44; New Tech B S 1992 p1102; NYTBR v98 S 26, 1993 p32)

Gleick, James. *Genius: The Life and Times of Richard Feynman.* Pantheon, 1992. 532 p., photos, biblio., index. 0–679–40836–3.

Feynman has been called the most brilliant scientist since Einstein. From his efforts on the Manhattan Project to his Nobel Prize winning work that gave physicists new techniques for modeling the interactions of subatomic particles, Feynman has left a giant mark on contemporary science. He was also a popular author, a social activist, and a truly eccentric character. Gleick, a first-rate science writer, recounts Feynman's life and explores the nature of his genius. (Choice v30 Mr 1993 p1114; Sci v259 Ja 22, 1993 p537; Time v140 D 7, 1992 p76)

Goodchild, Peter. *J. Robert Oppenheimer: Shatterer of Worlds.* Houghton Mifflin, 1981. 301 p., photos, index. 0–395–30530–6.

Portrays Oppenheimer's greatness and tribulations. This work is particularly strong on conveying the highly charged political environment of the war years and those in the aftermath, during which Oppenheimer was charged with "un-American activities." (BL v77 My 15, 1981 p81; KR v49 My 1, 1981 p613; NYTBR v86 Jl 12, 1981 p13)

Lanouette, William (with Bela Szilard). *Genius in the Shadows: A Biography of the Man Behind the Bomb.* Scribner's, 1993. 640 p., index. 0–684–19011–7.

Szilard was, perhaps more than any other scientist, the conscience of the Manhattan Project. He was as tireless a social activist as he was a brilliant theorist. This epic biography examines his papers and publications, as well as personal interviews, for insights into his thought and personality. (BL v89 D 15, 1992 p705; LJ v119 Mr 1, 1994 p54; Sci v261 S 10, 1993 p1461)

Pais, Abraham. *Subtle Is the Lord: The Science and Life of Albert Einstein.* Oxford University Press, 1982. 552 p., photos, index. 0–19–853907–x.

The most thorough scientific biography of Einstein available. Pais, a colleague of Einstein's, supplies great technical and historical detail. This book is perfectly readable, but the approach is technical in places; a less rigorous alternative is John Gribbin and Michael White's *Einstein: A Life in Science* (Dutton, 1994). (BL v79 Mr 15, 1983 p952; Choice v20 D 1982 p600)

Pflaum, Rosalynd. *Grand Obsession: Marie Curie and Her World.* Doubleday, 1989. 512 p. 0–385–26135–7.

Curie's life is one of the most fascinating in the history of science, from her Nobel Prize in 1903 for her work on radioactivity to the subsequent diseases that she contracted, largely as a result of that work. (KR v57 O 1, 1989 p1455; LJ v114 O 15, 1989 p88)

Rigden, John S. *Rabi: Scientist and Citizen.* Basic, 1987. 302 p., photos, 0–4650–6792–1.

I. I. Rabi recounted with pride that, as a youngster, his mother instructed him to ask at least one good question every day at school. Over the course of his career as a scientist, he asked a great many. Born in a tiny town in Galicia, he ascended to the highest ranks of American science, served as chair of the physics department at Columbia, and won the Nobel Prize in 1944. (BL v83 Ap 15, 1987 p1237; LJ v112 Ap 1, 1987 p155; Nature v326 Ap 30, 1987 p916)

Westfall, Richard. *Never at Rest: A Biography of Isaac Newton.* Cambridge University Press, 1980. 908 p., biblio., index. 0–5212–3143–4.

Newton's work concluded the scientific revolution that began with Copernicus and established the comprehensive model of the physical universe that dominated for nearly three centuries. He has been called the last man to know all that was to be known. For all of his greatness, though, he led what in many ways was a tortured personal life. Westfall's definitive biography shows both Newton's scientific genius and his personal frailty. A more popularized biography by the same author appeared in 1994. (Choice v18 Jl 1981 p1564; LJ v105 D 15, 1980 p2566; Sci v213 Ag 28, 1981 p998)

Also Recommended

Bernstein, Jeremy. *Hans Bethe: Prophet of Energy.* Basic, 1980. Covers Bethe's work on the hydrogen bomb and in the field of atomic energy. (LJ v105 O 1, 1980 p2075)

Blumberg, Stanley, and Louis Panos. *Edward Teller: Giant of the Golden Age of Physics.* Scribner's, 1990. Teller worked on the Bomb and was an architect of Reagan's "Star Wars." (SBF v26 S 1990 p54; New Tech B v75 S 1990 p1347)

Feynman, Richard. *Surely You're Joking, Mr. Feynman.* Norton, 1985. The best-selling misadventures of an eccentric genius. (BL v81 N 15, 1984 p407; Choice v22 Ap 1985 p1184; LJ v110 Mr 15, 1985 p56)

Segre, Emilio. *Enrico Fermi: Physicist.* University of Chicago Press, 1972. A very respectful biography by a colleague.

Segre, Emilio. *A Mind Never at Rest: The Autobiography of Emilio Segre.* University of California Press, 1993. The career and thoughts of a scientist never shy about expressing his opinions. (Phys W v7 Ap 1994 p62)

Wilson, David. *Rutherford: Simple Genius.* MIT Press, 1983. A thorough biography of the father of modern atomic physics and a pioneer in the study of radioactivity. (LJ v108 O 15, 1983 p1958; Nature v307 F 23, 1984 p761)

(*Note*: Listed here are several biographies about the lives of major physicists who worked on the Manhattan Project. All are worthwhile, but there is inevitable overlap. Also, Richard Rhodes's *The Making of the Atomic Bomb*, cited elsewhere in this chapter, is the definitive history of the project.)

(*Note*: See also entries in Chapter Six under "Autobiography and Biography" and in Chapter Eight under "Autobiography and Biography.")

History of Physics

Cohen, I. Bernard. *Birth of a New Physics*. Rev. ed. Norton, 1985. 258 p. 0–393–01994–2.

Since its original publication twenty-five years ago, this has remained one of the most popular histories of physics. This edition contains several new supplements. (BL v82 O 1, 1985 p176; Choice v23 N 1985 p467; LJ v110 Jl 1985 p80)

Crease, Robert P., and Charles Mann. *The Second Creation: Makers of the Revolution in Twentieth Century Physics*. Macmillan, 1986. 480 p., notes, index. 0025214403.

Over the course of this century, physicists have adopted and discarded numerous theories of matter and force, all the while building toward what is hoped will be a unified theory of the science. This is an account of progress toward that objective. Not limited to dry facts, this book tells who said what to whom, and why it was important. (BL v82 My 1, 1986 p1268; Choice v24 O 1986 p331; LJ v111 Je 1, 1986 p331)

Spielberg, Nathan, and Bryon D. Anderson. *Seven Ideas that Shook the Universe*. Wiley, 1987. 0–471–18477–6.

This book accepts the idea that science advances, not necessarily through linear progress, but rather through revolutionary episodes of discovery. (BL v82 My 15, 1987 p1393; New Tech B v71 N 1986 p624; S&T v74 S 1987 p263)

Also Recommended

Kevles, Daniel J. *The Physicists: The History of a Scientific Community in Modern Times*. Harvard University Press, 1987. A glimpse into the private and public worlds of science, from the late 1800s through World War II. (LATBR O 11, 1987 p14; NYTBR v92 D 13, 1987 p42; SBF v23 My 1988 p282)

Segre, Emilio. *From X-Rays to Quarks: Modern Physicists and Their Discoveries*. Freeman, 1980. Slightly more technical than most other books covering this subject, this would be a good choice for a reader with a modest background in physics. (BL v77 N15, 1980 p429; Choice v18 D 1980 p547; Sci v212 My 15, 1981 p782)

(*Note:* Although most of these histories focus on the twentieth century, they do so in a developmental context, and thus cover antecedent periods as well.)

General Topics

Bernstein, Jeremy. *Quantum Profiles*. Princeton University Press, 1991. 178 p. 0–691–08725–3.

Bernstein's *New Yorker* essays are read by an avid following. The strengths of the collection are, as the title suggests, the personal profiles, but his observations on topics such as beauty in physics and New Age mysticism also enlighten. (Choice v28 Je 1991 p1657; SBF v27 O 1991 p201)

Bethe, Hans. *The Road from Los Alamos* (from the Masters of Modern Physics series). Simon and Schuster, 1991. 283 p., index. 0–671–74012–1.

Collectively, the essays presented here are an eloquent statement on why scientists must put forward a responsible view of society and be active in public arenas where science plays a role. There are also many personal reminiscences on science and familiar scientists. (SBF v27 Ag 1991 p167)

Davies, Paul. *The Forces of Nature*. 2nd ed. Cambridge University Press, 1986. 175 p. 0–521–30933–6.

Beginning with classical mechanics, Davies builds to discussions of how atomic, nuclear, and quantum forces operate. His explanation of quantum chromodynamics is clever and understandable. (SBF v23 S 1987 p15; Nature v323 O 16, 1986 p588)

Gonick, Larry, and Art Huffman. *The Cartoon Guide to Physics*. Harper-Perennial, 1991. 213 p., illus. 0–0643618–6.

Gonick, an illustrator who has collaborated with several scientists in various fields, brings his formula for success to the discipline of physics. Humorous cartoon illustrations and an intelligent but informal look at fundamental and applied concepts in physics. A totally nonthreatening overview that could be used as an introduction to physics or to supplement class activities or other readings. (PW v238 Ja 4, 1991 p36; SBF v28 Ap 1992 p79)

Krauss, Lawrence M. *Fear of Physics: A Guide for the Perplexed*. Harper-Collins, 1993. Illus. index. 0–465–05745–4.

To many, modern physics, with its arcane and bewildering concepts, can seem impossible to understand. This book is written for those who have despaired of ever doing so. To break this barrier, Krauss concentrates, not on the esoteric theories, but on the tools and methods that physicists use in their daily work. A good first book to read, but so cleverly written that even those with some physics background will also enjoy it. (CSM v85 S 15, 1993 p15; KR v61 Jl 15, 1993 p914; LJ v119 Mr 1, 1994 p55)

Newton, Roger T. *What Makes Nature Tick?* Harvard University Press, 1993. 249 p. 0–674–95085–2.

The physical sciences utterly depend on mathematics. How and why that is so is the subject of this book. The mathematical foundations of several fields—

mechanics, thermodynamics, relativity, quantum physics, and so on—are discussed. Some familiarity with calculus is useful. (Nature v366 D 1993 p724; SBF v29 D 1993 p264)

Pais, Abraham. *Inward Bound: Of Matter and Forces in the Physical World.* Oxford University Press, 1986. 664 p., index. 0–19–851971–0.
A magnum opus chronicling what has been discovered and understood about the constituents of matter, the laws that govern them, and the forces that act upon them. Authoritative, but not easy reading. (BL v82 Ap 1, 1986 p1104; Choice v24 N 1986 p504; Nature v321 My 8, 1986 p123)

Trefil, James. *The Unexpected Vista: A Physicist's View of Nature.* Macmillan, 1985. 209 p., illus., index. 0–684–17869–9.
"It all depends on your point of view" is the title of one of the chapters in this book. Accordingly, Trefil propounds the viewpoints of physicists on how the world works and how we come to learn about it. (NYTBR v90 Mr 24, 1985 p40; PW v227 Ja 25, 1985 p92)

Wilczek, Frank, and Betsy Devine. *Longing for Harmonies: Themes and Variations from Modern Physics.* Norton, 1988. A husband and wife team of scientists combine physics and poetry in this integrated vision of culture. Although less strong on the literary side, a welcome perspective. (A Lib v19 Ja 1988 p12; BL v84 N 15, 1987 p523; LJ v112 D 1987 p121)

Also Recommended

Glashow, Sheldon. *The Charm of Physics.* Touchstone, 1991. Another in the Masters of Modern Physics series.

Leggett, A. J. *The Problems of Physics.* Oxford University Press, 1987. Problems to be solved in modern physics. (Choice v26 S 1988 p174; Nature v332 Mr 24, 1988 p315; SBF v24 Ja 1989 p140)

The New Physics. Paul Davies, ed. Cambridge University Press, 1988. Eighteen distinguished contributors offer articles on topics in pure and applied subjects. (Astron v17 S 1989 p102; Nature Je 1, 1989 p350)

Park, David. *The How and the Why: An Essay on the Origins and Development of Physical Theory.* Princeton University Press, 1988. On how science should be done and taught. College level. (Choice v26 Ap 1989 p1355; Nature v336 N 17, 1988 p290; SBF v24 Mr 1989 p233)

The Physics of Music. Freeman, 1978. Articles from *Scientific American.*

Rothman, Tony. *A Physicist on Madison Avenue.* Princeton University Press, 1991. Essays, meditations, and humor. (LJ v116 Je 15, 1991 p100; PW v238 Ap 19, 1991 p52)

Sherwood, Martin, and Christine Sutton. *The Physical World.* Oxford University

Press, 1988. Magnificent illustrations and intelligent text. (BL v84 Jl 1988 p1768; SBF v24 Mr 1989 p239)

Squires, E. *To Acknowledge the Wonder*. Taylor and Francis, 1985. Excellent at showing the connections between branches of physics. (Choice v23 Ja 1986 p768; New Tech B v71 My 1986 p47)

PHYSICS—SPECIAL TOPICS

Astrophysics

Davies, Paul. *The Cosmic Blueprint*. Simon and Schuster, 1988. 224 p., index. 0–671–60233–0.
  Is the universe self-ordering? Some astrophysicists believe that, over time, the physical forces of nature lead to higher states of order in all complex systems, including the universe. Davies synthesizes eclectic sources in forwarding this theory. (SBF v24 N 1988 p74)

Davies, Paul, and John Gribbin. *The Matter Myth: Dramatic Discoveries that Challenge Our Understanding of Reality*. Simon and Schuster, 1992. Biblio., index. 0–671–72841–5.
  Matter on Earth and throughout the universe is not always what it seems to be. Two of the very finest science popularizers collaborate to recount how science's perceptions have changed from the belief in a "clockwork" universe to one that embraces quantum uncertainties. (BL v88 Ja 15, 1992 p895; KR v59 D 1, 1991 p1510; LJ v118 Mr 1, 1993 p45)

Gribbin, John. *In the Beginning: After COBE and Before the Big Bang*. Little, Brown, 1993. 288 p., biblio., index. 0–316–32833–2.
  In 1992 NASA's Cosmic Background Explorer satellite produced evidence for the existence of so-called ripples in time, which provide the most compelling evidence yet for the veracity of the Big Bang theory. As with many new discoveries, however, this has fueled new speculations, which Gribbin puts forth with a sense of passion and wonder. (KR v89 Je 1, 1993 p1752; KR v61 Je 15, 1993 p760; PW v240 Je 21, 1993 p95)

Also Recommended

Chown, Marcus. *The Afterglow of Creation*. Arrow, 1993. Radiation in the universe and "ripples" at the edge of time. Chown writes of this discovery and the holes that it filled in physical theory. (New Sci v139 Jl 31, 1993 p42)

Herbert, Nick. *Faster than Light: Superluminal Loopholes in Physics*. New American Library, 1988. The possibilities of time travel through the universe.

(*Note:* See also entries in Chapter Six under "Cosmology.")

Chaos and Complexity Theories

Cohen, Jack, and Ian Stewart. *The Collapse of Chaos: Discovering Simplicity in a Complex World.* Viking, 1994. 0–670–84983–9.

First there was chaos theory, then complexity; perhaps the next inevitable scientific school of thought will be called simplicity. The authors freely admit that this book raises more questions than it answers, so nothing in here is gospel. But readers looking for a scientific smorgasbord will find it here. (*Note:* No reviews are available for this book.)

Gleick, James. *Chaos.* Viking, 1987. 303 p., illus., 0–670–81178–5.

The science of chaos emerged as common themes began to appear in disparate areas of inquiry. Gleick's very successful popularization has attracted a wide readership and remains the best book to read on the subject. (BL v84 O 1, 1987 p187; KR v55 Ag 1, 1987 p113; LJ v112 Ag 1987 p133)

Lewin, Roger. *Complexity: Life at the Edge of Chaos.* Macmillan, 1992. 320 p., biblio. 0–02–570485–0.

What has come to be called complexity theory is based on the observation that complex systems, from the behavior of molecules to the movements of whole societies, tend to organize themselves. Lewin writes this as a kind of scientific travelogue in which he visits with several of the researchers who have contributed to the development of theory at their laboratories around the world. (LJ v117 D 1992 p180; Nature v361 F 11, 1992 p507; Sci v259 Ja 15, 1993 p387)

Also Recommended

Waldrop, M. Mitchell. *Complexity: The Emerging Science at the Edge of Order and Chaos.* Simon and Schuster, 1992. About the Santa Fe Institute, established to develop complexity theory in physics, biology, and economics. (PW v239 Ag 17, 1992 p480)

Grand Unified Theories

Davies, Paul. *Superforce.* Simon and Schuster, 1984. 255 p. 0–671–60573–9.

"Everybody likes a good adventure story," Davies begins. Portraying the scientific search for a unified theory of physical forces as just that, Davies looks at symmetry, beauty, and an "ultimate plan" in nature. (Choice v22 F 1985 p843; LJ v109 N 15, 1984 p2152; Nature v312 D 20, 1984 p787)

Lindley, David. *The End of Physics: The Myth of a Unified Theory.* Basic, 1993. 275 p., refs., index. 0–465–01548–4.

A Grand Unified Theory that can make sense of the entire physical world has become something of a Holy Grail of modern physics, with highly respected researchers claiming that its discovery is at hand. Lindley, a senior editor at

Science, takes a more skeptical view. Among other things, he argues that such a theory could never be tested, so it would ultimately be of no use. (BL v89 My 15, 1993 p1663; LJ v119 Mr 1, 1994 p55; Nature v363 Je 17, 1993 p593)

Parker, Barry. *Search for a Supertheory.* Plenum, 1987. 292 p., illus., index. 0–306–42702–8.

This is a good first book to read about the subject of a unified theory and why it is presumed to exist. (BL v84 N 15, 1987 p522; SBF v23 My 1988 p282; S&T v75 Mr 1988 p278)

Weinberg, Steven. *Dreams of a Final Theory.* Pantheon, 1993. 334 p., notes, index. 0–679–41923–3.

The dream espoused in this book was dealt a temporary setback in 1994 when the U.S. Congress voted to suspend funding on the supercolliding particle accelerator. Still, the vision persists, and Weinberg does an excellent job of articulating it in terms for general readers. (BL v89 D 1, 1992 p638; Choice v30 Jl 1993 p86; New Sci v138 My 29, 1993 p39)

Nuclear and Particle Physics—General

Asimov, Isaac. *Atom: A Journey Across the Subatomic Cosmos.* Dutton, 1991. 0–525–24990–7.

So many "new" particles have been found in recent years that one physicist remarked that the next Nobel Prize should go to whoever does *not* discover one. With his signature clarity, Asimov explains the bewildering world of particle physics and shows how there is order to it after all. (LJ v117 Mr 1, 1992 p47; New Tech B v77 Mr 1992 p328)

Close, Frank. *The Particle Explosion.* Oxford University Press, 1987. 239 p., illus., 0–19–951965–6.

Rutherford discovered the atom's nucleus in 1911. Since then, there has occurred in explosion (in the figurative sense) of particle discovery and identification. Well illustrated. (BL v83 Je 1, 1987 p1482; Choice v24 Jl 1987 p1723; SciTech v11 Jl 1987 p14)

Ne'eman, Yuval, and Yoram Kirsh. *The Particle Hunters.* Cambridge University Press, 1986. 272 p., illus., index. 0521301947.

Ne'eman, a major figure in theoretical physics, collaborates with Kirsh to write an almost entirely nonmathematical account of the years in physics between Rutherford's experiments and the discovery of the W and Z particles at CERN, the European research center. (Choice v24 Mr 1987 p1092; Nature v323 O 30, 1986 p766; Sci v236 My 22, 1987 p999)

Sutton, Christine. *Spaceship Neutrino.* Cambridge University Press, 1992. 244 p., photos, index. 0–521–36404–3.

Neutrinos are the most mysterious particles, so amorphous that they can pass through the Earth without a trace. However evanescent they may be, understanding how these particles work is important to several theories of the universe. (Choice v30 Je 1993 p1665; S&T Jl 1993 p60)

Von Bayer, Hans Christian. *The Taming of the Atom: The Emergence of the Visible Microworld.* Random House, 1992. 223 p., photos, 0–679–40039–7.

"The first time I saw an atom," the author writes, "it blinked." Through scanning tunneling microscopy, atoms can actually be imaged. Von Bayer notes that "even as the atom emerges into public view like a jewel of exquisite design, it will open up and allow, for the first time since the dawn of science, a glimpse inside." (BRpt v11 Mr 1993 p55; LJ v118 Mr 1, 1993 p45; Nature v361 Ja 21, 1993 p215)

Weinberg, Steven. *The Discovery of Subatomic Particles.* 2nd ed. Freeman, 1990. 222 p., index. 0–71767–2121–x.

The historical and mathematical treatment in this, the second edition of Weinberg's classic, makes this a good intermediate work on the subject for advanced high school readers and up. (SBF v27 Ja 1991 p7)

Also Recommended

Fritsch, Harald. *Quarks: The Stuff of Matter.* Basic, 1983. Dated, but holding its own in terms of popularity among general readers.

Pickering, Andrew. *Constructing Quarks: A Sociological History of Particle Physics.* University of Chicago Press, 1986. The social sciences and physical sciences meet in this book. (AJS v91 My 1986 p1479; CS v15 Ja 1986 p120)

Riordan, Michael. *The Hunting of the Quark.* Simon and Schuster, 1987. A moderately technical rendering. (KR v55 S 1, 1987 p1301; NYTBR v92 S 27, 1987 p42)

Watkins, Peter. *The Story of W and Z.* Cambridge University Press, 1986. An account of the experiments leading to the discovery of two particles, which confirmed the electroweak theory. (SBF v23 S 1987 p17; SciTech v10 Jl 1986 p13)

Nuclear and Particle Physics—History

Rhodes, Richard. *The Making of the Atomic Bomb.* Simon and Schuster, 1987. 886 p., illus., 0–671–44133–7.

Many lesser books about the Manhattan Project had been written before this painstakingly researched and brilliantly written book was published. The politics, the science, the patriotism, the fears, and the personalities are all depicted in a way that makes the history come to life. Won the Pulitzer Prize. (BL v83 Ja 15, 1987 p730; Sci v236 My 22, 1987 p974; Time v129 Mr 23, 1987 p84)

Also Recommended

Szasz, Ferenc Morton. *The Day the Sun Rose Twice.* University of New Mexico Press, 1984. The first detonation of an atomic bomb. (Choice v22 Ap 1985 p1220; LATBR Mr 31, 1985 p8; LJ v109 D 1984 p2291)

Optics

Bova, Ben. *The Beauty of Light.* Wiley, 1988. 350 p., illus. 0–471–62580–9.

Bova, a science fiction writer, forays into science fact with this review of the artistic, scientific, and biological dimensions of light. There is even a chapter entitled "Lust." (BL v85 O 1, 1988 p606; KR v56 S 1, 1988 p1288; PW v234 S 16, 1988 p72)

Hecht, Jeff. *Laser Pioneers.* Academic, 1992. 298 p. 0–123–36030–7.

The history of lasers and interviews with some main figures. (SciTech v16 Ap 1992 p26)

Laurence, Clifford L. *The Laser Book: A New Technology of Light.* Prentice-Hall, 1986. 208 p., plates. 0–135–23622–3.

A "brief and concise introduction to the subject of lasers for people with non-technical backgrounds." Gives guidance to those contemplating careers in the field. From the Frontiers of Science series. (Choice v24 Ap 1987 p1250)

Also Recommended

Hecht, Jeff. *The Laser Guidebook.* TAB, 1992. 498 p. 0–070–27737–0. A reference guide and tutorial. See also the author's *Understanding Lasers: An Entry Level Guide* (IEEE Press, 1992). (SBF v29 Ja 1993 p10; SciTech v15 N 1991 p26)

Optics Today. John N. Howard, ed. American Institute of Physics, 1986. 344 p. 0–88318–4990. Published articles from *Physics Today.* Chapters have good introductions and are well organized.

Philosophical Topics

Capra, Fritjof. *The Tao of Physics.* Rev. ed. Bantam, 1984. 346 p. 0–553–26379–x.

Love it or hate it: there's no in-between with this book. Capra argues that there are deep metaphysical convergences between the doctrines of some Eastern religions and the concepts of modern theoretical physics. Many scientists have denounced the book stridently, but it has generated debate and spawned several spinoffs and imitations. For a counterpoint perspective, see *Physics and Psychics* by Victor Stegner (Prometheus, 1990). (Choice v13 Je 1976 p541; LJ v100 O 15, 1975 p1933)

Gregory, Bruce. *Inventing Reality: Physics as Language.* Wiley, 1988. Refs., biblio., index. 0–471–61388–6.

Because much of what physicists study exists outside of the realm of common experience, there are no words to describe it. New languages must thus be created. Gregory tells of the history of physics as the development of new terminologies, which in turn have changed the way we think about basic physical realities. In stressing his point, the author quotes the poet Muriel Rukeyser: ''The universe is made of stories, not atoms.'' (BL v85 O 1, 1988 p207; KR v56 S 15, 1988 p1380)

Lederman, Leon, with Dick Teresi. *The God Particle: If the Universe Is the Question: What Is the Answer?* Houghton Mifflin, 1993. 434 p., index. 0–393–55849–2.

Lederman, former director of Fermilab and a Nobelist, asks the same question as the Greek philosopher Democritus: What is the world made of? In a historical context, he discusses how, repeatedly, elementary particles were found to consist of even more elementary particles, and with each new discovery their properties became more strange and complex. Lederman believes that the ultimate Grand Unified Theory that scientists are currently stalking might be so succinct that it could fit on a t-shirt. (BL v89 F 1, 1993 p954; Choice v31 S 1993 p169; LJ v119 My 1, 1994 p55)

Morris, Richard. *The Edges of Science: Crossing the Boundary from Physics to Metaphysics.* Prentice-Hall, 1990. 244 p. 0–13–235029–7.

Contemporary theoretical physics has strayed into many areas of thought considered metaphysics. Morris has an impeccable knack for sorting out the philosophical consequences of scientific concepts. This is a solid, objective book on a topic too often distorted. (LJ v116 Mr 1, 1991 p63; SBF v27 Ap 1991 p69)

(*Note:* See also entries in Chapter Five under ''Social Aspects of Science— Science and Religion.'')

Quantum Physics

Davies, P.C.W. *The Ghost in the Atom: A Discussion of the Mysteries of Quantum Physics.* Cambridge University Press, 1986. 157 p. 0521307902.

Although quantum physics challenges many commonsense notions, Davies does a good job of explaining it. He describes that which is well established in the field, as well as some of the more fanciful areas of speculation. This book is a spinoff of a BBC radio program, and included are some interview transcripts. (Choice v24 Je 1987 p1582; Nature v324 D 4, 1986 p420; SBF v23 S 1987 p15)

Feynman, Richard. *QED: The Strange Theory of Light and Matter.* Princeton University Press, 1985. 158 p. 0691083886.

From a series of popular lectures given by the author, the sections in this book take on the challenge of explaining quantum electrodynamics without mathematics. (Choice v23 My 1986 p1424; LJ v111 Mr 1, 1986 p49; Nature v320 Ap 17, 1986 p661)

Gribbin, John. *In Search of Schrodinger's Cat: Quantum Physics and Reality.* Bantam, 1984. 302 p. 0553341030.

Physicist Erwin Schrodinger once explained quantum physics by using an analogy in which an unobserved cat in a black box could be both alive and dead *at the same time.* As Gribbin shows, this is no trick. He explains quantum theory in a historical context beginning with seventeenth century mechanics and, along the way, introduces readers to Einstein, Bohr, Planck, Schrodinger, and others. (BL v80 Ag 1984 p1580; KR v52 Je 15, 1984 p565; LATBR v109 Jl 1984 p1337)

Han, M. Y. *The Probable Universe: An Owner's Guide to Quantum Physics.* TAB, 1992. 146 p., gloss., index. 0–8306–4191–2.

Many think of quantum physics as incomprehensible to ordinary humans and irrelevant to daily life. Han, an expert in the field, proves both presumptions wrong. This remarkably lucid book is a fine first choice for anybody who wishes to read a nonmathematical explanation of the subject. (SBF v29 Ja 1993 p13)

Also Recommended

Herbert, Nick. *Quantum Reality.* Doubleday, 1987. Another introductory view. (CSM v90 Je 24, 1988 pB4)

Hey, Tony, and Patrick Walters. *The Quantum Universe.* Cambridge University Press, 1987. A review of quantum theory for those with some calculus. (SBF v23 Ja 1988 p147)

Pagels, Hans. *The Cosmic Code.* Penguin, 1982. An earlier but still popular book on the subject. Can be found in many libraries. (BL v78 Mr 1, 1982 p837; LJ v107 F 15, 1982 p442)

Relativity Theory

Einstein, Albert. *The Meaning of Relativity.* Crown, 1952. 164 p., index.

Einstein's own popularization of relativity theory is important for libraries to own, although it is not the first book to recommend on the subject. Still, there are some who prefer to go straight to the source.

Mook, Delo, and Thomas Vargish. *Inside Relativity.* Princeton University Press, 1987. 306 p., illus., gloss., index. 0–691–08472–6.

This collaboration between a physicist and a literature professor gives a thorough, balanced overview of relativity theory. Strongest on special relativity and its consequences. Not the most entertaining read, but well done for more serious students. Good use of illustrations. (Choice v25 Ap 1988 p1277; LJ v113 Ja 1988 p91; SBF v24 S 1988 p14)

Schwartz, Joseph, and Michael McGuinness. *Einstein for Beginners*. Pantheon, 1979. 173 p., illus., biblio. 0394505883.
   The title says it all. (PW v216 Ag 6, 1979 p90)

Will, Clifford. *Was Einstein Right? Putting Relativity to the Test*. Basic, 1986. 274 p. 0–465–09088–5.
   When Einstein first published relativity theory, many reacted by acknowledging that it worked mathematically, but by also questioning if it could be proven experimentally. Will shows how the theory has passed the experimental muster time and again. The combined theoretical/experimental approach of this book makes it one of the best for readers seeking introductory information. (BL v83 S 1, 1986 p12; LJ v111 S 15, 1986 p94; NYTBR v91 O 5, 1986 p46)

Also Recommended

Calder, Nigel. *Einstein's Universe*. Viking, 1979. Written for Einstein's centennial. (PW v217 Mr 28, 1980 p48; SBF v15 Mr 1980 p198)

Einstein, Albert. *Three Complete Books*. Wings, 1993. A nice reissue that would fill out any collection lacking these writings. Includes *Relativity, Ideas and Opinions*, and *Out of My Later Years*.

Russell, Bertrand. *The ABC of Relativity*. 4th ed. New American Library, 1985. Better books exist for general readers, but this was one of the most influential in its time.

Superconductivity

Hazen, Robert M. *The Breakthrough: The Race for the Superconductor*. Summit, 1988. 271 p., index. 0–671–65829–8.
   Research into superconductivity is breaking new ground in applied technologies, and the field has high hopes for the future. Just a tiny bit behind the times, this remains the best available book on the subject for general readers. (SBF v24 Mr 1989 p239; BL v84 My 15, 1988 p1555)

(*See also* entries in Chapter Thirteen under ''Superconductivity.'')

Symmetry

Pagels, Heinz. *Perfect Symmetry: The Search for the Beginning of Time*. Bantam, 1986. Photos, index. 0671465481.
   The origin of the universe and the Big Bang theory, and their relationships with time, which, the author believes, will be understood through the appreci-

ation of the "perfect symmetries" of the universe. (LJ v111 Ja 1986 p49; SBF v22 S 1986 p47; S&T v71 Ap 1986 p361)

Zee, Anthony. *Fearful Symmetry: The Search for Beauty in Modern Physics.* Macmillan, 1986.

Time and again, physical reality has proven to possess qualities that human beings perceive as "beautiful" and "elegant." Accordingly, physicists believe that a unified theory should appeal to human aesthetic values, which are founded in the symmetries, or regularities, of nature. (BL v83 Ja 15, 1987 p42; Choice v24 Jl 1987 p1724; PW v230 N 21, 1986 p42)

Thermodynamics

Atkins, Peter W. *The Second Law.* 2nd ed. Freeman, 1994. 216 p., photos, index. 0–7167–6006–1.

The second law of thermodynamics governs entropy—hot objects cool, but cool objects do not heat up. Much of this book deals with energy and entropy in chaotic systems. (Nature v312 N 15, 1984 p215)

Goldstein, Martin, and Inge F. Goldstein. *The Refrigerator and the Universe: Understanding the Laws of Energy.* Harvard University Press, 1993. 424 p., gloss., index. 0–674–75324–0.

Nature's laws of energy conservation and entropy are made clear in this unique book. The scope is impressive; the reader will find, first, a discussion of the molecular processes of heat production and dissipation, along with an introduction to probability calculations, then a far-ranging historical and speculative account of how energy works in our lives, our world, and the universe. Familiar examples aid comprehension. (SBF v30 Ap 1994 p70)

Also Recommended

Finn, C.B.P. *Thermal Physics.* Routledge, 1986. A college text for nonengineers.

Time

Coveney, Peter, and Roger Highfield. *The Arrow of Time: A Voyage Through Science to Solve Time's Greatest Mystery.* Fawcett, 1991. 378 p., index. 0–449–90630–2.

Time moves forward—doesn't it? Actually, that is the subject of a fair amount of scientific debate. The authors draw upon the second law of thermodynamics, which governs entropy, to argue that our commonsense perception of time is indeed true. Not easy stuff, but accessible to a reader willing to stick with it. (BL v87 Je 1, 1991 p1848; LJ v117 Mr 1, 1992 p47; Nature N 22, 1990 p356)

Gribbin, John. *Unveiling the Edge of Time.* Harmony, 1992. 051758591x.

General relativity makes possible a variety of strange consequences. This book explores some of them. Much of the text covers black holes and other exotic objects, and how space-time operates in and around them. (Choice v 30 Mr 1993 p1179; KR v60 Ag 1, 1992 p963; PW v239 Ag 24, 1992 p67)

Also Recommended

Davies, Paul. *The Physics of Time Asymmetry.* University of California Press, 1985. Discusses the difficulties inherent in explaining physical laws in a universe where time is relative.

OTHER RESOURCES

Periodicals for General Readers

American Journal of Physics. American Association of Physics Teachers, m., revs. 0002–9595.
   Devoted to the instructional and cultural aspects of physics. College level.

Physics Teacher. American Association of Physics Teachers, m. 0031–921x.
   Although written for teachers, the coverage is current, topical, nontechnical, and of broad general interest. Conveys a good sense of how physics can stimulate the imagination.

Physics Today. American Institute for Physics, m., revs. 0031–9228.
   The general publication of the Institute. Covers all topics of interest to physicists in a minimally technical way. Editorials and commentaries are also insightful.

Physics World. Institute of Physics, m., revs. 0953–8585.
   The general interest periodical of the British Institute of Physics contains more news clips and short, popular articles than can be found in its American counterpart, *Physics Today.*

Audiovisual Materials

The Atom—Future Quest. Hawkhill Associates, 1991 (27 minutes).
   From the series *Future Quest*, which brings several issues on science and technology to action video, this installment goes into the research centers at Fermilab and IBM, where viewers meet research scientists and see them at work. Covers various pure and applied subject areas. (SBF v27 Ag 1991 p185)

Atomic Physics and Reality. Jorlund Film–Denmark, 1985 (40 minutes, color).
   Albert Einstein and Niels Bohr had an irreconcilable difference of opinion regarding the quantum nature of atomic phenomena. Although over the years Bohr's quantum theory gained acceptance, it remained somewhat in question until 1982, when experimental proof settled the debate. This superior video

recounts the controversy and its resolution. For all viewers high school and up. (SBF v2 N 1987 p116)

Crystals and Lasers. Films for the Humanities, 1989 (58 minutes each).

Filmed as part of the prestigious Christmas lecture series at the British Royal Institution, these videos cover broad topics in optics, crystallography, catalysis, and applied physical sciences. The strength of the presentation lies in the demonstrations of laboratory techniques and technologies that are used in this field of research. (SBF v27 Ja 1991 p25)

A Lightly Story. Landmark Films, 1993 (25 minutes each).

This four-part series, produced by a Western Australia television station, covers the history and science of optics. Within the historical context of the presentation, there are some excellent demonstrations of optic phenomena, as well as the biological apparatuses by which we perceive light. Titles include "A Glimmer of Understanding," "Rainbow and Red Skies," "In a Darkened Room," and "Light Pressure for Some Heavy Ideas." The Australian accents and terminology take some getting used to, but do not detract from the fine content and production. (SBF v30 Ap 1994 p86)

The Mechanical Universe . . . and Beyond. Annenberg/CPB Collection, 1987 (30 minutes each, color).

The Mechanical Universe series is the fruit of a major, privately funded initiative to produce a complete video series covering the span of physics. Owning the entire twenty-six segments would be ideal, especially for college libraries, but the following installments are particularly recommended: "Atoms to Quarks," "Engine of Nature," "Entropy," "Magnets," "Optics," and "Velocity and Time." (SBF v23 N 1987 p116)

Physics in Action. Films for the Humanities, 1990 (20 minutes each).

Ten videos are included in the British series. Especially recommended are "The Laws of Motion" and "Relativity." (Phys T v31 Mr 1993 p174; SBF v27 N 1991 p249)

Also Recommended

The Day after Trinity. Pyramid Films, 1981 (two films, 88 minutes total). Oppenheimer and the detonation of the first atomic bomb.

Explorers of the Nucleus: Cyclotron Experiment #190. Indiana University, 1987 (60 minutes, color). This is a case study of a cyclotron experiment and the nuclear physicists who designed and implemented it. (SBF v23 Ja 1988 p181)

How Low Can You Go, I and II. Media Guild, 1987 (25 minutes). Extreme low temperature physics. From the *Physics of Matter* series. (SBF v24 N 1988 p113)

Laser Holography. Thomas Edison Foundation, 1987 (15 minutes, color). A short but informative introduction. (SBF v23 Ja 1988 p181)

CD-ROMs

Physics Infomall. Physics Teacher's CD-ROM Toolkit, 1993 (one disc).

Designed to meet the needs of high school physics teachers, this collection contains articles, handbooks, data sources, and various other instructional materials deemed essential.

Chapter Thirteen

Technology and Applied Sciences

The engineer, inspired by the law of economy and governed by mathematical calculation, puts in accord with universal law. He achieves harmony.

Le Corbusier, Toward a New Architecture

The way to solve the conflict between human values and technological needs is not to run away from technology, that's impossible. The way to resolve the conflict is to break down the barriers of dualistic thought that prevent a real understanding that technology is not an exploitation of nature, but a fusion of nature and the human spirit into a new kind of creation that transcends both.

Robert Pirsig, Zen and the Art of Motorcycle Maintenance

If a man . . . make a better mouse-trap than his neighbour, tho' he build his house in the woods, the world will make a beaten path to this door.

Ralph Waldo Emerson

Science and technology are two distinct cultural phenomena. A common but erroneous view of technology is that it is the end product of a scientific quest. In this view, technology borrows ideas from the pure, theoretical sciences and puts that knowledge to work in practical ways. In fact, technology often progresses parallel to, but independent of, science. It can be argued that, historically, technology is older than science, for even ancient humans achieved magnificent feats of engineering, but science, as we now define it, was not born until approximately the seventeenth century, and then chiefly in Europe. Our species's use of technology is so primordial that Homo habilis (literally, "handy man") is so named because of its use of stone and flint tools.

Therefore, being technologically literate is not exactly the same thing as being scientifically literate. Even the greatest of theoretical scientists might be stymied by the complexities of programming a VCR. As technology—especially electronic technology–increasingly and ineluctably changes the familiar ways we do things, there is confusion, sometimes resistance, and in a great many cases, even when acceptance is won, the inner workings of new devices remain mysterious to people who use them daily. One aspect of technological literacy involves breaking that barrier of ignorance.

For example, many people use computers regularly in their jobs, but have no glimmer of knowledge of what makes them work. That is fine—until something changes. When asked to make informed, evaluative assessments of system capabilities or to solve problems, such people are ill equipped to meet the demands of an evolving situation. More technologically literate computer users are not likely to put technical support services out of business, but they can become more efficient and articulate when discussing system problems and needs. Toward that goal, several of the books listed in this chapter are of the how-to or user education variety.

At a broader level, technological literacy is critical for our society. "Technophobia" is a cultural pathology resulting from the collective frustrations, anxiety, and ignorance of many people who fear that the potential risks of high technology are greater than any possible benefits. At the other extreme, "technophilia" is the utopian passion of zealots who advocate unhesitating development in the name of progress. Sound management of technology must find a middle ground that is cautious, but also forward-thinking. Technology cannot make those decisions for us.

TECHNOLOGY—GENERAL

Reference

Applied Science and Technology Index. H. W. Wilson, 1958– . q.
Widely used and recognized in libraries, this index covers all fields of technology, including aerospace, chemical engineering, computers, construction, general engineering, petroleum science, robotics, and telecommunications. A good place to start a research paper or search for consumer information. (RSSEMA p119)

Blackburn, David, and Geoffrey Holister. *G. K. Hall Encyclopedia of Modern Technology.* G. K. Hall, 1987. 248 p. 0–8161–9056–9.

The greatest attribute of this general encyclopedia is its splendid illustrations. Topically arranged for the interested layperson. (RSSEMA p121)

Dictionary of Computing. 3rd ed. Oxford Science Publications, Oxford University Press, 1990. 510 p. 0–19–853825–1.

The third edition contains 550 new entries for terms in computing and the associated fields of electronics, mathematics, and logic. Contributions come from experts in America and Europe. (SBF v27 Ap 1991 p72)

Encyclopedia of Building Technology. Henry Cowan, ed. Prentice-Hall, 1988. 322 p. 0132755203.

Cowan, an editor of *Architectural Science Review*, has compiled 210 short articles that span the entire field of building technology. The contributors are experts in their fields. The bibliographies and reading lists accompanying each article are highly useful. (Choice v25 Ja 1988 p744; SciTech v12 Mr 1988 p27)

Lesko, Matthew. *Lesko's New Tech Sourcebook: A Directory to Finding Answers in Today's Technology-Oriented World.* Harper and Row, 1986. 726 p. 0–060–96036–1.

Compiled with the layperson specifically in mind, this directory meets its goal of connecting people who need answers with professionals who can help them. Nearly 700 technical subject areas are covered. A new edition of this very valuable reference source would be welcome. (BL v82 Jl 1986 p1598; Choice v23 Je 1986 p1523; LJ v111 Ja 1986 p72)

McGraw-Hill Dictionary of Engineering. Sybil Parker, ed. McGraw-Hill, 1984. 659 p. 0070454124.

This standard dictionary is an effective roadmap to the concepts and vernaculars of all fields of engineering.

Steinberg, Mark, and Sharon Cosloy. *The Facts on File Dictionary of Biotechnology and Genetic Engineering.* Facts on File, 1993. 195 p. 0–8160–1250–4.

In twenty-five to fifty word definitions, this dictionary codifies the "basic modern vocabulary" of the field for a wide audience, which includes students and others with a personal or professional involvement. In addition to new, specialized topics, it also covers fundamental concepts in the biological sciences.

Also Recommended

Bianchina, Paul. *Illustrated Dictionary of Building Materials and Techniques.* TAB, 1986. Defines over 4,000 terms and includes sixty pages of appendices with helpful information.

Encyclopedia of Computer Science and Engineering. Anthony Ralston and Edwin D. Reilly, Jr., eds. Van Nostrand, 1992. Superb coverage of all technical fields associated with computing. (Choice v30 My 1993 p1438; SciTech v17 Ja 1993 p5)

Jane's Encyclopedia of Aviation. Portland House, 1989. 948 p. 0517691868. The source for information on airplanes and air history.

Raymond, Eric S. *The New Hacker's Dictionary.* MIT Press, 1993. The definitive lexicon of computer hackers. This book is so witty and so cleverly organized that it can actually be read for pleasure as well as used for reference. (ARBA v24 1993 p699; LJ v119 Mr 1, 1994 p56)

World of Invention. Gale, 1993. Over 1,100 alphabetical entries on both famous and little-known inventions. Also has good biographical coverage. (LJ v119 F 1, 1994)

(*Note*: Other reference style resources are listed in various sections of this chapter, as noted.)

Autobiography and Biography

Hodges, Andrew. *Alan Turing: The Enigma.* Simon and Schuster, 1983. 587 p. 0–671–49207–1.
   Turing forwarded early computer science as no other; his work is still core to scholars in the artificial intelligence field. He was, as this book's title suggests, an enigma: a homosexual, the object of prejudice and abuse, he was also entrusted with government secrets. A fascinating work that serves for lay readers and scholars. (Choice v21 Ap 1984 p1157; KR v51 N 1, 1983 p1161; LJ v108 D 1, 1983 p2246)

Josephson, Matthew. *Edison: A Biography.* Wiley, 1992. 0–471–548065–5.
   This biography does equal justice to Edison's creative genius and his entrepreneurial drive. The author captures the era admirably; several major figures from history appear as important players in Edison's life story. (NYTBR v97 Mr 22, 1992 p28)

Philip, Cynthia Owen. *Robert Fulton: A Biography.* Watts, 1985. 384 p., illus. 0–531–09756–0.
   Known for his invention of the steam-powered riverboat, Fulton was a versatile technologist as well as an aspiring, often frustrated artist. The author scoured primary source materials in writing this biography. (Atl v256 O 1985 p106; BL v82 S 15, 1985 p98; LJ v110 Ag 1985 p90)

Poundstone, William. *Prisoner's Dilemma.* Doubleday, 1992. 288 p. 0–385–41567–2.
   A creative and stimulating introduction to game theory and computer mathematics through a biographical survey of the life of John Von Neumann. (BL v88 F 15, 1992 p1077; Byte v17 My 1992 p370; LJ v117 Ja 1992 p144)

Rabinow, Jacob. *Inventing for Fun and Profit.* San Francisco Press, 1990. 0911302646.
   An octogenarian when he wrote this, his first book, Rabinow put a lifetime's

worth of creativity and perspicacity into it. The title, which sounds a bit kitschy, is not really representative of the content; what the reader finds is an autobiographical exploration of an inventive mind. During his career, Rabinow applied his insatiable curiosity to a number of tasks, and never seemed willing to dismiss a problem as unsolvable. (Sci My 18, 1990 p892)

Also Recommended

Ada, the Enchantress of Numbers. Betty Toole, ed. Strawberry, 1992. Byron's daughter, a gifted mathematician and proto-programmer.

The Biographical Dictionary of Scientists, Engineers and Inventors. David Abbott, ed. Blond Educational, 1985. A collection of short biographical entries following a chronological introduction. (TES S 20, 1985 p57)

Lacey, Robert. *Ford: The Men and the Machine.* Little, Brown, 1986. The life of Ford and his automotive industry. (Bus W D 15, 1986 p10; Choice v24 Ap 1987 p1186; LJ v111 Mr 1, 1986 p46)

Niske, W. Robert. *The Life of Wilhelm Conrad Roentgen: Discoverer of the X-Ray.* University of Arizona Press, 1971. A readable biography; contains translations of some original works. (BRD 1971 p1009; Choice v8 O 1971 p1055; Sci v173 S 24, 1971 p1225)

Ritchie, David. *Computer Pioneers.* Simon and Schuster, 1986. The origins and development of computers and those who pioneered the technology. (LJ v112 Mr 1, 1987 p31; SciTech v10 O 1986 p5)

History of Technology

Garrison, Ervan. *A History of Engineering and Technology: Artful Methods.* CRC Press, 1991. 275 p., illus., index. 0–8493–8836–8.
This introductory history covers people, concepts, and events that contributed significantly to the history of engineering. The author emphasizes ingenuity and artfulness in engineering methods, and he uses fine, detailed photos and drawings to illustrate visually how different engineering problems have been solved. Concludes with a good chapter on technology and the future. (Choice v30 S 1992 p1243)

Inkster, Ian. *Science and Technology in History: An Approach to Industrial Development.* Rutgers University Press, 1991. 391 p. 0813516803.
The dynamics of industrial progress are interwoven with society, culture, and economy. Inkster investigates these relationships within a historical continuum. Although written to stand up to scholarly scrutiny, no technical knowledge is necessary. (Choice v29 Mr 1992 p1100; SciTech v15 O 1991 p25)

Pacey, Arnold. *Technology in World Civilization: A Thousand Year History.* MIT Press, 1990. 238 p., illus., maps, index. 0–262–16117–6.

Covering the period from 700 to 1970, this sweeping survey takes a global view of technological discovery and development. Pacey, a British historian, demonstrates how many key inventions resulted from technological "dialogues" between cultures. Since most general histories tend to be Eurocentric, the international scope of this book makes it unique. (LJ v116 Mr 1, 1991 p64; PW v237 Mr 16, 1990 p59; New Sci v128 O 6, 1990 p56)

Also Recommended

De Camp, L. Sprague. *The Ancient Engineers.* Ballantine, 1974. A perennial favorite on the marvels of ancient technology. (BRD 1963 p254; LJ v88 Mr 1, 1963 p1019; Sci v140 Je 28, 1963 p1387)

Encyclopedia of the History of Technology. Ian McNeil, ed. Routledge, 1990. Parts include "Materials," "Power and Organization," "Transportation," "Communications," and "Technology and Society." (RSSEMA p123)

Williams, Trevor. *The History of Invention.* Facts on File, 1987. 352 p. A sweeping introduction to human creativity and inventiveness. Lavish illustrations give this book a powerful visual appeal. (Choice v25 F 1988 p928; LJ v113 Mr 1, 1988 p33; SBF v23 My 1988 p301)

General Topics

Brennan, Richard. *Levitating Trains and Kamikaze Genes: Technological Literacy for the Nineties.* Wiley, 1990. 262 p., illus., index.
   Reading one book does not make a person technologically literate, but this is perhaps the best crash course available. In nine chapters, the author delineates new realms of technology in engineering, computer science, and biomedicine. Each chapter includes a summary of key concepts and a self-test. (LJ v116 Mr 1, 1991 p64; New Sci v128 O 13, 1990 p49; SBF v26 S 1990 p52)

Introduction to Appropriate Technology. R. J. Congdon, ed. Rodale Press, 1977. 205 p. 0878571884.
   The idea of using "appropriate" technology—just the right equipment for the job, no more or less—is at the heart of many conservation plans and has become an influential movement in contemporary engineering. This concept is intended to humanize technology. (SBF v14 S 1978 p75; SLJ v24 Mr 1978 p142)

Lewis, H. W. *Technological Risk.* Norton, 1990. 353 p., index. 0-393-02883-6.
   Risk is associated with the introduction of every new technology. Often, the perceived risk is greater than the actual. Lewis provides a realistic assessment of various types of hazards and how to make sense of how serious the risks are. (KR v58 S 1, 1990 p1230; LJ v115 S 15, 1990 p96; PW v237 Ag 24, 1990 p50)

McCauley, David. *The Way Things Work*. Houghton Mifflin, 1988. 384 p., illus., index. 0–395–42857–2.

Nobody makes learning about technology more enjoyable than McCauley. Here, he looks at four areas—movement, elements, light and sound, and electricity and automation—and, in his characteristically ebullient style, shows what makes things tick. The drawings are magnificent. (LJ v114 Ja 1989 p99)

Nye, David. *American Technological Sublime*. MIT Press, 1994. Illus., index. 0–262–14065–x.

In America, perhaps more than any other country on Earth, technology is glorified as monuments to national pride. The "sublime" is an expression of ideal cultural identity. Focusing on certain monuments, such as the Empire State Building, and events, such as the detonation of the Atomic Bomb, Nye examines the aesthetic of the sublime and how it has changed over time. (*Note*: No reviews are available for this book.)

Pacey, Arnold. *The Maze of Ingenuity: Ideas and Idealism in the Development of Technology*. MIT Press, 1992. 306 p., index. 0–262–66075–x.

Any great engineering feat begins with a vision. That is not to say, however, that all visions are equally worthy. Pacey argues for a technological ethic based on need and functionality. (Nature v356 Ap 9, 1992 p488; SBF v28 Ag 1992 p166)

Penzias, Arno. *Ideas and Information: Managing in a High-Tech World*. Norton, 1989. 224 p., index. 0–393–02649–3.

This book by a Nobel laureate and vice president of Bell Laboratories examines some of the ways that the tools and machines that human beings create empower them to reach greater cognitive heights. (Am Sci v77 My 1989 p312; BL v85 Ja 15, 1989 p823; LJ v114 F 1, 1989 p80)

Petrowski, Henry. *The Evolution of Useful Things*. Knopf, 1992. 288 p. 0679412263.

Tools like knives, pencils, paper clips, aluminum cans, and telephones are so familiar that it's easy to take them for granted. In fact, though, they evolved into their present forms after trial and error. Petrowski shows why the adage "form follows function" is true. (BL v89 D 1, 1992 p639; Choice v30 Ap 1993 p1335; Sci v260 My 21, 1993 p1166)

Postman, Neil. *Technopoly: The Surrender of Culture to Technology*. Knopf, 1992. 222 p. 0394582721.

A "technopoly," according to the author, is a system in which society embraces technology as a solution to all cultural problems. In such a culture, technology overwhelms other areas of thought and expression, such as art, philosophy, and morality. Postman claims that, in America today, we are close to living in such a state. (BL v88 D 1, 1991 p657; LJ v117 Ja 1992 p169; NYTBR v97 My 10, 1992 p20)

Technology and the Future. 6th ed. Albert Teich, ed. St. Martin's, 1993. 383 p. 0–312–06747–x.

As director of technology at the American Association for the Advancement of Science, Teich knows the forces that shape technology and society very well. Twenty years have passed since publication of the fifth edition. Within these articles there are diverse viewpoints and disagreements, but rationality prevails. (SBF v30 Ja 1994 p10)

Also Recommended

Barbour, Ian. *Ethics in an Age of Technology.* Harper, 1993. Ethics, Christianity, and appropriate technology.

Cone, Robert J. *How the New Technology Works: A Guide to High Tech Concepts.* Oryx Press, 1991. In seventeen brief chapters, major high-tech concepts are defined and described. (ARBA v23 1992 p689; LJ v117 Mr 1, 1992 p47)

Copp, Newton H., and Andrew Zanella. *Discovery, Innovation and Risk.* MIT Press, 1993. A series of case studies designed to give students insight into technological and engineering reason. (Choice v31 S 1993 p148; New Sci v138 My 22, 1993 p43; SBR v29 Ja 1993 p143)

Flatow, Ira. *They All Laughed . . . From Light Bulbs to Lasers: The Fascinating Stories Behind the Great Inventions that Have Changed Our Lives.* Harper-Collins, 1992. The host of a popular National Public Radio program brings his entertaining style to this book, in which he revels in the irony and the ingenuity behind great discoveries. (BL v88 Jl 1992 p1908; LJ v117 Jl 1992 p115; SBF v29 Mr 1993 p41)

Machines and Inventions. Time-Life Books, 1993. Devices found at home and at the office, and how they work. (SBF v30 Ap 1994 p78)

Technology 2001. Derek Leebaert, ed. MIT Press, 1991. Twelve visions of technology in the twenty-first century by major corporate researchers. (Choice v29 O 1991 p313; SBF v27 My 1991 p104)

TECHNOLOGY AND APPLIED SCIENCE—SPECIAL TOPICS

Agriculture

Rodale, Robert. *Save Three Lives: A Plan for Famine Prevention.* Sierra Club, 1991. 256 p., biblio., index. 0–87156–621–4.

Rodale, the famous naturalist who completed this book just prior to his death, contends that the recent waves of famine in African countries are due largely to unsound agricultural practices that undermined delicate local ecologies. His solution is to abandon large-scale relief projects in favor of local initiatives designed to "save three lives." (BL v88 S 1, 1991 p12; KR v59 Ag 1, 1991 p996; LJ v117 Mr 1, 1992 p42)

Solbrig, Otto T., and Dorothy Solbrig. *So Shall You Reap: Farming and Crops in Human Affairs*. Island Press, 1994. Illus., biblio., index. 0–55963–308–5.

Historians note that the development of agriculture was one of the monumental accomplishments of ancient civilizations. Today, we are on the verge of a new revolution in crop science. The authors offer insights on how agricultural sustainability can be achieved. (LJ v119 Ap 1, 1994 p122)

Soule, Judith D., and Jon K. Piper. *Farming in Nature's Image: An Ecological Approach*. Island Press, 1992. 286 p., index. 0–933280–89–0.

A hallmark principle of sustainable agriculture is that farming practices must imitate the ecological conditions under which a crop grows naturally. This commonsense approach has too often been ignored by corporate agricultural entities, with sometimes calamitous results. The authors contend that an ecological approach to farming can bring better harvests to a hungry world. (SBF v28 My 1992 p104)

Also Recommended

Erickson, Jonathan. *Gardening for a Greener Planet*. TAB, 1992. Environmentally conscious home gardening. (SBF v28 Je 1992 p142)

Lacey, Richard. *Hard to Swallow: A Brief History of Food*. Cambridge University Press, 1994. An anecdotal history of food and the food industry, with comments on human diets.

Millichap, J. Gordon. *Environmental Poisons in Our Food*. PNB, 1993. A consumer guide to food chemicals. (SBF v30 Mr 1994 p37)

National Research Council. *Alternative Agriculture*. National Academy Press, 1989. Popularized version of a committee report on alternative farming. (Choice v27 Ap 1990 p1342)

Architecture and Buildings

Alexander, Christopher. *The Timeless Way of Building*. Oxford University Press, 1979. 552 p.; *A Pattern Language*. Oxford University Press, 1977. 1171 p. 0195019199.

These two books have influenced a school of architectural thought in which cognitive patterns, which we commonly perceive and intuitively look for in our environment, can be incorporated into buildings to create a "timeless" style. (LJ v102 D 1, 1977 p2424; LJ v103 Je 15, 1978 p1260)

Brand, Stewart. *How Buildings Learn: What Happens after They're Built*. Viking, 1994. Photos. 0–670–83515–3.

Economically and aesthetically, it often makes sense to adapt an existing building for new purposes than to build a new one. "Evolutionary design" might be the next wave in architecture. (BL v90 Je 1, 1994 p175; KR v62 S15, 1994 p513; LJ v119 My 1, 1994 p98)

Mark, Robert. *Light, Wind and Structure: The Mystery of Master Builders.* MIT Press, 1990.

Written as a companion to NOVA's *The Mystery of Master Builders* production, this book analyzes the architecture in three watershed eras: (1) Ancient Rome, (2) High Gothic, and (3) Renaissance/Baroque. The author shows how understanding historical structures is relevant to the tasks of architects today. (Choice v28 S 1990 p94; LJ v115 My 15, 1990 p76; SBF v27 Ja 1991 p8)

Also Recommended

National Audubon Society. *Audubon House: Building the Environmentally Responsible, Energy Efficient Office.* Wiley, 1994. A team of architects, engineers, designers, and contractors renovated a decaying building to create an environmentally responsible prototype office.

Aviation

Captain X. *Unfriendly Skies.* Berkeley Books, 1989. 236 p. 0425121828.

Captain X, a veteran pilot who remains anonymous for fear of reprisals from his employer, documents horrifying conditions in the air travel industry. Near misses, deferred maintenance, overworked controllers—these are just some of the unsafe situations the captain claims are common. (BL v85 My 15, 1989 p1579; LATBR Ap 30, 1989 p6; LJ v114 Je 1, 1989 p140)

Irving, Clive. *Wide Body: The Triumph of the 747.* Morrow, 1992. Photos, biblio., index. 0–688–09902–5.

In the early 1960s, the struggling Boeing aviation company took a tremendous gamble by developing the 747, a plane that many believed would never fly. The plane not only flew, it changed the history of commercial air travel. (LJ v118 Mr 1, 1993 p43)

Kaplan, James. *The Airport.* Morrow, 1994. 278 p., photos, index. 0–688–09247–0.

To travelers, New York's JFK Airport can seem like a dizzying maze. Managing so busy, large, and vital an institution can sometimes be no less confusing. This is a look at the culture and organization of a big city airport, which is almost like a city unto itself.

Reiss, Bob. *Frequent Flyer: One Plane, One Passenger, and the Spectacular Feat of Commercial Flight.* Simon and Schuster, 1994. 304 p. 0–671–77650–9.

Experienced flyers sometimes take air travel for granted. What goes on before, during, and after a flight, however, is never taken for granted by the crew and mechanics. Here is a look at the small but critical details of commercial aviation. A good book to read on the plane. (LJ v119 Ja 1994 p156)

Rich, Ben, and Leo Janus. *Skunk Works*. Little, Brown, 1994. 370 p., index. 0–316–74330–5.

The Skunk Works is the Lockheed Corporation's research and development wing for aviation defense technologies. As the nickname implies, the engineers and pilots who worked there possessed gritty resilience, dogged determination, and a strong sense of mission. Rich headed the Skunk Works and writes with pride of its accomplishments. (*Note*: No reviews are available for this book.)

Also Recommended

Batchelor, John, and Christopher Chant. *Flight: The History of Aviation*. McGraw-Hill, 1990. 192 p. A concise overview of the story of the dream and reality of human flight.

Bergman, Jules, and David Bergman. *Anyone Can Fly*. Doubleday, 1986. For those who wish to learn the rudiments of aviation, or for others who would just like to know something about how it is done. (LJ v112 Mr 1, 1987 p36)

Bryan, C.D.B *The National Air and Space Museum*. Abrams, 1988. A visual and textual documentation of this popular museum. (SBF v25 S 1989 p19)

Dorsey, Gary. *The Fullness of Wings: The Making of a New Daedalus*. Viking, 1990. 350 p., photos. 0–670–82444–5. The construction and flight of a light-weight, human-powered aircraft. (LJ v116 Mr 1, 1991 p60)

Biotechnology and Genetic Engineering

Fox, Michael. *Superpigs and Wondercorn: The Brave New World of Biotechnology and Where It May Lead*. Lyons and Burford, 1992. 209 p. 1558211829.

A good general overview of the sociological and technical aspects of genetic engineering. Very readable and entertaining. (BL v89 S 1, 1992 p6; KR v60 Jl 15, 1992 p894)

Kimbrell, Andrew. *The Human Body Shop: The Engineering and Marketing of Life*. Harper, 1993. 348 p. 0062505246.

Whether in the sale of organs or the design of drugs to enhance bodily attributes, a market mentality has come to influence how we view our bodies. The author argues that we must respect the sanctity of the human body lest it become just another commodity. (BL v89 Ap 15, 1993 p1474; Nature v365 S 1993 p304; NYTBR v98 S 12, 1993 p26)

Smith, George. *The New Biology: Law, Ethics, and Biotechnology*. Plenum, 1989. 303 p. 0898854482.

Biotechnology has advanced so rapidly that it has left the law far behind. The legal system is just beginning to grapple with the unique and sometimes odd issues brought to bear by genetic technologies. (Choice v27 Ap 1990 p1343)

Witt, Steven C. *BriefBook: Biotechnology, Microbes, and the Environment.* Center for Science Information, 1990. 219 p. 0912005033.

Biotechnologically engineered microorganisms are "infinitely small, historically huge." The Center for Scientific Information published this as a citizen's primer to biotechnology and related issues. The style is a bit breezy; this works best for beginners. (Choice v28 F 1991 p956)

Also Recommended

Siddhanti, Smita. *Multiple Perspectives on Risk and Regulation: The Case of Deliberate Release of Genetically Engineered Organisms into the Environment.* Garland, 1991. Environmental risks, problems, and solutions.

Teitelman, Robert. *Gene Dreams: Wall Street, Academia, and the Rise of Biotechnology.* Basic, 1989. Genetic engineering as big business. (LJ v 115 Mr 15, 1990 p45; SBF v25 My 1990 p238)

(*Note*: See also entries in Chapter Seven under "Molecular Biology" and in Chapter Eight under "Biochemistry.")

Computer Sciences—General

Henle, Richard, and Boris Kuvshinoff. *Desktop Computers in Perspective.* Oxford University Press, 1992. Illus., index. 0–19–507031–3.

Personal computer books go out of date rapidly, but this one should retain its shelf life longer than most. Its superior organization, simple but not patronizing language, and effective illustrations and figures make it valuable to novices; additional information on special topics such as modems, windows, and Unix further enhances its value. (Choice v30 Mr 1993 p1192; SBF v29 Ja 1993 p5)

Kidder, Tracy. *Soul of a New Machine.* Little, Brown, 1981. 293 p. 0–316–49170–5.

This Pulitzer Prize winner tells of the researchers who developed a new supermini computer at Data General Corporation. An enduring favorite, especially among hackers, this book captures the zest and idiosyncrasies of an R&D clan in the computer industry. (BL v78 O 15, 1981 p274; Bus W S 7, 1981 p8; LJ v106 Ag 1981 p1558)

Levy, Steven. *Insanely Great: The Life and Times of Macintosh, the Computer that Changed Everything.* Viking, 1994. 292 p. 0–670–85244–9.

The Mac did indeed change the personal computer industry. This is the story of its development, which began with the stated goal of creating a PC that would be "insanely great." (LJ v119 Ja 1994 p150)

Mandell, Stephen. *Dr. Mandell's Ultimate Personal Computer Desk Reference.* Rawhide, 1994. 573 p., index. 0–9637426–12.

Billed by the author as the computer user's equivalent of *Physicians' Desk*

Reference, this is a remarkably complete and comprehensible resource with over 300 topics and reviews. A good book to keep next to your PC.

Mungo, Paul, and Bryan Clough. *Approaching Zero: The Extraordinary Underworld of Hackers* . . . Random House, 1992. 246 p., biblio. 0–679–40938–6.

Cyberspace, the territory that is connected by computer networks, is also the frontier of a new band of outlaws who go by names like Captain Zap and Acid Phreak. Although they are unknown and unseen, these individuals mirthfully undermine computer security and spawn viruses that can wreak havoc on systems upon which many of us depend. (BL v89 F 1, 1993 p960; Byte v18 Jl 1993 p49; LJ v118 Mr 1, 1993 p98)

Pickover, Clifford A. *Computers, Pattern, Chaos and Beauty*. St. Martin's, 1990. 394 p., illus., gloss., index. 0–312–04123–3; *Computers and the Imagination*. St. Martin's, 1992. 419 p., illus., gloss., index. 0–312–06131–5.

These companion volumes integrate computer science with chaos theory, fractal mathematics, and a large measure of graphic artistry in exploring the stunning ways in which computers can visually represent worlds both real and imagined. For a reader willing to invest some time and mental energy, these books are vastly rewarding. Even those who merely wish to sample the subject will be dazzled by the computer simulations. (Byte v16 N 1991 p447; Choice v29 Jl 1992 p1714; LJ v116 Mr 1, 1990 p61)

Reid, T. R. *The Chip: How Two Americans Invented the Microchip and Launched a Revolution*. Simon and Schuster, 1985. 320 p. 0–671–45393–9.

A look at the creative routes taken in the development of microchip technologies, essential to modern computing. (LJ v111 Mr 1, 1986 p46)

Also Recommended

Computers Before Computing. William Aspray, ed. Iowa State University Press, 1990. Counting machines and computational devices before the advent of computers. (LJ v116 Mr 1, 1991 p61)

Forester, Tom, and Perry Morrison. *Computer Ethics: Cautionary Tales and Ethical Dilemmas in Computing*. MIT Press, 1990. The only accessible book available addressing issues related to the ethical and legal aspects of computer security, privacy, and so on. (Choice v28 Mr 1990 p1171; LJ v116 Mr 1, 1991 p61; SBF v26 N 1990 p101)

Hafner, Katie, and John Markoff. *Cyberpunk: Outlaws and Hackers on the Computer Frontier*. Simon and Schuster, 1991. This book is as much about the minds of renegade hackers as it is about the technology that they have mastered. (BL v87 Jl 1991 p2015; Byte v16 Jl 1991 p351; LJ v117 Mr 1, 1992 p45)

McGraw-Hill Personal Computer Programming Encyclopedia. McGraw-Hill, 1989. Information on major and hard to find programming languages. (ARBA v21 1990 p717)

Vassiliou, M. S., and J. A. Orenstein. *Computer Professional's Quick Reference.* McGraw-Hill, 1992. For people who work in a multicomputer environment and need an authoritative cross-reference. (BL v88 Jl 1992 p1958; SBF v28 Je 1992 p133)

Yourdon, Edward. *The Decline and Fall of the American Programmer.* Prentice-Hall, 1992. How hubris and aggressive overseas competition led to the alleged decline of America's computer programming supremacy. (Byte v17 Jl 1992 p284)

(*Note*: Although not strictly necessary, and even though they become outdated at a notoriously rapid rate, a core of computer guidebooks will be popular additions to any science literacy collection. These might include beginners' guides, handbooks and manuals, and reference sources for major operating systems, software packages, programming languages, applications technologies, and so on.)

Computer Sciences—Artificial Intelligence and Robotics

Crevier, Daniel. *AI: The Tumultuous History of the Search for Artificial Intelligence.* Basic/HarperCollins, 1993. 386 p., index. 0–465–02997–3.
    While artificial intelligence (AI) is the wave of the future in computer science, its history dates back some thirty years. Crevier interviewed many of the leading researchers in the field for this book. (BL v89 Ap 1, 1993 p1388; Byte v18 Jl 1993 p49; LJ v118 My 1, 1993 p112)

Jubak, Jim. *In the Image of the Brain: Breaking the Barrier Between the Human Mind and Intelligent Machines.* Little, Brown, 1992. 348 p., biblio., index. 0–316–47555–6.
    Neural networks, computer circuitry which mimics human neural structure, represent a great advance in artificial intelligence technology. The ultimate goal is to make a computer that works like the human brain. (BL v88 My 1, 1992 p1564; KR v60 Ap 1, 1992 p444; LJ v118 Mr 1, 1993 p43)

Kurzweil, Raymond. *The Age of Intelligent Machines.* MIT Press, 1990. 565 p., biblio., photos, index. 0–262–11121–7.
    This book provides the technical background for understanding what AI can and cannot do, and for assessing its human, social, and philosophical implications. Kurzweil's approach is enthusiastic and his scope is broad. (BL v87 S 15, 1990 p117; LJ v116 Mr 1, 1991 p61; NYTBR v95 S 9, 1990 p36)

Moravec, Hans. *Mind Children.* Harvard University Press, 1988. 214 p. 0674576160.
    One of the most fanciful technological excursions in recent publishing history. The author of this book envisions a future in which computer science and robotics have become so advanced that a human mind can be downloaded into an

automaton. This is not science fiction—remember, Harvard University Press is the publisher. (BL v85 O 1, 1988 p202; KR v56 S 1, 1988 p1307)

Penrose, Roger. *The Emperor's New Mind: Concerning Computers, Minds and the Laws of Physics.* Oxford University Press, 1989. 466 p., illus., biblio. 0–19–851973–7.

Penrose, a renowned theoretical physicist and colleague of Stephen Hawking, ponders the nature of human and machine thought and puts forward the view that some facets of human cognition can never be emulated by computer. He foresees the development of a mature theory of intelligence as being critical to the future of science. (Choice v27 Ap 1990 p1341; LJ v115 Mr 1, 1990 p45; Nature v341 O 5, 1989 p393)

Also Recommended

Asimov, Isaac, and Karen Frenkel. *Robots: Machines in Man's Image.* Harmony, 1985. Although written prior to many AI innovations, this gives a good history and summary of topics up to the time of publication. (BL v81 My 15, 1985 p1282; LJ v110 Je 1, 1985 p136)

Campbell, Jeremy. *The Improbable Machine.* Simon and Schuster, 1989. What AI research can tell us about how the brain works. (LJ v115 Mr 1, 1990 p43; NYTBR v94 D 24, 1989 p12)

Caudill, Maureen, and Charles Butler. *Naturally Intelligent Systems.* MIT Press, 1989. A slightly more sophisticated book on neural nets. (Choice v28 S 1990 p158; LJ v115 Ja 1990 p144)

Haugeland, John. *Artificial Intelligence: The Very Idea.* MIT Press, 1985. Psychological and philosophical views on AI. (Choice v23 Ap 1986 p1243; LJ v111 F 1, 1986 p88; SBF v25 N 1989 p59)

Robotics. Marvin Minsky, ed. Doubleday, 1986. Ten essays on robotics, especially AI topics. (LJ v111 Mr 1, 1986 p31)

Staugaard, Andrew. *Robotics and AI: An Introduction to Applied Machine Intelligence.* Prentice-Hall, 1987. When AI and robotics research came together, a new breed of machine was created. This book gives the technical background. (Choice v24 My 1987 p1434; SciTech v11 Mr 1987 p36)

Thro, Ellen. *Robotics: The Marriage of Computers and Machines.* Facts on File, 1993. For high school and public libraries. (BL v89 Jl 1993 p1956; SBF v29 Je 1993 p145)

Computer Science—Artificial Life

Emmeche, Claus. *The Garden in the Machine.* Princeton University Press, 1994. c. 188 p. 0–691–03330–7.

A technical and philosophical journey into artificial life. The approach is interdisciplinary. Originally published in 1992 in France.

Levy, Steven. *Artificial Life: The Quest for a New Creation.* Pantheon, 1992. 390 p., index. 0–679–40774–x.
Computer viruses behave so much like biological viruses that some argue that they are indistinguishable. Computers are capable of creating emergent systems that do things it was once believed only living things could do, and that fact raises questions about just what defines a living thing. (Bus W Jl 27, 1992 p12; LATBR S 13, 1992 p15; LJ v117 Je 1, 1992 p164)

Also Recommended

Kelly, Kevin. *Out of Control: The Rise of Neo-Biological Civilization.* Addison-Wesley, 1994. Emerging computer technologies that are modelled after natural-istic systems. (KR v62 My 1, 1994 p608; NYTBR v99 J 24, 1994 p26; PW v241 Je 6, 1994 p53)

Computer Sciences—Networking

Krol, Ed. *The Whole Internet User's Guide and Catalog.* O'Reilly and Asso-ciates, 1992. 1–56592–025–2.
A bestseller among Internet books, this ranks high in terms of usefulness to all levels of users. Explains all basic commands and advanced features. (A Lib v24 Ap 1993 p334; Am Sci v81 Mr 1993 p194; LJ v117 D 1992 p193)

LaQuey, Tracey. *The Internet Companion: A Beginner's Guide to Global Net-working.* Addison-Wesley, 1993. 0–201–62224–6.
Internet connectivity has changed the ways people communicate and revolu-tionized the information environment. To many unfamiliar with the Net, how-ever, getting connected and learning its protocols can be intimidating. This book is recommended as a practical introduction that contains information beneficial to more experienced networkers, too. (LJ v118 F 1, 1993 p41)

Rheingold, Howard. *The Virtual Community: Homesteading on the Electronic Frontier.* Addison-Wesley, 1993. 336 p., index. 0–201–60870–7.
In the "virtual" community, you might share some of your most personal thoughts and opinions with "neighbors" whom you have never met. Electronic mail, bulletin boards, and discussion groups connect networkers of like minds in forums for open exchange and unprecedented freedom of expression. This may be the ultimate democratization of information, and Rheingold helps us understand the social bonds that make these connections a true community. (Bus W D 13, 1993 p15; Fortune v129 F 7, 1994 p157; SBF v30 Mr 1994 p36)

Also Recommended

Braun, Eric. *The Internet Directory.* Ballantine, 1994. At present, probably the most comprehensive directory available.

Dern, Daniel. *The Internet Guide for New Users.* McGraw-Hill, 1993. Another good introductory treatment. (SBF v30 Ja 1994 p6; SciTech v17 D 1993 p40)

Kehoe, Brendan. *Zen and the Art of the Internet.* Prentice-Hall, 1993. The first edition of this favorite guide was published on-line; the print version obviously expanded its audience. (Byte v18 Mr 1993 p185; LJ v118 F 1, 1993 p41)

Computer Science—Virtual Reality

Heim, Michael. *The Metaphysics of Virtual Reality.* Oxford University Press, 1993. 175 p., index. 0–19–508178–1.
What does it mean, philosophically, if our senses cannot distinguish between the ''real'' and ''virtual'' worlds? One need not be a philosopher to be intrigued by the questions raised in this book. Heim sees virtual reality as a hopeful realm for humankind, but one we should enter, literally and figuratively, with eyes wide open. (LJ v118 Je 1, 1993 p128; LJ v119 Mr 1, 1994 p55)

Rheingold, Howard. *Virtual Reality.* Summit, 1991. Index. 0–69363–8.
Technologically simulated reality has been with us for many years, in the form of television and video games. Today, virtual reality represents an extension of that technology, which is rapidly entering the public mainstream. Rheingold tours high-tech laboratories worldwide to try some of the alternative realities we can experience through computer technologies. (LJ v117 Mr 1, 1992 p45; SLJ v118 Je 1, 1993 p128)

Also Recommended

Pimentel, Ken, and Kevin Teixeira. *Virtual Reality: Through the New Looking Glass.* Intel, 1993. Worlds generated by computers, with applications in aviation, health sciences, architecture, entertainment, and other fields. (BL v89 Ja 1, 1993 p772; Choice v30 My 1993 p1504)

Engineering—General

Adams, James L. *Flying Buttresses, Entropy, and O-Rings: The World of an Engineer.* Harvard University Press, 1992. Index. 0–674–30688–0.
''This book is not about gee-whiz devices and software,'' the author tells us in the introduction; ''we want to focus on the *process* of engineering.'' This process approach gets at the heart of how engineers think through and solve problems. Adams looks at engineering successes and failures, and comments on what can be learned from both. (Nature v356 Ap 9, 1992 p488; SBF v28 Mr 1992 p46; SciTech v16 Ap 1992 p25)

The Builders: The Marvels of Engineering. National Geographic, 1992. 288 p. 0870448366.

The Eiffel Tower, the Golden Gate Bridge, the Channel Tunnel, and other large-scale civil engineering projects that stand as monuments to building technology. Graphs, figures, and photos give perspective from every angle. (Choice v31 S 1993 p160; SciTech v17 Mr 1993 p32)

Ferguson, Eugene S. *Engineering and the Mind's Eye.* MIT Press, 1992. 241 p., illus., biblio., index. 0–262–06147–3.

Ferguson argues that good engineering design is the product not only of technological expertise, but also of intuition and an aesthetic approach to problem solving. Failure to adopt this holistic view of engineering results in piecemeal design and can have disastrous consequences. (Choice v30 F 1993 p991; LJ v118 Mr 1, 1993 p45; Sci 258 N 27, 1992 p1504)

Florman, Samuel. *The Civilized Engineer.* St. Martin's, 1987. 258 p., index. 0312001142.

The author wrote these essays with the belief that ''as engineering becomes increasingly central to the shaping of society, it is more important that engineers become introspective.'' Accordingly, Florman reflects upon his career and offers his opinions. Particularly strong on human topics like ethics and traditions. (BL v83 Ap 1, 1987 p1163; LJ v112 Je 1, 1987 p123)

Norman, Donald. *The Psychology of Everyday Things.* Basic, 1988. 297 p., illus., index. 0–465–06709–3.

Have you ever bumped your nose as a result of pushing a door that was meant to be pulled? The fault, according to Norman, may not be yours, but the designer's who failed to provide clear cues about the proper method for opening the door. Such misunderstandings can be avoided, he contends, if designers simply adhere to some of the basic principles of cognitive psychology. (Choice v26 O 1988 p400; KR v56 Ap 1, 1988 p520; LJ v113 O 15, 1988 p97)

Petroski, Henry. *The Pencil: A History of Design and Circumstances.* Knopf, 1990. 434 p. 0–394–57422–2.

Petroski can legitimately lay claim to the title of reigning poet laureate of engineering. In this, one of his most popular books, he looks at the pencil, turning this simple device into a metaphor for engineering and the process of design. (BL v86 Ja 15, 1990 p961; LJ v115 Mr 1, 1990 p112; Sci v24 My 18, 1990 p894)

Wright, Paul H. *Introduction to Engineering.* Wiley, 1994. 260 p., index. 0–471–57930–0.

A text that could serve in introductory engineering courses at the freshman level. Gives good case studies. (SBF v30 Ap 1994 p78)

Also Recommended

Baynes, Ken, and Francis Pugh. *The Art of the Engineer.* Overlook, 1981. Thoughts on technology and quality of life. (LJ v106 O 1, 1981 p1939; PW v220 O 28, 1981 p102)

Florman, Samuel. *The Existential Pleasures of Engineering.* St. Martin's, 1977. Essays on the engineering life and profession. (LJ v102 Mr 1, 1977 p548; NYTBR Je 26, 1977 p41)

McPhee, John. *The Control of Nature.* Farrar, Straus, Giroux, 1989. Large-scale civil-engineering projects aimed at controlling relentless natural forces. (BL v85 Mr 1, 1989 p1053; LJ v114 Ap 1, 1989 p108; Time v134 Ag 7, 1989 p60)

Petroski, Henry. *Beyond Engineering: Essays and Other Attempts to Figure Without Equations.* St. Martin's, 1986. Essays on assorted thoughts and insights. (BL v82 Ap 15, 1986 p1169)

Engineering—Electrical and Electronic

Amdahl, Ken. *There Are No Electrons.* Clearwater, 1991. 322 p. 0962781592.
   Yes, there really are electrons, but the somewhat flippant title of this book is meant to assure readers that they need not possess an engineer's understanding of electronics to grasp the basic principles. A good first source of information for readers with no prior knowledge on the subject. Light and slightly whimsical, but the content is informative. (Choice v29 N 1991 p486; WCRB v16#1, 1991 p46)

Levine, Sy. *A Library on Basic Electronics.* 3 vols. Electro-Horizons, 1986–1989. 0–939527–01–42.
   There are three volumes in this set: (1) *Basic Concepts and Passive Components*, (2) *Discrete Semiconductors and Optoelectronics*, and (3) *Integrated Circuits and Computer Concepts.* In all, they cover the field quite well. (SBF v25 Mr 1990 p209)

Also Recommended

Lenk, John D. *Lenk's Video Handbook* and *Lenk's Audio Handbook.* McGraw-Hill, 1991. Repair and maintenance of those devices that confound so many people. (SciTech v15 Je 1991 p4; SciTech v15 Ja 1991 p31)

Nye, David. *Electrifying America: Social Meanings of a New Technology, 1880–1940.* MIT Press, 1991. A history of the technology that changed night into day. (SBF v27 My 1991 p102)

Rains, Darell. *Major Home Appliances: A Common Sense Repair Manual.* TAB, 1987. A fix-it guide to large electrical home devices. (LJ v113 Mr 1, 1988 p33)

Veley, Victor F. C. *The Benchtop Electronic Reference Manual.* TAB, 1990.

Basic to intermediate electronics, with all the mathematical calculations. (R&R Bk N v6 Je 1991 p36)

Wyatt, Alan. *Electric Power: Challenges and Choice.* Book Press, 1986. Issues in electrical generation and power resources. (CG v107 Ap 1987 p342)

Engineering—Structural

Levy, Matthys, and Mario Salvadori. *Why Buildings Stand Up: The Strength of Architecture.* Norton, 1990. 311 p., illus., index. 0393306763; *Why Buildings Fall Down: How Structures Fail.* Norton, 1992. 352 p., illus., index. 0–393–03356–2.

These two volumes make nice bookends in any science literacy collection. In a literary and informative style, they elucidate basic principles of structural engineering. Will appeal to a wide variety of readers. (BL v88 Je 15, 1992 p1795; LJ v117 My 1, 1992 p112)

Petroski, Henry. *To Engineer Is Human: The Role of Failure in Successful Design.* St. Martin's, 1985. 247 p. 0312806809.

When a bridge collapses, the first thing a civil engineer wants to know is why. The reason is much more than morbid curiosity. Although nobody wants to see structural design failures, sometimes they can give information to the engineer that cannot be obtained in any other way. (BL v81 Je 15, 1985 p1425; KR v53 Je 15, 1985 p574; LJ v110 S 1, 1985 p206)

Energy and Environmental Technology

Ecological Economics: The Science and Management of Sustainability. Robert Costanza, ed. Columbia University Press, 1991. 525 p., index. 0–231–07562–6.

Until recently, ecology and economics were quite separate subjects. This volume represents a large step toward their synthesis, a necessary prerequisite for dealing with a host of environmental problems. Academic and larger public libraries. (Choice v29 F 1992 p936; Nature v357 Je 4, 1992 p371; SBF v28 Ap 1992 p69)

Gipe, Paul. *Wind Power for Home and Business.* Chelsea Green, 1993. 414 p., index. 0–930031–64–4.

A pragmatic approach to helping individuals decide if the power of the wind can be put to work for them. Quite comprehensive. (SBF v29 N 1993 p232; SciTech v17 S 1993 p50)

Golob, Richard, and Eric Brus. *The Almanac of Renewable Energy.* Holt, 1993. 348 p., photos, index. 0–8050–1948–0.

Valuable and practical information on hydroelectricity, biomass energy, geothermal energy, solar thermal energy, photovoltaic energy conversion, wind power, tidal power, and innovative energy storage methods. Gives sources for additional information and resources. Probably the handiest and most complete single source available on the subject. (Choice v31 S 1993 p161; LJ v118 Jl 1993 p66; SBF v29 Je 1993 p142)

The McGraw-Hill Recycling Handbook. Herbert Lund, ed. McGraw-Hill, 1993. Gloss., index. 0–07–039096–7.

Everything that you could possibly need to know about recycling can be found in this book's thirty-five chapters. The information could help you with managing domestic recycling, but is really aimed at those wishing to initiate larger-scale projects. (BL v89 Ap 1993 p1457; Choice v30 Jl 1993 p1800; SBF v29 Ag 1993 p173)

National Research Council. *Fuels to Drive Our Future.* National Academy Press, 1990. 223 p., gloss., index. 0–309–04142–2.

A bottom-line cost and technology assessment of the viability of petroleum alternatives. This is the report of a federally funded committee charged with studying energy consumption past, present, and future. (Choice v28 O 1990 p339)

Planet Management. Michael Williams, ed. Oxford University Press, 1993. 256 p., index. 0–19–520945–1.

From Oxford's Illustrated Encyclopedia of World Geography series, this handsome, informative tome gives regional assessments of environmental problems and solutions. Works great in high school libraries and up. (LJ v117 Ag 1993 p98; SBF v29 D 1993 p268)

Also Recommended

Carless, Jennifer. *Renewable Energy: A Concise Guide to Green Alternatives.* Walker, 1993. The author, a zealous environmental advocate, discusses the prospects of the major forms of renewable energy. (SBF v29 Ag 1993 p172)

Energy and the Environment in the Twenty-First Century. Jefferson Tester et al., eds. MIT Press, 1991. Tipping the scales at 1006 pages, this voluminous guide is based on the proceedings of an international conference. Best in academic and larger public libraries. (Choice v29 Mr 1992 p1098; SciTech v15 S 1991 p34)

Feshbach, Murray, and Alfred Friendly, Jr. *Ecocide in the USSR: Health and Nature under Seige.* Basic, 1992. The environmental policies of the former Soviet Union and their catastrophic results. (LJ v118 Mr 1, 1993 p43; Nature v357 Je 11, 1992 p451)

Shelton, Mark L. *The Next Great Thing: The Sun, the Stirling Engine, and the Drive to Change the World.* Norton, 1994. Trials of a solar energy device currently in the developmental stage. Especially interesting are accounts of frustrations of dealing with the Department of Energy. (LJ v119 Mr 15, 1994 p98)

(*Note*: See also entries in Chapter Eleven under "Environmental Topics.")

Home Construction

Locke, Jim. *The Well Built House*. Rev. ed. Houghton Mifflin, 1992. 302 p. 0–395–62951–9.

"Cobby" is the word that the author uses to describe substandard building practice and the use of inferior materials. Ergo, this book could be dedicated to the elimination of cobbiness in home construction. Also addresses legal and financial aspects of home projects. (BL v85 O 1, 1988 p208; PW v234 S 16, 1988 p88)

Wing, Charles. *The Visual Handbook of Building and Remodeling*. Rodale Press, 1990. 498 p. 087857901x.

A veteran author of how-to manuals shares many years of experience and knowledge in this hefty work. Unlike most manuals, this one places figures and illustrations at the center of the explanation, with text employed for secondary clarification. This technique allows for quick assimilation of construction principles. (LJ v115 Ag 1990 p135)

Also Recommended

Ferguson, Myron. *Build It Right!* DIMI, 1994. A guide for home buyers who wish to ensure the quality of the building's design and construction.

Litchfield, Michael. *Renovation: A Complete Guide*. 2nd ed. Prentice-Hall, 1990. The ins and outs of renovation in this second edition of what is widely considered a classic. (LJ v115 Ag 1990 p135)

Information Sciences

Lubar, Steven. *Infoculture: The Smithsonian Book of Information Age Inventions*. Houghton Mifflin, 1993. 408 p., illus., index. 0–395–57042–5.

The author says, "American culture can best be understood and interpreted as an *infoculture* . . . where what we see, hear and think about is shaped to an ever increasing extent by our information machines." This book looks at those machines in words and pictures. (BL v90 N 1, 1993 p489; LJ v118 N 1, 1993 p144)

Tufte, Edward R. *Envisioning Information*. Graphics Press, 1990. 126 p., illus., index. 0–961–3921–18.

A favorite among librarians, Tufte's work explores techniques for the visual design and display of information. In over 400 illustrations, the author demonstrates how raw information can be presented in ways that leap off a flat page and assume vivid visual dimensions. (LJ v115 O 1, 1990 p108; NYTBR v95 Ag 5, 1990 p19; SciTech v14 Jl 1990 p9)

Inventions

Brown, Kenneth. *Inventors at Work: Interviews with Sixteen Notable American Inventors.* Microsoft, 1988. 408 p. 1–55615–042–3.

The inventors featured here are from aviation, space sciences, computer science, telecommunications, and other fields. The author's questions are intelligent and intriguing, and pull no punches. (BL v84 D 15, 1987 p662; LJ v113 Mr 15, 1988 p64; PW v232 N 6, 1987 p63)

Freidel, Robert D., Paul Israel, and Bernard Finn. *Edison's Electric Light: A Biography of an Invention.* Rutgers University Press, 1986. 263 p. 0813511186.

The authors pored over Edison's notes, correspondence, and other documents in researching this balance and in-depth study of the man, his thoughts, and the factors that influenced this invention. Throughout, short clarifying essays help put points in context. (Choice v23 My 1986 p1408; LJ v111 Jl 1986 p99; Sci v232 My 23, 1986 p1018)

Inventors and Discoveries: Changing Our World. National Geographic Society, 1988. 320 p., illus., index. 0–87044–751–3.

An exciting book that captures the novelty and wonders of discovery. Sections on electricity, materials, computers, and several other subjects feature lucid, sometimes anecdotal text and some rare, captivating photos. (BL v85 Mr 15, 1989 p1231; LJ v114 Mr 15, 1989 p83; SBF v25 S 1989 p41)

Also Recommended

Hughes, Thomas P. *American Genesis: A Century of Invention and Technological Enthusiasm, 1870–1970.* Viking, 1989. The period of time covered in this book was one of astonishing technological inventiveness in America. (Am Sci v77 Jl 1989 p416; LJ v114 Mr 15, 1989 p83; SBF v25 Ja 1990 p135)

McCracken, Calvin. *Handbook for Inventors: How to Protect, Patent, Finance, Develop, Manufacture and Market Your Ideas.* Macmillan, 1983. Tips from a professional inventor. (LJ v108 Je 15, 1983 p1264)

Materials Science

Cotterill, Rodney. *The Cambridge Guide to the Material World.* Cambridge University Press, 1985. 05121246407.

Although now almost ten years old, there is no better book for general readers that can serve as an introduction to the complex world of materials science. A sense of genuine awe is conveyed in the great variety, beauty, and usefulness of the thousands of materials that can be built from the ninety-two basic elements. (Choice v22 Jl 1985 p1650; Nature v314 Ap 25, 1985 p686; TES My 17, 1985 p50)

Also Recommended

Images of Materials. David B. Williams, Alan Pelton, and Ronald Gronsky, eds. Oxford University Press, 1992. Proceedings of the World Materials Conference; the chapters of this book were written to provide an introductory treatment to the subject, but parts are fairly technical. (New Sci v134 Je 6, 1992 p48; SBF v28 Ag 1992 p168; Sci v257 S 25, 1992 p1976)

Military Technology

Anderberg, Bengt, and Myron Wolbarsht. *Laser Weapons: The Dawn of a New Military Technology.* Plenum, 1992. 250 p., index. 0–306–44329–5.
"Star Wars" this isn't. Low energy laser weapon technology is under development by several world powers, the United States among them. Certain laser devices were used successfully in the Persian Gulf conflict. (PW v239 O 5, 1992 p60; SBF v29 Mr 1993 p42; SciTech v17 Ja 1993 p39)

Jackson, Robert. *High Tech Warfare.* Sterling, 1992. 152 p., illus. 0806987340.
Featured are infantry, air, and naval weapons, both actual and in development. The glossy photos and intricate line drawings are eye-catching. A British publication, this book has a NATO-wide focus that lends a valuable internationalist perspective. (BL v89 Ja 15, 1993 p858)

Nanotechnology

Drexler, Eric K., et al. *Unbounding the Future: The Nanotechnology Revolution.* Morrow, 1991. 304 p., gloss., index. 0–688–09124–5.
Nanotechnology is perhaps the final phase in technological miniaturization, where matter can actually be manipulated at the molecular level. Drexler, an unabashed enthusiast of nanotechnology, sees its applications as almost limitless, from curing diseases to cleaning up toxic wastes. (LJ v117 Mr 1, 1992 p47)

Nuclear Technology

Gould, Jay M., and Benjamin Goldman. *Deadly Deceit: Low Level Radiation, High Level Cover Up.* Four Walls Eight Windows, 1990. 222 p. 0941423352.
Gould, who has served as an expert witness in several environmental health litigations, and Goldman, author of *Hazardous Waste Management: Reducing the Risk*, meticulously examine death and disease statistics for populations living near nuclear power plants and give their realistic assessment of related health risks. (BL v86 Ap 15, 1990 p1591; Nature v346 Jl 26, 1990 p325; Sierra v75 S 1990 p93)

Herman, Robin. *Fusion: The Search for Endless Energy.* Cambridge University Press, 1990. 267 p., index. 0–521–38373–0.

Harnessing the energy created by nuclear fusion in an extremely difficult but enormously promising endeavor. Even some of fusion's most ardent enthusiasts have been skeptical about whether it can become a practical means of producing energy. Herman recounts the hopes and disappointments of the three generations of scientists that have studied fusion technology. (KR v58 O 1, 1990 p1369; LJ v116 Mr 1, 1990 p63)

League of Women Voters. *The Nuclear Waste Primer: A Handbook for Citizens.* Lyons and Burford, 1993. 1–55821–226–4.
    NIMBY, an acronym for "Not In My Back Yard," characterizes the typical and understandable response of locals whose community is near a real or proposed nuclear waste disposal site. This book should be required reading for anybody who falls in that category. Concise, factual, and nonsensationalized. (SciTech v18 Ja 1994 p47)

Nuclear Power: Policy and Prospects. P.M.S. Jones, ed. Wiley, 1987. 416 p. 0–471–90732–4.
    Reviews public opinion, risk analyses, technological capabilities, and industrial experiences as they relate to nuclear power throughout the world. (Choice v25 My 1988 p1424)

Rhodes, Richard. *Nuclear Renewal: Common Sense about Energy.* Viking, 1993. 126 p. 0–670–85207–4.
    From the author of *The Making of the Atomic Bomb*, this is a succinct and rational endorsement of nuclear energy in the United States. Granting that problems exist and mistakes have been made, Rhodes nonetheless maintains that nuclear technology is safe and, further, will soon become necessary as finite natural fuels are used up. What is needed, he suggests, is a plan that will not replicate the mismanaged and misguided initiatives of the past. The simplicity of his arguments is compelling. (BL v898 Ag 1993 p2019; KR v61 Ag 1, 1993 p988; PW v240 Jl 12, 1993 p62)

Also Recommended

Cohen, Bernard L. *The Nuclear Energy Option.* Plenum, 1990. This was one of the first post-Chernobyl books to urge reconsideration of the viability of nuclear power. (PW v237 O 12, 1990 p52; SBF v27 Mr 1991 p36)

Gallagher, Carole. *American Ground Zero: The Secret Nuclear War.* MIT Press, 1993. A fine photographic history of the victims of government nuclear testing in the 1950s. (BL v89 Ap 1, 1993 p1393; LJ v118 Ap 15, 1993 p122; New Sci v138 Ap 24, 1993 p43)

Medvedev, Zhores. *The Legacy of Chernobyl.* Norton, 1990. An insider's account of the 1986 disaster. (LJ v116 Mr 1, 1991 p64)

Superconductivity

Schechter, Bruce. *The Path of No Resistance: The Story of the Revolution in Superconductivity.* Simon and Schuster, 1989. 0–671–65785–2.
The problem with conventional conducting materials is that they waste energy. With superconductors, however, energy that might otherwise be lost can be harnessed for applications as diverse as medical imaging and high-speed public transportation. (LJ v115 Mr 1, 1990 p46; SBF v25 Ja 1990 p132)

Also Recommended

Langone, John. *Superconductivity: The New Alchemy.* Contemporary Books, 1989. Complete with a science fiction introduction and examples of what this technology will mean to all of us. (LJ v115 Mr 1, 1990 p46)

(*Note*: See also entries in Chapter Twelve under "Superconductivity.")

Telecommunications

Oslin, George. *The Story of Telecommunications.* Mercer University Press, 1992. 507 p. 0865544182.
Telecommunications, from the earliest telegraphs to the satellites that direct our calls today, has evolved as a very reliable and essential technology. The author, who worked for many years in the industry, tells its history from the social and technological sides. (Choice v30 Je 1993 p1649; SciTech v17 Mr 1993 p2)

Pierce, John R., and Michael Noll. *Signals: The Science of Telecommunications.* Freeman, 1990. 247 p. 0–7167–5026–0.
A revolution in telecommunications is upon us. Tracing a general history of the technology, from Morse to Graham Bell, the authors survey what is currently available and what future inventions might bring. (Choice v28 F 1991 p959; LJ v116 Mr 1, 1991 p64)

Truxal, John G. *The Age of Electronic Messages.* MIT Press, 1990. 487 p., illus., index. 0262–20074–0.
From the Sloan Foundation's New Liberal Arts series, this book was written as a comprehensible and comprehensive survey of telecommunications for high school science students or undergraduate nonscience majors. The breadth of coverage is impressive; there are chapters on all major telecommunications technologies. (LJ v115 Je 1, 1990 p169)

Also Recommended

Bernstein, Jeremy. *Three Degrees above Zero: Bell Labs in the Information Age.* Scribner's, 1984. No fewer than seven Nobel laureates have been affiliated with Bell. Much more than a corporate history, this book is about public and private aspects of basic research. (LJ v109 S 15, 1984 p1747)

Singleton, Loy. *A Global Impact: The New Telecommunication Technologies.*
Harper Business, 1989. Amazing new technologies on the telecommunications
horizon.

Transportation and Automotive Technology

Boyne, Walter J. *Power Behind the Wheel: Creativity and the Evolution of the
Automobile.* Stewart, Tabori and Chang, 1988. 240 p. 1–55670–042–3.

The history of the automobile is shown in words and splendid pictures. A
strength, often overlooked in other books, is the discussions of how the auto's
component parts were developed. (BL v85 Ja 1, 1989 p738; LJ v114 F 1, 1989
p80; PW v234 S 2, 1988 p94)

Flink, James J. *The Automobile Age.* MIT Press, 1988. 456 p., photos, index.
0–262–06111–2.

Society and economics changed dramatically with the rise of the automobile.
Several themes are explored, but the unifying one is how the automobile transformed personal and institutional American "lifeways." (CSM v80 Jl 6, 1988
p17; KR v56 Mr 1, 1988 p338)

Hawkes, Nigel. *Vehicles.* Macmillan, 1991. 239 p., illus., index. 0–02–549106–
7.

Vehicular technologies from the steamboat to the dirigible. The focus is on
great historical achievements, like Lindbergh's Atlantic crossing and Yeager's
breaking the sound barrier. The oversize format of this book and its quality
illustrations and cutaway diagrams are appealing. (Bus W D 16, 1991 p134; LJ
v117 F 15, 1992 p134; SBF v28 Mr 1992 p46)

Lay, M. G. *Ways of the World: A History of the World's Roads and the Vehicles
that Used Them.* Rutgers University Press, 1992. 402 p., illus., index. 0–8135–
1758–3.

Travel comes naturally to human beings. Some of the earliest technologies of
the human species were developed in order to facilitate itinerant lifestyles. This
is an ambitious and readable history of human migration and movement technologies. (BL v84 N 15, 1992 p568; Choice v30 Je 1993 p1681)

Potter, Stephen. *On the Right Lines? The Limits of Technological Innovation.*
St. Martin's, 1987. 0–312–00488–5.

Unlike in Europe, the railways in America have been in decline for some
years. This history examines the birth, maturity, and subsequent fall of the rail
industry, and outlines an agenda for restoring it to economic vitality. Much is
tied to new high-speed train technologies.

Also Recommended

Bobrick, Benson. *Labyrinths of Iron: Subways in History, Myth, Art, Technology
and War.* Holt, 1994. The world of underground travel from the Romans to
modern cities.

Duffy, James E. *Modern Automotive Technology*. Goodheart-Wilcox Co., 1994. A good contemporary guide to the workings and maintenance of automobiles.

Sikorsky, Robert. *Drive It Forever: Your Key to Long Automobile Life*. Mc-Graw-Hill, 1989. Auto maintenance for anybody. (LJ v116 Je 1, 1991 p94)

OTHER RESOURCES

Periodicals for General Readers

Air and Space. Smithsonian Institution, bi-m. 0886–2257.
The next best thing to a visit to the National Air and Space Museum is a subscription to this journal. Shows fascinating aviation technology and air history, with an emphasis on American innovations.

Architectural Record. McGraw-Hill, m., revs. 0003–858x.
America's leading journal of architecture is aimed at practicing architects and students, but many general readers also appreciate it for the news, descriptions, and photographs of the inspiring projects it features.

Building Design and Construction. Cahners Publishing, m. 0007–3407.
A source of information on projects, economic and industry trends, and new building technologies. Many articles give insider views of building endeavors.

Bulletin of the Atomic Scientists. Educational Foundation for Nuclear Science, b-mi., revs. 0096–3402.
Born in the early days of the nuclear era, this popular magazine has kept track of the "doomsday clock" for many years. Articles offer penetrating observations on nuclear science and its social impacts. A voice of reasoned conscience.

Byte. McGraw-Hill, m., revs. 0360–5280.
Microcomputer and small systems enthusiasts in schools and businesses read this magazine regularly; hackers consider it almost scripture. The critical reviews are technical, but even readers with minimal computer knowledge will find many of the articles appealing.

Electronics Now. Gernsback Publishing, 6/yr. 0737–1692.
For hobbyists, with features on the various technologies: video, audio, computers, and so on.

Engineering Education—ASEE Prism. American Society for Engineering Education, bi-m. 0022–0809.
Articles by and for educators. Although the focus is specialized, the discipline-wide perspective and interplay of opinions and techniques make this of broader general interest.

ENR: Engineering News Record. McGraw-Hill, w. 0891–9526.

The most widely read magazine of the construction industry, *ENR* reports news and current events, notable new projects, business ventures, and other professional issues.

EPRI Journal. Electric Power Research Institute, 8/yr. 0362–3416.

Although a professional specialty publication, this magazine reaches out to general readers with its popularized research summaries. Also covers environmental topics and consumer issues.

Home Mechanix. Times Mirror, m. 8755–0423.

Covers all areas of home technology, with a noticeable recent emphasis on environmental consciousness. Each issue gives a "Source List" of where to find additional information on products and subjects featured within it.

Motor Trend. Peterson Publishing, m. 0027–2094.

Of the various popular automobile magazines, *Motor Trend* is probably the most appropriate for a science literacy collection because of the technical nature of its vehicle assessments and its coverage of the industry.

PC Novice. Peed Publishing, m. 0033–7862.

"The magazine for computer newcomers" is just the algorithm for computer illiterates looking for a painless entry into the world of automation.

Popular Mechanics. Hearst Corp., m. 0032–4558.

How-to articles for tinkerers and inventors of every type. Glossy, detailed photographs assist the do-it-yourselfers, and the brief news snippets keep them up to date.

Robotics World. Communications Channels, Inc., 6/yr. 0737–7908.

This is a popular trade journal for a field that attracts people from all walks of life, not just specialists.

Technology and Culture. University of Chicago Press, q., revs. 0040–165x.

Don't let the scholarly look and feel of this journal put you off. The articles are readable, the subjects fascinating, and the reviews in-depth. Published under the auspices of the Society for the History of Technology.

Technology Review. MIT Press, 8/yr., revs.

An important journal for technological literacy. Covers developments and breaking news, with an emphasis on practical applications. Moderately technical, this title is edited for the attentive general public. Covers biological and medical technologies, too.

Audiovisual Resources

COMPUTERS

Artificial Intelligence: Fulfilling the Dream. TV Ontario, 1989 (30 minutes each, color).

The four videos in this series look at the history, technology, and potential applications of machine intelligence. Each segment builds on the prior one, beginning with the inception of the idea of AI in the 1950s and concluding with frontiers of computers and robotics. Titles include "In Search of the Thinking Machine" (history), "Cloning the Experts" (expert systems), "Making the Quantum Leap" (machine communications), and "Beyond the Vision" (machine vision). (SBF v25 Mr 1990 p217)

Data as Power. Films for the Humanities, 1990 (26 minutes, color).
It has been said that "information is power." This is especially true in the computer age. This video, from the *Computer Revolution* series, shows how computers store and manage information and suggests possibilities. (SBF v30 Ja 1994 p22)

Subject of Matter. Films for the Humanities, 1990 (26 minutes each, color).
The first video in this set, "The Brain Machine," discusses the workings of the human brain, and the second, "Simulated Intelligence," focuses on how computers can replicate brain functions. (SBF v29 AP 1993 p88)

Virtual Reality. Films for the Humanities, 1993 (26 minutes, color).
Virtual Reality is a dazzling, mind-boggling excursion into the worlds that exist in computer circuitry, but which we perceive as real with our senses. Also discusses uses and abuses. (SBF v30 My 1994 p118)

Also Recommended

The Search for Realism: Computer Graphics, Animation and Design. Media Guild, 1990 (47 minutes, color). Computer graphics and animation technologies give viewers an appreciation of the power and potential they possess. Experts who are interviewed convey enthusiasm in their work. (SBF v27 N 1991 p245)

ENERGY

Energy Alternatives: A Global Perspective (series). Filmmakers Library, 1991 (52 minutes each).
As these videos show, alternative energy technologies exist, they are practical, and they can be implemented now. There are three parts: "Changing the Way the World Works" (problems and solutions), "The Rich Get Richer" (affluent nations and energy consumption), and "Power to the People" (Third World energy issues). One reviewer said, "This series should be required viewing for all who consider themselves parts of the human race." (SBF v28 Je 1992 p150)

Energy Efficiency and Renewables. Video Project, 1993 (22 minutes, color); *Recycling.* Video Project, 1993 (18 minutes, color).
Over history, human beings have successfully converted from one primary fuel source to another several times. Today, with finite fossil fuels being depleted rapidly, the need to do so again is nearly upon us. This video features work

being done at the Rocky Mountain Institute, a leading research center. (SBF v30 Ja 1994 p23)

Wind: Energy for the Nineties and Beyond. Video Project (24 minutes, color).

Harnessing the power of the wind is an ancient idea with great promise for the future. Wind-generated electricity has the advantages of being totally clean and endlessly renewable, and in many localities it is reliable. (SBF v30 Ap 1994 p85)

Also Recommended

The Generation of Electricity and *Commercial Generation and Transmission of Electricity.* Films for the Humanities, 1990 (20 minutes each, color). These two companion videos are from the *Physics in Action* series. The first begins with Michael Faraday's experiments with electromagnetism and concludes with a visit to a coal fueled power plant. The second demonstrates how that electricity is generated and distributed, and includes a visit to a nuclear power plant. (SBF v O 1991 p 216)

ENGINEERING AND TECHNOLOGY—GENERAL

Exploring Technology. Bergwall Productions, 1989.

A five-part series that shows how technology moves from the research and development stage to implementation. High school and college freshman level. (SBF v25 Ja 1990 p161)

The Mystery of the Master Builders. Coronet Film and Video–WGBH TV (Boston), 1988 (59 minutes).

An intelligent look at the symbiotic relationships among architecture, structural engineering, and building construction in a historical context. Shown on the *Nova* program.

To Engineer Is Human. Films Inc.–BBC, 1987 (50 minutes).

Based on Henry Petroski's very popular book of the same title, this video shows some graphic footage of engineering design failures and discusses what lessons can be learned from them.

Also Recommended

The Electromagnetic Spectrum. Films for the Humanities, 1990 (20 minutes, color). A British production, part of which was filmed in a high school physics laboratory, discusses the entire range of the spectrum—microwaves, X-rays, and ultraviolet, infrared, and visible light.

One Small Step. Media Guild, 1989 (25 minutes). Computer systems at NASA and the various reliability checks that control and safeguard systems.

Out of the Fiery Furnace, into the Machine Age. Landmark Films, 1991 (52 minutes). Metallurgy and heavy industry's roles in business and technology in the United States.

THE ENVIRONMENT

The Big Spill. Coronet–MTI Film and Video, 1990 (produced by Nova). (58 minutes, color).
Gives an overview of the ecological consequences of the Exxon *Valdez* oil spill and the immense difficulties involved in the clean-up. Also discusses the bureaucratic complexities and finger-pointing that played roles in the unfortunate affair. (SBF v26 N 1990 p136)

Cleaning Up Toxics at Home and *Cleaning Up Toxics in Business.* Video Project, 1990 (30 minutes each, color).
"Think globally, act locally" is an oft-used adage of environmentalists. These videos show steps that individuals can take in the safe use, storage, and disposal of toxic substances that they commonly encounter. (SBF v27 My 1991 p118)

Also Recommended

Acid Rain: A North American Challenge. National Film Board of Canada, 1990 (16 minutes, color). The images of forests damaged by acid rain convincingly drive home the need to take remedial measures. This short but informative video benefits from its Canadian perspective on the acid rain issue. Enlightening to American viewers. (SBF v26 N 1990 p136)

Garbage: The Movie—An Environmental Crisis. Churchill Films, 1990 (16 minutes, color). The early nineties voyage of the New York garbage barge that had no place to dump its cargo has become a symbol of the solid waste disposal problem in America. This video assesses alternative trash-handling and recycling measures. (SBF v27 Ja 1991 p23)

MILITARY TECHNOLOGY

The Killing Machines. WGBH TV (Boston) and KCET TV (Los Angeles)–Coronet Films, 1990 (58 minutes).
This production, originally aired on *Nova*, is on United States military policy and high-tech weapons development.

TRANSPORTATION

The Entrepreneurs, Expanding America. PBS Video, 1987 (60 minutes).
Robert Mitchum appears in this program on how transportation technology and vehicular mechanization figured in the conquest of the North American continent.

Fastest Planes in the Sky. Films for the Humanities. 1993 (60 minutes, color).
A *Nova* segment converted to VHS for a new audience, this historical documentary ranges from the first crude propeller planes, which were considered

fast in their day, to modern experimental aircraft that can exceed twice the speed of sound. (SBF v29 N 1993 p248)

Transportation 2000: Moving Beyond Auto America. Video Project, 1993 (20 minutes, color).

Taking for granted that, in the future, reliance on the gasoline automobile for personal transportation cannot continue, this brief video urges development of alternatives and suggests some possibilities. (SBF v29 N 1993 p246)

Also Recommended

America by Design: The Street. PBS video–WTTW TV (Chicago), 1986 (56 minutes). City planning and road engineering in the United States.

History of Aviation. World Aviation Video, 1988 (50 minutes). A World Aviation and Video City production, this video concentrates in particular on the period of time from 1903 to 1940.

CD-ROMs

CASPAR: Computer Aided Science Policy and Research Data. Quantum Research Corp., 1991 (one disc).

Retrieval and analysis of statistical data on science and engineering resources from government offices.

Jane's All the World's Aircraft. Janes Information Group, Avionics, 1992.

Information and analysis on all civil and military aircraft in service in the world today. For students, scholars, professionals, and zealous aviation hobbyists.

Longman Microinfo World Research Database. Longman, Microinfo, 1991 (one disc).

This database compiles and arranges data from several world research center directories and who's whos.

Materials Safety Data Sheets. National Safety Data Corp.

A government information database on hazards, toxicity, physical properties, and special handling procedures for materials. Mostly for professionals, but could also be used in learning laboratory situations.

Book Review Source Abbreviations

A&S Sm	*Air and Space/Smithsonian*
AF	*American Forests*
AHR	*American Historical Review*
AJS	*American Journal of Sociology*
Am Ent	*American Entomologist*
A Lib	*American Libraries*
Am Sci	*American Scientist*
ARBA	*American Reference Books Annual*
Astron	*Astronomy*
Atl	*Atlantic*
Aud	*Audubon*
BAS	*Bulletin of the Atomic Scientists*
BioSci	*Bioscience*
BL	*Booklist*
BRD	*Book Review Digest*
Brit Bk N	*British Book News*
BRpt	*Book Report*
Bus W	*Business Week*
BW	*Book World*
BWatch	*Bookwatch*

Byte	*Byte*
CG	*Canadian Geographic*
Choice	*Choice*
Comw	*Commonweal*
CS	*Contemporary Sociology*
CSM	*Christian Science Monitor*
Ecol	*Ecology*
Fortune	*Fortune*
Fut	*Futurist*
GJ	*Geographical Journal*
Gov Pub R	*Government Publications Review*
HB	*Horn Book*
JAL	*Journal of Academic Librarianship*
J Ch Ed	*Journal of Chemical Education*
Kliatt	*Kliatt Young Adult Paperback Book Guide*
KR	*Kirkus Reviews*
LATBR	*Los Angeles Times Book Review*
LJ	*Library Journal*
Nature	*Nature*
New R	*New Republic*
New Tech B	*New Technical Books*
NH	*Natural History*
New Sci	*New Scientist*
NW	*Newsweek*
NYTBR	*New York Times Book Review*
Phys T	*Physics Today*
Phys W	*Physics World*
PT	*Psychology Today*
PW	*Publisher's Weekly*
R&R Bk N	*Reference and Research Book News*
RSSEMA	*Reference Sources in Science, Engineering Medicine and Agriculture*
SA	*Scientific American*
S&T	*Sky and Telescope*
SBF	*Science Books and Films*
Sch Lib	*School Librarian*
Sci	*Science*
SciTech	*SciTech Book News*

Sierra	*Sierra*
SLJ	*School Library Journal*
T&C	*Technology & Culture*
TES	*Times Educational Supplement*
Time	*Time*
TLS	*Times Literary Supplement*
Variety	*Variety*
WCRB	*West Coast Review of Books*
WLB	*Wilson Library Bulletin*
WSJ	*Wall Street Journal*

Author Index

Title Index

Subject Index

illustrations, scientific, 41
Immense Journey, The, 53
inductive method. *See* scientific method
"information explosion," 25
International Association for the Evaluation of Educational Achievement, 11
International Center for Advancement of Science Literacy, 5
International Mathematics Study, 11
internet, 61–62; "gopher," 61; World Wide Web, 62

"knowledge acquirers," 33–35
"knowledge intermediaries," 33–34
"knowledge producers," 33
Konner, Melvin, 36, 43–44
Kunh, Thomas, 22

LaFollette, Marcel, 26
Lapp, Ralph, 26
"levels of relevance," 34, 40, 45, 59
librarians, 16–17, 29, 31, 34, 40, 46, 49
library collections, 34, 46; evaluations of materials (general), 46–48, (books), 49–55, (periodicals), 55–57, (video), 58–60, (new technologies), 60–62
"lifetime learning," 5, 16, 31, 49

Making Science Our Own, 26
Manhattan Project, 26
Mathematical Principles of Natural Philosophy. See Principia, the
mathematics, information, 32, 36
Miller, Jon, 5, 14
Muller, Johannes, 25
multimedia, 61
Myers, Greg, 43

narrative documentary (videos), 58
National Center for Educational Statistics, 7
National Institute of Health, 14
National Science Foundation, 11–14
Nelkin, Dorothy, 43
Nation at Risk, 7
New Hacker's Dictionary, 3
New Priesthood, 26
Newton, Isaac, 23–24

1990 Science Report Card, 7
"nonattentive public," 6

Old Man's Toy, 39–40
On the Connection of the Physical Sciences, 25
On the Revolutions of the Heavenly Spheres, 22
Origin of Species, 25

"pacing." *See* "staging"
Paulos, John Allen, 16
periodicals, 55–57
Poincare, Jules-Henri, 7
"Poles of Technical Communication" model, 44, 56; "Zones of Authority," 56–57
Pollett, Miriam, 51
Prewitt, Kenneth, 6
Principia, the, 23–24
producers (videos), 60
publishers (books), 51–53

Raymond, Eric, 3
reference books, 55
"religion of science," 25
"representational" writing, 33
Revolutionibus Orbium Caelestium. See On the Revolutions of the Heavenly Spheres
Roy, Rustrum, 27
Royal Society, 15, 24, 33

Sagan, Carl, 15, 27
Scholastic Apptitude Test, 11
Science: communications (general), 21–28, 32–34, 36; journalism, 35, 40, 42, 44, 56–57; popularization of, 15–16, 23–29, 32–45; "primary literature," 33–34, 41; social aspects, 3–5, 10–11, 15–16; writing, 32–44, 49–53
Science and Engineering Pipeline, 12–13
Science as Writing, 32
science education, 4, 7, 8–9, 14; benchmarks, 7–8, 10; college, 12; high school, 8; social factors, 9–11; surveys, 7–11
science literacy, 4–7, 12, 22, 32; defini-

About the Author

GREGG SAPP, head of Access Services at the University of Miami's Richter Library, writes the annual *Library Journal* column "Best Sci-Tech Books for General Readers." He has studied issues related to science literacy, science education, and information resources for 10 years. He is coauthor of *Notable Historical Figures in Fiction* (1994) and *Access Services in Libraries: New Solutions for Collection Management* (1992) and currently preparing a work entitled *Primary Sources: Original Scientific Texts and Their Publics*, a series of original essays in the history of science.